Surface Treatments for Biological, Chemical, and Physical Applications

Surface Treatments for Biological, Chemical, and Physical Applications

Edited by Mehmet Gürsoy and Mustafa Karaman

WILEY-VCH

The Editors

Mehmet Gürsoy
Selcuk University
Department of Chemical Engineering
Alaaddin Keykubat Kampüsü
Merkez/Konya 42075
Turkey

Prof. Mustafa Karaman
Selcuk University
Department of Chemical Engineering
Alaaddin Keykubat Kampüsü
Merkez/Konya 42075
Turkey

Cover
Pond image - fotolia_© Kalle Kolodziej
and image of droplets - fotolia_© fotofuerst.

Library of Congress Card No.: applied for

British Library Cataloguing-in-Publication Data
A catalogue record for this book is available from the British Library.

Bibliographic information published by the Deutsche Nationalbibliothek
The Deutsche Nationalbibliothek lists this publication in the Deutsche Nationalbibliografie; detailed bibliographic data are available on the Internet at http://dnb.d-nb.de.

Print ISBN: 978-3-527-34083-5
ePDF ISBN: 978-3-527-69880-6
ePub ISBN: 978-3-527-69882-0
Mobi ISBN: 978-3-527-69883-7
oBook ISBN: 978-3-527-69881-3

Cover Design Adam Design, Weinheim, Germany
Typesetting SPi Global Private Limited, Chennai, India
Printing and Binding Markono Print Media Pte Ltd, Singapore

Printed on acid-free paper

Contents

List of Contributors

Aydin Cihanoğlu
Department of Chemical Engineering
İzmir Institute of Technology
Gülbahçe-Urla
35430 İzmir
Turkey

Daryoush Emazadah
Islamic Azad University
Department of Chemical Engineering
Gachsaran Branch
Gachsaran
Iran

Beyhan Erdoğan
DYO Paints Manufacturing & Trading Company INC
Atatürk Organize Sanayi
Bölgesi 10003 Sok. No: 2
35620 Çiğli – İZMİR
Turkey

Gökhan Erdoğan
Gediz University
Department of Biomedical Engineering
35665 Seyrek – Izmir
Turkey

Ahmad Fauzi Ismail
Universiti Teknologi Malaysia
Advanced Membrane Technology Research Centre (AMTEC)
Skudai
Johor 81310
Malaysia

Susana C.M. Fernandes
Division of Glycoscience
School Biotechnology
Royal Institute of Technology (KTH)
Roslagstullsbacken 21
Stockholm SE-10691
Sweden

Nilay Gazel
Selcuk University
Department of Chemistry
Alaaddin Keykubat Campus
42075 Konya
Turkey

Günnur Güler
Ege University
Center for Drug Research and Development and Pharmacokinetic Applications (ARGEFAR)
35100 Bornova – Izmir
Turkey

Mehmet Gürsoy
Selcuk University
Department of Chemical Engineering
Alaaddin Keykubat Kampüsü
Merkez/42075 Konya
Turkey

Rasoul Jamshidi Gohari
Islamic Azad University
Department of Chemical Engineering
Bardsir Branch
Bardsir
Iran

Mustafa Karaman
Selcuk University
Department of Chemical Engineering
Alaaddin Keykubat Kampüsü
Merkez/42075 Konya
Turkey

Duygu O. Kiliç
Center for Materials Research
Izmir Institute of Technology
35430 Urla – Izmir
Turkey

Tuğba Kiliç
École polytechnique fédérale de
Lausanne (EPFL)
Institute of Bioengineering
Lausanne CH 1015
Switzerland

Hilal D. Kivrak
Yuzuncu Yil University
Faculty of Engineering
Department of Chemical Engineering
65080 Van
Turkey

Mahmut Kuş
Selcuk University
Department of Chemical Engineering
Alaaddin Keykubat Kampüsü
42075 Konya
Turkey

Woei-Jye Lau
Universiti Teknologi Malaysia
Advanced Membrane Technology
Research Centre (AMTEC)
Skudai
Johor 81310
Malaysia

Epameinondas Leontidis
University of Cyprus
Department of Chemistry
Kallipoleos 75
1678 Nicosia
Cyprus

Nadzirah Mohd Mokhtar
Universiti Teknologi Malaysia
Advanced Membrane Technology
Research Centre (AMTEC)
Skudai
Johor 81310
Malaysia

Nik Abdul Hadi Md Nordin
Universiti Teknologi Malaysia
Advanced Membrane Technology
Research Centre (AMTEC)
Skudai
Johor 81310
Malaysia

Chi-Siang Ong
Universiti Teknologi Malaysia
Advanced Membrane Technology
Research Centre (AMTEC)
Skudai
Johor 81310
Malaysia

Fatih Özcan
Selcuk University
Advanced Technology Research and Application Center
Alaaddin Keykubat Kampüsü
42075 Konya
Turkey

Faruk Özel
Karamanoglu Mehmetbey University
Department of Materials Science and Engineering
65080 Karaman
Turkey

Gizem Payer
Department of Chemical Engineering
İzmir Institute of Technology
Gülbahçe – Urla
35430 İzmir
Turkey

Diego Hernán Quiñones-Murillo
Universidad del Atlántico
Department of Chemical Engineering
Barranquilla 080020
Colombia

Özlem G. Şahin
Selcuk University
Department of Chemical Engineering
Alaaddin Keykubat Kampüsü
42075 Konya
Turkey

Nur Aimie Abdullah Sani
Universiti Teknologi Malaysia
Advanced Membrane Technology Research Centre (AMTEC)
Skudai
Johor 81310
Malaysia

Zahide Tosun
Selcuk University
Department of Physics
Alaaddin Keykubat Kampüsü
42075 Konya
Turkey

Uğur Türkan
Gediz University
Department of Biomedical Engineering
35665 Seyrek – Izmir
Turkey

Esma Yenel
Selcuk University
Advanced Technology Research and Application Center
Alaaddin Keykubat Kampüsü
42075 Konya
Turkey

Huseyin B. Yildiz
KTO Karatay University
Department of Materials Science and Nanotechnology Engineering
Alaaddin Kap Street
42020 Konya
Turkey

Preface

Solids are composed of a bulk material covered by a surface. The surface acts as an interphase to the surrounding and it is usually the surface phase that determines if a material is suitable for an intended application or not. Hence, treatment of a surface through either chemical or physical routes has been an important topic that has attracted the attention of both researchers and industry because of the endless application possibilities. Surface modification techniques are now being used extensively in the textile, aerospace, automotive, biomedical, defense, chemical, tooling, construction industries, and many more as well. The purpose of this book is to give the reader some generalized concepts on the modification and characterization, as well as some insight on the recent applications of the surfaces. Other than a few old books there is no complete book that describes the whole picture. In addition, we want to get the attention of a broad range of readership including those of undergrad students, technicians, and even individuals that are interested in this topic. The language of the book is clear and concise and provides many excellent illustrations to make their point. The book covers the most basic concepts without complicated analyses so that an individual who is not familiar with the subject can also benefit. We have used as many visuals as possible to show the importance of surface treatments on many applications, including surfaces that are naturally engineered. The first three chapters of the book include basic parts: (1) surfaces in nature, (2) surface modification techniques, and (3) surface characterization techniques. The remaining chapters deal with some emerging chemical and physical applications: (4) surface modification of polymeric membranes for various separation processes, (5) Langmuir–Blodgett films: sensor and biomedical applications and comparisons with the layer-by-layer (LbL) method, (6) surface modification of biopolymer-based nanoforms and their biological applications, (7) enzyme-based biosensors in food industry via surface modifications, and (8) heterogeneous catalysis.

The most remarkable examples of surface engineering are abundant in nature. Since the beginning of the life, all kinds of living organisms from one-celled creatures to plants and animals have adapted to their environments for survival. By long trial and error processes, nature itself has generated a great number of outstanding living creatures with tested and proven sustainable biological functions. There have been numerous studies in materials science, based on mimicking natural materials. In Chapter 1, examples of biomimetic materials especially the related surfaces are given.

Following the increasing demand for high performance materials from the past century, the field of materials surface modification by various synthetic strategies has undergone enormous expansion. Significant numbers of studies have been carried out to develop efficient techniques for adding new functionalities to a materials surface as well as to understand the fundamental aspects of the various techniques. In Chapter 2, the basics of most significant surface modification processes are given. Although both dry- and wet-coating techniques are involved, emphasis is given to vapor deposition techniques.

The chemical, physical, and morphological features of surfaces including wettability, optical, adhesive, mechanical, and so on play a very crucial role in material properties. Therefore, characterization of a surface is very important for materials science. The properties of natural and fabricated material surfaces can be determined using various characterization techniques. In Chapter 3, some of the most common characterization methods are given.

Starting from Chapter 4, emerging applications of surface science are given. Membranes are industrially important materials that are used extensively in fluid separation processes. The membrane separation process is mainly governed by the characteristics of the membrane top layer. Thus, modifications of membrane surface properties is the practical and effective approach to achieve desired separation efficiency. In general, membrane surface modification can be performed by techniques such as interfacial polymerization (for thin film composite membranes), organic–inorganic blending (for mixed matrix membranes), LbL assembly, photo-initiated polymerization, and conventional spray coating. Chapter 4 aims to overview the surface modification methods available for polymeric membranes development and discusses the importance of surface-modified membranes for various applications, covering aqueous, solvent, and gas phases.

Chapter 5 focuses on Langmuir–Blodgett films that has been an active area of research for more than a century. Many of the methods for the production of thin organic films have been adopted by nanotechnology for the production of hybrid organic–inorganic films, containing molecular layers, nanoparticles, sheet-like inorganic materials, and biopolymers. The chapter focuses mainly on Langmuir–Blodgett films and compares them to LbL films.

Chapter 6 addresses chemical surface modifications of biopolymers-based nanoforms namely nanocellulose, including bacterial cellulose, and nanochitin, for biological applications. These biopolymers provide sustainable solutions to the need for new (bio)materials in biological applications due to their unique biological properties like low toxicity, biocompatibility, biodegradability, and bioactivity. Nonetheless, further chemical functionalization has been advanced in order to optimize their intrinsic properties and/or generate novel functions to develop new materials, in particular functional bionanocomposites.

Biotechnological applications of enzyme-based biosensors have become an important tool for the detection of chemical and biological components for food monitoring due to their exceptional performance capabilities toward food materials, which include high specificity and sensitivity, rapid response, low cost, relatively compact size, and user-friendly operations. In Chapter 7, detailed information is shared about surface modifications of enzymes, immobilization of enzymes, main characteristics of enzyme based biosensors, and their usage in food analysis.

In Chapter 8, the importance of heterogeneous catalysis is discussed in terms of solid surface science. Considering this purpose, general information about heterogeneous catalysis and surface science including their industrial and economic importance, reaction mechanisms and kinetics in heterogeneous catalysis systems, and preparation and characterization methods for heterogeneous catalysts are given.

Finally, we wish to express sincerely our gratitude to all authors and co-authors of the book, because of their great effort during the preparation of this book.

22 September 2016 | *Mehmet Gürsoy*
Konya, Turkey | *Mustafa Karaman*

1

Surfaces in Nature

Mehmet Gürsoy and Mustafa Karaman

"Natura nihil frustra facit" (Nature does nothing in vain).
Aristotle (384–322 BC)

1.1 Introduction

Human beings have used nature in order to meet their needs. Early inhabitants took advantage of natural resources and materials just for their fundamental needs such as food, shelter, and clothing. Over the past several centuries, science and technology have developed exponentially in the world. Conventional materials and methods such as self-cleaning surfaces, new generation optical devices, biomaterials, and so on are not enough to meet the requirements of high-technology applications.

Thanks to improvements in surface analysis techniques, scientists can look deep into nature. In this way, the relationships between the structure and functions of living organisms can be investigated.

Since the beginning of life, all kinds of living organisms from one-celled creatures to plants and animals have adapted to their environments for survival. By extensive trial and error processes, nature itself has created a great number of outstanding living creatures with tested and proven sustainable biological functions. As a result of the long adaptation process (spanning millions of years), organisms have developed impressive features that have equipped them better to compete for limited resources, defend themselves against their predators, and live longer. Therefore, it can be easily said that nature is the best materials scientist ever. Because of this, copying or mimicking of biological systems is an effective way to produce desired high-technology materials. This approach is called as *biomimicry*.

The word biomimicry is a combination of two Ancient Greek words, "bios" meaning "life," and "mimesis" meaning "to imitate." Biomimicry can be defined as that branch of science that seeks to imitate processes or structures existing in nature. The main philosophy behind biomimicry is "If Nature can do it, so can we." With this approach, nature is used as a guide to tackle problems using biomimetic materials and processes.

Surface Treatments for Biological, Chemical and Physical Applications, First Edition.
Edited by Mehmet Gürsoy and Mustafa Karaman.

Biomimetic innovation is based on the observation and mimicking of nature. Indeed, observation and learning from nature began in the initial days of mankind. That is why, biomimicry has been accepted as an ancient discipline. Throughout history, nature has inspired us to invent many tools.

During the Renaissance Period, numerous impressive bio-inspired designs were produced by Leonardo Da Vinci. For instance, he designed a flying machine by investigating bird anatomy. His words on nature show why he applied bio-inspired idea on his designs: "Although human genius through various inventions, makes instruments corresponding the same ends, it will never discover an invention more beautiful, nor more ready, nor more economical than does nature, because in her inventions nothing is lacking and nothing is superfluous [1]."

In 1969, the term *biomimetics* was first used by bioengineer Otto Herbert Schmitt in his paper at the Third International Biophysics Congress in Boston [2]. However, biomimicry was popularized by Janine Benyus in her book Biomimicry: Innovation Inspired by Nature (1997) [3]. Subsequently, biomimetic approaches have become more and more popular and important during the past decades in almost all research fields.

There have been a lot of studies in materials science, based on mimicking natural materials. In this chapter, examples of biomimetic materials, especially the related surfaces, are given.

1.2 Inspiring Natural Surface Structures

1.2.1 Self-Cleaning Surfaces

Self-cleaning surfaces are in great demand for fundamental research and various industrial applications. These types of surfaces must have two important criteria: very high contact angle and low contact angle hysteresis.

Contact angle can be defined as a measure of the wettability of a solid surface by a liquid drop (Figure 1.1a). Theoretically, the contact angle values must be between 0° and 180°. If the contact angle is less than 90°, these surfaces are classed as hydrophilic. If the wetting angle is higher than 90°, these kind of surfaces can be accepted as water repellent. When a contact angle approaches 180°, the surface is considered superhydrophobic. Adhesion of droplets on the surface can be determined by dynamic contact angle measurements. Dynamic contact angle is referred as *advancing contact angle* (the maximum value of the contact angle) and *receding contact angle* (the minimum value of the contact angle). Contact angle hysteresis is the difference between them. There is a strong relation between sliding angle and contact angle hysteresis. Sliding angle is the required minimum angle to move the droplet on surfaces (Figure 1.1b). As the contact angle hysteresis decreases the drops can easily roll off from the surfaces.

In nature, self-cleaning properties have been observed in various plant leaves. On a rainy day, the raindrops do not spread on the plant leaf and these drops completely roll off the leaf. And thus, undesirable particles on the leaves are easily removed by rolling water drops. It is also known that pathogenic microorganisms

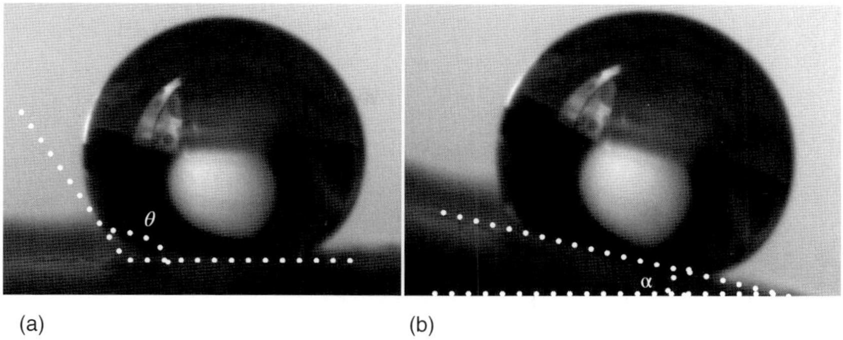

Figure 1.1 (a) θ is the water contact angle and (b) α is defined as the inclination angle at which a water drop rolls off the surface.

cannot germinate and infect leaves without water. For this reason, self-cleaning helps prevent the occurrence of plant diseases [4].

Among the plants that are self-cleaning, lotus is one of the most popular examples due to its very high contact angle and very low hysteresis. These values are 164° and 3°, respectively [5]. That is why, the term *Lotus Effect* is also used in place of "Self-Cleaning Effect" in the literature [6]. The lotus grows in an aquatic environment such as lakes, and shallow and muddy water. It always achieves to remain clean even in dirty waters [7, 8]. For this reason, the lotus is considered as a symbol of purity.

The plants are covered by a cuticular surface except for their roots. The cuticle layer is a natural composite that is the interface between plants and their environment [9]. This composite consists mostly of two parts, soluble lipids and bio-polyesters [10, 11]. Because of the chemical structures of these components, the cuticle layers usually exhibit hydrophobic properties. Lotus leaf surface is also covered by low surface energy cuticular surface, which contains mainly $—CH_2—$ groups [12]. The relation between the contact angle and surface energy was formulated in Young's Eq. (1.1) describing wetting phenomena in terms of thermodynamic equilibrium [13] (Figure 1.2).

$$\cos\theta = \frac{\gamma_{SV} - \gamma_{SL}}{\gamma_{LV}} \tag{1.1}$$

where θ is the contact angle of the liquid, γ_{SL} is the interfacial surface tensions between the solid and the liquid, γ_{SV} and γ_{LV} are the solid and liquid surface free energy, respectively. According to Eq. (1.1), decreasing γ_{SV} should increase the contact angle value.

However, it is well known that the lower surface energy of $—CH_3$ groups or fluorocarbons do not exist in any biological systems [12]. So, the contact angle of

Figure 1.2 The schematic representation of a liquid drop with the contact angle and tension vectors.

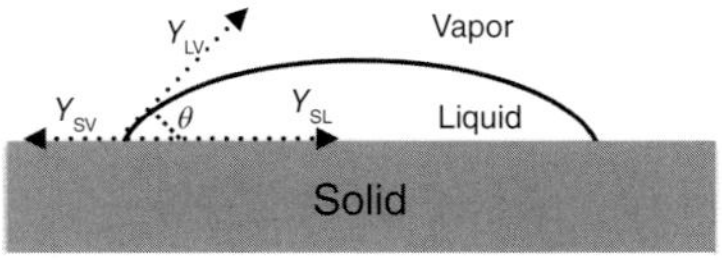

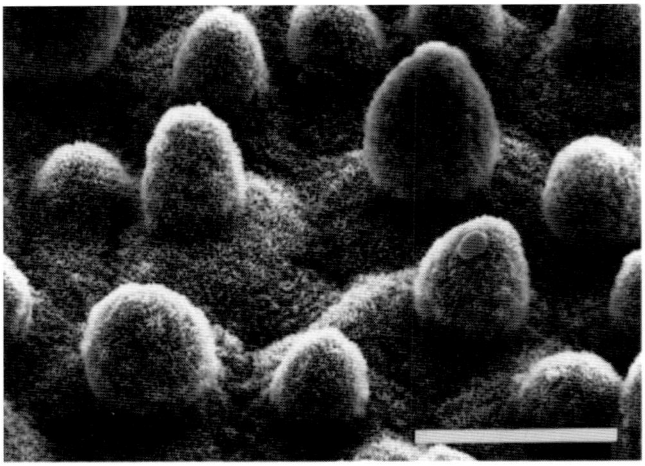

Figure 1.3 The SEM image of adaxial lotus leaf surface, scale bar = 20 μm. (Barthlott 1997 [9]. Reproduced with permission of Springer.)

a planar cuticular surface can be a maximum of about 110° [14]. Thus, in nature only having low surface energy surface is not enough to be superhydrophobic. By means of scanning electron microscope (SEM), micrometer scale bumps and nanometer scale wax crystals were observed on the lotus leaf surface [9]. The SEM image of adaxial lotus leaf surface structure is given in Figure 1.3.

The effect of surface roughness on the wettability can be explained with the help of the Wenzel model [15, 16]. The Wenzel model describes the following Eq. (1.2).

$$\cos\theta = R_f \cos\theta_0 \tag{1.2}$$

In which, θ is the contact angle of a rough surface, θ_0 is the contact angle of a smooth surface, and R_f is the surface roughness factor. The roughness factor is defined as the ratio of the actual surface area to the geometric surface area. If the surface is flat, $R_f = 1$; however, this value must be higher than 1 for rough surfaces. For a hydrophilic surface, θ_0 must be lower than 90°, roughness decreases the contact angle. On the other hand, for hydrophobic surfaces, as in lotus leaf, θ_0 is greater than 90°. Therefore, according to Wenzel equation, it is expected that increasing the surface roughness increases hydrophilicity [17].

In brief, water repellent surfaces can be produced in two different ways: changing the surface morphology and decreasing the surface free energy. In the former method, the underlying principle is to create micro/nanoscale rough structures on the surfaces. In the latter method, the surfaces are usually coated with hydrophobic functional groups. Only having low surface energy or only having rough surfaces may not be sufficient to be superhydrophobic. Thus, the production of superhydrophobic surfaces mostly requires the combination of two methods.

The extraordinary surface morphology of lotus leaf minimizes the contact area between its surface and water drops. The hierarchical micro/nanostructures provide that air is trapped underneath the water drop. That is why, very high contact angle values are observed on the lotus leaf. The unique structure of the lotus leaf can be directly imitated in order to produce self-cleaning materials.

Roughening fluorinated polymers and silicones directly leads to superhydrophobic surfaces, because of the inherent hydrophobic nature of these kinds of polymers. For example, Barshilia and Gupta [18] treated polytetrafluoroethylene (PTFE/Teflon) surfaces with argon and oxygen plasma to obtain a superhydrophobic surface. The highest average water contact angle and the maximum surface roughness were found for 4 h plasma treatment. During plasma treatment, inspite of changing the surface morphology, functional groups were preserved. The contact angle value increased from 102° to 158°. After 10 months, the contact angle was again measured from the modified surface and almost the same values were found. According to the results, the superhydrophobicity of obtained surface remains unchanged even after a long time. Similarly, Jin *et al.* [19] created a polydimethylsiloxane (PDMS) surface containing micro-, submicro-, and nanostructures using a one-step laser etching method. Etched PDMS surface with these special structures showed high water contact angle (162°) and low sliding angle ($< 5°$).

Coating rough or hierarchical surfaces with low surface energy materials is also a common method to produce self-cleaning materials. Ma *et al.* [20], produced superhydrophobic fabrics by a two-step process. In the first step, poly(caprolactone) (PCL) fibers were generated by the electrospinnig method. And then, the fiber mat surfaces were coated with perfluoroalkyl ethyl methacrylate (PPFEMA) by Initiated Chemical Vapor Deposition (iCVD). These PFEMA-coated electrospun fibers exhibited very good self-cleaning properties with contact angle of 175° and sliding angle of lower than 2.5°. These results are attributed to the combination of the inherent roughness of the electrospun mats and the low surface energy of the PPFEMA coating. Liu *et al.* [21] fabricated micro-nanoscale binary structured composite particles of silica/fluoropolymer using an emulsion-mediated sol–gel method to mimic the surface microstructures on the lotus surface. With this method, super hydrophobic surfaces with water contact angle larger than 150° were obtained. Grewal *et al.* [22] investigated the effect of different micro- and nanopatterned surfaces on their wettability and tribological surfaces. The hierarchical patterns were designed, imitating the topography of the adaxial surface of lotus leaf. The advancing and receding contact angle of PTFE-coated hierarchical structure surface was found to be similar to those of the lotus leaf.

The casting method (soft molding) is another simple and effective way to replicate leaf surface structures. Sun *et al.* [23] successfully applied this method for lotus leaf at the first time. They cast PDMS on the lotus leaf, and then the PDMS layer was peeled off. After that, this negative PDMS layer was used to make a positive PDMS layer. According to SEM results, the positive replica and the original lotus leaf showed the same surface structures. The positive template also exhibited similar superhydrophobic properties as the fresh lotus leaf. The contact angle of the positive replica was found as 160° and the water droplets could easily roll off this surface.

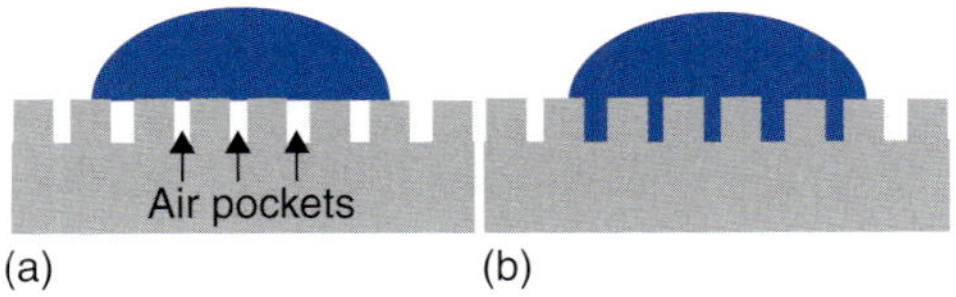

Figure 1.4 (a) Cassie model wetting regime and (b) Cassie impregnating wetting regime.

1.2.2 Adhesive Hydrophobic Surfaces

Similarly to lotus leaves, the hierarchical nano- and microstructures on the red rose petal surface provide a high contact angle. However, differently from lotus leaves, water droplets do not slide off the surface of a red rose petal.

Basically, two main hypotheses are used to explain superhydrophobicity on rough surfaces: Wenzel and Cassie wetting regimes. The former regime has been already mentioned in the beginning of this chapter. The Cassie model was developed after the Wenzel's state. According to the Cassie model [24], air is trapped in micro- and nanostructures underneath the water droplets, as seen in Figure 1.4. Because of these air pockets, the liquid cannot wet the surface. It is thus expected that water drops easily roll off this kind of surface, just as in the lotus leaf. The wetting mechanism of lotus leaf is an excellent example for Cassie wetting regime. On the other hand, spherical water droplets usually stick on the red rose petal surfaces. This different behavior can be attributed to difference in the surface topography between the red rose petal and lotus leaf. It is observed that the sizes of the structures on the red rose petal are larger than those of lotus leaves [25]. While water drops cannot enter into the grooves of the lotus surface, they can enter into the grooves of the red rose petal surface.

This phenomenon is known as the *Cassie impregnating wetting regime* as seen in Figure 1.4 [26]. Due to the high adhesion between the water drops and petal surface, the petal surface exhibits a high contact angle. Therefore, water droplets do not fall off even if the petal surface is tilted to 180°. The unique wettability behavior of red rose petal surface was defined as the "petal effect" by Feng *et al.* [27] for the first time in the literature. Recently, there has been a great deal of interest in fabrication of artificial "petal effect" coating and surfaces that mimic the original red rose petal.

Karaman *et al.* produced a thin "petal effect" polymer sheet using a combination of casting and iCVD methods [28]. Firstly, they poured poly vinyl alcohol (PVA) solution on the surface of a fresh red rose petal. The obtained PVA negative mold was placed in an iCVD reactor, then coated by poly(glycidyl methacrylate) (PGMA) and poly(1*H*,1*H*,2*H*,2*H*-perfluorodecyl acrylate) (PPFDA). This schematic duplication process of the red rose petal surface is presented in Figure 1.5. The contact angle of the positive replica was found as $152 \pm 3°$ and the water did not roll off even when the biomimetic polymer sheet was turned upside down.

Fluorinated polyimide was synthesized by the electrospinning method by Guangming *et al.* [29]. The surface of nanofibers consisted of dented nano- and/or micro bowl-like structures. The air is trapped in bowl-like particles below the water droplets; this situation provides a very high contact angle. Moreover, when the droplet is lifted, the air pocket expands, which creates a negative pressure, and thus the adhesion between water droplet and surface is increased.

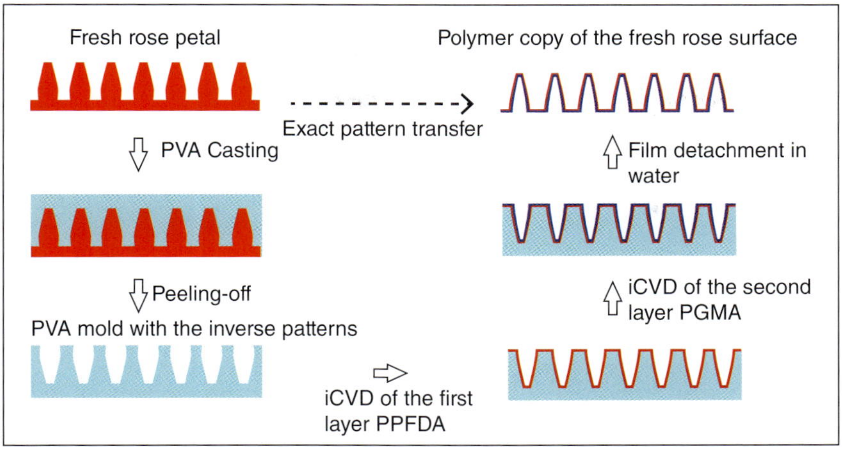

Figure 1.5 The schematic duplication process of the red rose petal surface. (Karaman 2012 [28]. Reproduced with permission of Elsevier.)

The maximum adhesion force of the petal effect surface was measured as 127 μN and the contact angle was found as 153°.

Recently, the tunable adhesive superhydrophobic surfaces have been produced by controlling the reaction parameters, such as reaction time, particle sizes, surface roughness, and so on. In other words, superhydrophobic surfaces having tunable water adhesion capability allow choosing wetting behaviors, "lotus effect" or "petal effect." For instance, Liu *et al.* [30] developed a one-step electrodeposition process in that, superhydrophobic surfaces with controlled adhesion can be easily produced by just changing the reaction time. When the reaction time is 10 min, a petal effect surface was obtained and the maximum contact angel was found as 155.1°. When the reaction time increased further up to 30 min, a self-cleaning surface was obtained. Its static contact angle and contact angel hysteresis were found to be 161.7° and 3°, respectively.

Another fabrication of superhydrophobic surfaces with tunable water droplet adhesion was carried out by Xie *et al.* [31]. They used both O_2 plasma etching and plasma deposition of thin films to create a superhydrophobic wood surface. Firstly, wood substrates were exposed to O_2 plasma, and then were coated with pentafluoroethane (PFE) films. The obtaining surfaces showed lotus effect properties with high contact angle (161.2° ± 1.5°) and low sliding angle (~15°). When wood samples were coated with diamond-like carbon (DLC) after the etching property, thees surfaces also exhibited high contact angle (153.7° ± 2.7°). However, differently from PFE-coated wood, DLC-coated samples showed a petal effect.

1.2.3 Unidirectionally Superhydrophobic Surfaces

In superhydrophobic structures that show lotus effect property, having low contact angle hysteresis is not enough for some applications, the unidirectional movement of water is also important. The water droplets move in all directions on the lotus leaf surfaces. On the other hand, the water drops on rice leaf surfaces

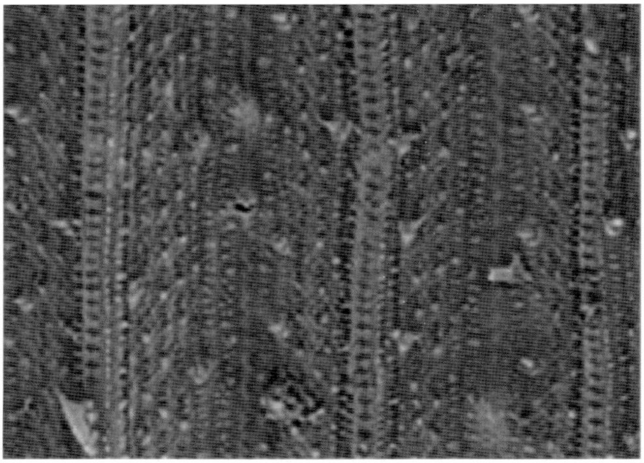

Figure 1.6 The SEM image of adaxial rice leaf surface [35]. (Yao http://link.springer.com/article/10.1007/s11434-012-5220-1. Used under CC BY 4.0 https://creativecommons.org/licenses/by/4.0/.)

roll along just the long-axis direction [32]. The difference between lotus leaf and rice leaf can be attributed to surface topography. Actually, there are similar structures on both these natural surfaces. However, while these microstructures are randomly located on a lotus leaf, they are set on a one-dimensional order (parallel to the leaf edge) on the rice leaf [33]. This observation is consistent with the difference between the parallel and perpendicular direction sliding angles on the rice leaf. These values of parallel direction and perpendicular to the leaf edge were found as 3–5° and 9–15°, respectively [34]. Rice leaf surface structure is an excellent example for the fabrication of unidirectionally superhydrophobic surfaces. The hierarchical structures can be seen in Figure 1.6 [35].

Yang *et al.* used the combination of lithography- assisted electrochemical etching, anodic oxidation, and fluoridation methods to fabricate artificial rice leaf structures [36]. They achieved to fabricate a three-level microstructure (macro/micro/nano) of rice leaves on aluminum. This biomimetic structure shows superhydrophobicity and anisotropic sliding behavior.

Zhu *et al.* fabricated the large area surface with ordered binary structure arrays by mimicking the rice leaf surface structure [37]. The underlying pattern on the substrate can be easily modified by changing the polymer solution concentration, which provides the fabrication of various topographies. The obtaining surfaces demonstrate anisotropic wettability similarly to rice leaf.

In another study, Yao *et al.* developed a two-step soft transfer to produce an artificial rice leaf structure [35]. The obtained biomimetic rice surface exhibited superhydrophobicity and anisotropic sliding properties that were similar to those of natural rice. Parallel and perpendicular sliding angles were found as 25° and 40°, respectively.

Wu *et al.* used improved laser interference lithography to fabricate micropearl arrays for adjusting two-directional unidirectional wetting structures [38]. They systematically investigated the effect of laser beam intensity ratio and resin thickness on anisotropic wetting behavior. According to appropriate parameters, micropearl arrays were designed and modifying them with fluoroalkylsilane created biomimetic surfaces exhibiting wettability properties very similar to those of rice leaf.

1.2.4 Fog Harvesting Surfaces

Access to safe and sufficient water is of vital importance to people. However, water scarcity is one of the major environmental issues, in today's world. According to World Health Organization (WHO) reports, an estimated 2.5 billion people have no access to improved sanitation and, unfortunately, each year hundreds of thousands children die from water-related diseases such as diarrhea [39–41]. Learning from living creatures that live in arid conditions is an efficient way to obtain clean water. For instance, Namib Desert beetles (*Stenocara*) overcome the lack of water by collecting moisture from air. Actually, collecting fine fog droplets is not easy in the heat and breeze of the desert. The fog collecting of mechanism of Stenocara was revealed by Parker and Lawrence. They discovered an array of hydrophilic bumps on the beetle backs, which are surrounded by hydrophobic waxy lines [42]. The fine moisture droplets collect on hydrophilic bumps and start to grow. When the collecting droplets reach sufficient weight, they detach and roll down the tilted hydrophobic beetle's back surface to the mouth. The hierarchical structures on the Namib Desert beetle's back can be seen in Figure 1.7 [42].

Zhai *et al.* successfully mimicked the back of the Stenocara beetle creating hydrophilic patterns on superhydrophobic surfaces [43]. Garrod *et al.* used a two-step plasma chemical method to produce the hydrophilic–hydrophobic pattern, which is similar to the Stenocara beetle's back [44]. Another study that is inspired by the fog harvesting surface structure of the Stenocara beetle's back was carried out by Dorrer and Rühe [45]. They fabricated various superhydrophobic surfaces patterned with smooth, circular patches of hydrophilic domains. According to the results, it was found that the pinning force for a given pump was constant and independent of the drop volume. Except for the Namib Desert beetle, some other organisms have water collection ability. Cactus is one of these organisms. It can survive in extremely arid conditions because of its efficient fog collection mechanism. A cactus consists of conical spines

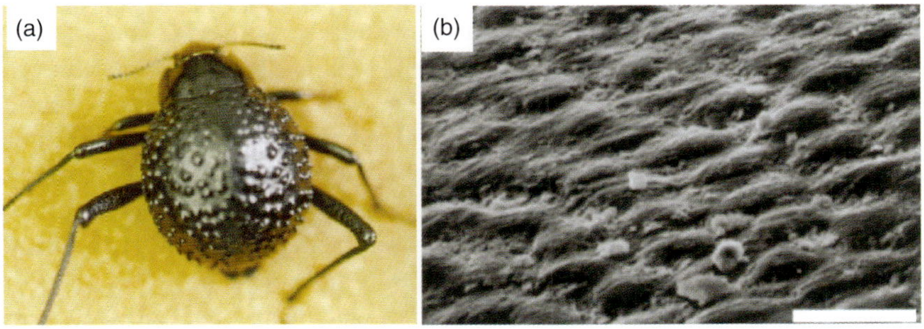

Figure 1.7 (a) The photo of an adult female *Stenocara* sp. and (b) the SEM image of the *Stenocara* sp. dorsal surface, scale bar = 10 μm. (Parker 2001 [42]. Reproduced with permission of Nature Publishing Group.)

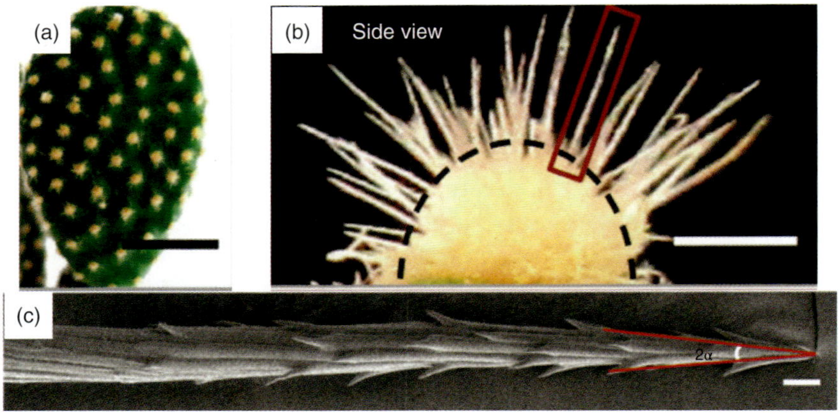

Figure 1.8 (a) The photograph of the cactus *Opuntia microdasys* scale bar = 5 cm, (b) the photography of a single cluster containing a lot of spines, scale bar = 100 mm, and (c) the SEM image of the spine, scale bar = 20 μm. (Ju 2012 [46]. Reproduced with permission of Nature Publishing Group.)

and hierarchically hydrophilic/hydrophobic structures [46]. The fog collection system of cactus is based on the Laplace pressure gradient and the wettability difference [47–49]. The hierarchical structures of the cactus' spines can be seen in Figure 1.8 [46]. Cao *et al.* produced biomimetic microtip arrays, which are similar to those of cactus, using a modified magnetic particle-assisted molding method [49]. The morphology of the tips is adjusted by changing the weight ratio of PDMS to magnetic particles. The optimal ratio of PDMS to magnetic particles was found to be 2 : 1. Andrews *et al.* reported that the *Cotula fallax* plant can also collect water from moisture due to its unique hierarchical 3D arrangement formed by its leaves and the fine hairs covering them [50].

1.2.5 Anti-reflective Surfaces

Anti-reflective coatings are used in many technological applications such as glasses computer screens, solar cells, military hardware, and so on. The purpose of using anti-reflective coating is to maximize the transmission of light through an optical surface [51].

In nature, anti-reflective structures are found in some insects. These structures can be found in their eyes, which make it possible to see objects in even low-light environments. These natural anti-reflective structures can also be found on their body surface that helps in hiding from enemies by reducing reflections [52]. For example, the eyes of the moth consist of hexagonal arrays that reduces optical losses. The dimensions of this structure are smaller than the wavelength of the light, so the reflection of light is effectively suppressed [53]. The hierarchical structure of the moth eye is shown in Figure 1.9 [54].

Moth eyes inspired anti-reflective structures that have been fabricated by various methods. One of these studies was carried out by Raut *et al.* [55]. They produced anti-reflective structures using "sacrificial layer mediated nanoimprinting." For wavelengths between 400 and 1000, while non-anti-reflective

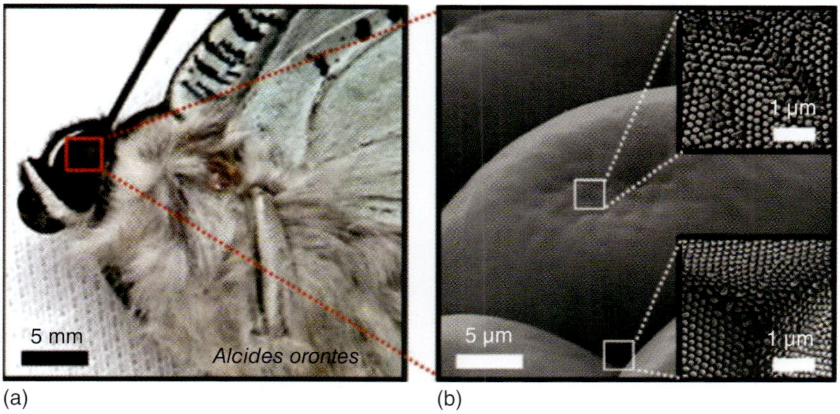

Figure 1.9 (a) The photograph of the moth *Alcides orontes* and (b) SEM images of real moth-eye structures of the moth. (Kwon 2016 [54]. Reproduced with permission of American Chemical Society.)

structures reflected 8.7% of light, the moth-inspired arrays reflected just 4.8%. The minimum reflectance was found to be 1.4% from 400 to 1000 nm in wavelength, when the arrays were designed on both sides of the substrate. Oh *et al.* also replicated the structure of moth eyes using thermal imprinting processes and plasma treatment methods [56] and with their study, the sub-wavelength structures were obtained. Except for moth eyes, other organisms also inspire researchers to design anti-reflective structures. For instance, Li *et al.* mimicked the surface of the eyes of the butterfly, *Euploea mulciber*, which consists of the hierarchical nipple array structure, in order to obtain anti-reflective structures [57]. They fabricated biomimetic amorphous carbon structure using a one-step vacuum sintering method. This structure exhibited a reflectance of 2–3% in visible light; this value for an amorphous carbon plate (without hierarchical nipple array structure) was 11% reflectance. Xu *et al.* were inspired from mosquito eye structures to produce anti-reflective structures [58]. With this aim, they used a combination of self-assembled polymer spheres and nanoimprint lithography. It was found that the topography of this biomimetic surface is similar to that of mosquito eyes. Because of the hierarchical structures, the surface reflection was considerably decreased.

1.2.6 Structural Color

Color is perhaps the most diverse property in biological creatures. The coloration in the animal kingdom provides adaptation to the surrounding environment for misleading their enemies [59]. It can also be used for sexual interactions [60]. Basically, the color source can be classified into three groups: pigments, bioluminescence, and structural colors [61, 62]. Pigmental color is known as *chemical color*, and is obtained by selective absorption of visible light by pigments [61]. Bioluminescence is produced by chemical reactions in the living organisms. Structural color is known as *physical color* that is highly related to surface structure, and is based on the nano- and microstructures on surfaces.

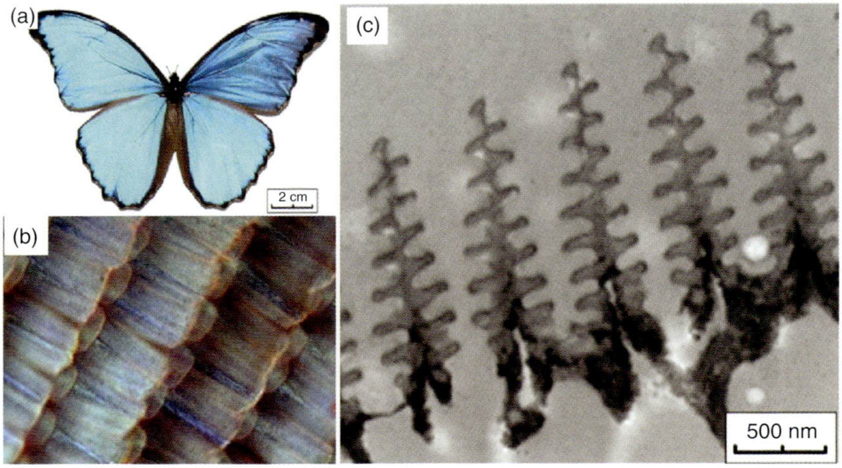

Figure 1.10 (a) The photograph of *Morpho didius* butterfly, (b) the magnified image of wing, and (c) the cross-section TEM image of wing scales. (Jiang 2014 [66]. Reproduced with permission of Elsevier.)

In contrast to chemical colors, structural colors show high resistance to discoloration because of chemical and thermal changes. According to archeological finds, structural colors in fossils are preserved as long as the structure details are maintained [63]. Moreover, when compared to pigment color, structural color is more efficient in terms of energy consumption [59]. Differently from the pigment-based color materials, physical colors can be produced without hazardous chemicals, which makes fabrication of structural colors safe and eco-friendly [64]. Due to all these advantages, the structurally based colors have drawn great interest in the past years.

In nature, there are many bright and vivid structural colors in living creatures. *Morpho* butterflies, which belong to the Nymphalidae family, are one of the excellent examples of structural color [65]. Figure 1.10 shows (a) The photograph of *Morpho didius* butterfly, (b) the magnified image of wing, and (c) the cross-section TEM image of wing scales [66].

The iridescent metallic blue color of morpho butterflies is a result of the microscopic structures on their wings, which reflect the light in order to produce this color – without pigments. The electron microscope was first used to observe the wing surface of Morpho, and ordered microstructures were found [67]. Inspiring this unique nanostructure that nature has created, biomimetic materials are produced. For example, Watanabe *et al.* used ion beam chemical vapor deposition (FIBCVD) method to mimic the structure of the Morpho wing [68]. The obtained structure morphology is almost the same as that of Morpho. Both Morpho butterfly and the replica exhibited very similar reflection intensity spectra for the various incidence angles. Another fabrication of morpho blue color was performed by Saito *et al.* [69]. They coated TiO_2/SiO_2 layers on stepped quartz using a combination of electron-beam lithography and dry etching.

Apart for insects, structural colors were also observed in some birds. Peacock is one of them, the beautiful colors of its tail feather are based on physical color. Cong *et al.* fabricated crystal thin film composed of poly(styrene-methylmethacrylate-acrylicacid) (P(St-MMA-AA)) on an inclined silicon substrate [70]. The different colors were obtained from the different layers of stair-like thin film. With this study, the colors of the peacock's tail feathers were successfully mimicked.

1.2.7 Drag Reduction and Antifouling Surfaces

Nature has created various organisms that have special shapes and surfaces in order to reduce drag in air and water. By this means, animals can move faster by consuming low energy [71, 72]. Shark is one of this type of animals. Besides their aerodynamic shapes, they have unique skin structures. During long-term evolution, their skin structures have been optimized in order to minimize the frictional resistance between the water and their body. This makes sharks one of the fastest animals in the ocean. Shark skin is covered with tooth-like scales also called *dermal denticles*, which are aligned along the direction of water flow [73–75]. These microstructures on shark skins reduce not only water friction during swimming but also prevent bacterial growth on their bodies [76, 77]. This self-cleaning mechanism is quite different from those of the lotus and the rose petal effect. Shark skin is not superhydrophobic; in fact, it is even hydrophilic [76]. The reason for having a clean surface can be attributed to the rough shark skin structures that reduce the contact area for adhering and fouling marine organisms. In addition to surface area, the contact time is also decreased because of the accelerated flow rate on the shark's body surface [78, 79].

Inspired by the unique surface structures of shark skin, there are many studies in the literature to obtain antifouling coatings. One of these studies was carried out by Carman *et al.* [80]. They fabricated various patterns (pillars, pits, ridges, channels) on polydimethylsiloxane elastomers (PDMSe) surfaces using photolithography method. They reduced the settlement of Ulva spores by 86% when compared to smooth PDMS. Wen *et al.* used 3D printing to produce thousands of artificial shark denticles on membrane [81]. According to the results, while swimming speed is increased by 6.6%, energy cost-of-transport was reduced by 5.9%. Han *et al.* created biomimetic shark skin by direct replication of the shark skin structure [82]. The skin of *Carcharhinus brachyurus* was used as a template. This artificial sharkskin structures demonstrated a drag reduction efficiency of 8.25%. The SEM images of the natural and biomimetic shark skin structures are given in Figure 1.11a,b, respectively.

Not only marine creatures, but also birds have excellent drag reduction properties. Because of their body shape and feather structures, the birds minimize air drag and thus, they exhibit excellent flying performance. That is why researchers investigate birds in order to reduce air friction. For instance, Chen *et al.* mimicked the herringbone riblets of pigeon feathers [83]. In this study, the drag reduction efficiency of herringbone riblets was found to be 16%.

1.2.8 Adhesive Surfaces

Adhesive tapes have been widely used since their in 1845 [84]. They can be easily used without the need for any solvent or heat, these properties make them very

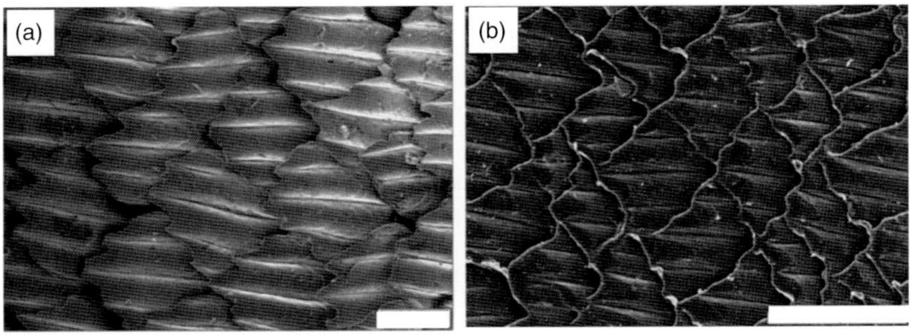

Figure 1.11 (a) The SEM image of the shark skin template and (b) the SEM image of the biomimetic shark skin. Scale bars, (a) 100 μm and (b) 200 μm . (Han 2008 [82]. Reproduced with permission of Springer.)

versatile and practical for various purposes. However, the traditional adhesive tapes cannot be used for hanging heavy objects. Moreover, they do not work under vacuum conditions [85]. Researchers seek for a solution in nature in order to produce alternative adhesive tapes without these handicaps.

The adhesive types in nature can be roughly categorized into two groups: (i) wet adhesion and (ii) dry adhesion. The former is based on secreted body fluids between the animal surface and the object [86]. This adhesion mechanism is widely observed in insects, for instance the wet adhesion was found in *Hemisphaerota cyanea* beetle [87].

In dry adhesion, mainly van der Waals bonding plays an important role and this force is generated as a consequence of the interaction between animal adhesive pads and objects [76]. The gecko lizard is a famous example of this kind of adhesion mechanism. The skin of gecko pads consists of well-aligned hairs (setae), which end in spatulae [88, 89]. The hierarchal nano- and microstructures provide enough van der Waals force to overcome gravity and, thus, geckos climb vertical surfaces and can stick to them upside down [90]. In fact, the similar hierarchal structures and micro spatulae are also observed in other animals such as insects and arachnids [91]. As can be seen in Figure 1.12, the diameters of the setae decrease with the body weight of the creature, in other words, the amount of setae per unit area increases with the weight of the animal. The gecko displays the highest density of setae, and it is also the biggest creature that generates dry adhesion [78].

Therefore, the gecko is one of the most spectacular living being for researchers to mimic its adhesive properties. One of these studies was carried out by Qu *et al.* [92]. They used a combination of PECVD and fast heating method to fabricate vertically single-walled carbon nanotubes (VA-SWNTs). It was reported that the VA-SWNTs show the highest achievable force ($29.0\,N\,cm^{-2}$) among all of the synthetic and natural gecko feet. Cho *et al.* used anodic aluminum oxide (AAO) membrane with the controllable pore channels as a replication template to fabricate gecko-inspired hairy hard PDMS films with nanopillars [93]. The obtained structures showed high adhesion and superhydrophobicity. Therefore, when a water droplet is placed on this structure, it does not roll even if the biomimetic polymer sheet is turned upside down. The reason for this "petal

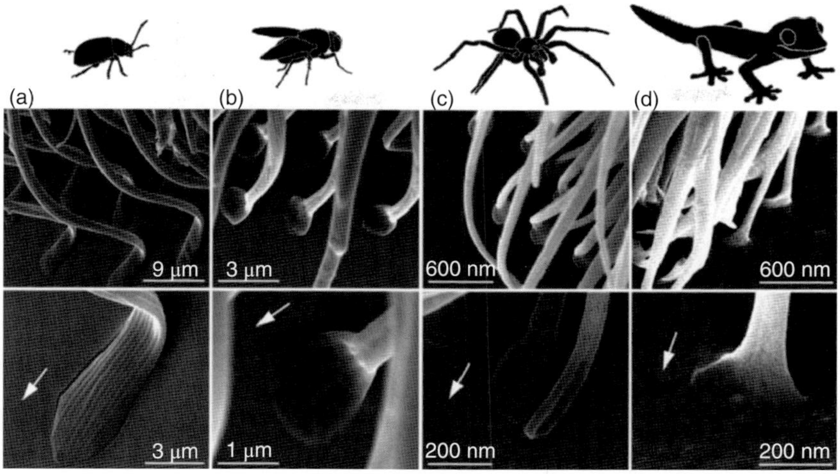

Figure 1.12 SEM images of the spatula-shaped terminal elements of various animals which have adhesive foot. (a) The beetle *Gastrophysa viridula*, (b) the fly *Calliphora vicina*, (c) spider *Cupiennius salei*, and (d) the gecko *Gecko gekko*. Arrows point in distal direction. (Varenberg 2010 [91]. Reproduced with permission of Royal Society of Chemistry.)

effect" can be attributed to van der Waals forces between the water droplets and the molded surface composed of densely packed hairy PDMS nanopillars. Davies *et al.* produced artificial PDMS-based gecko hair arrays using photolithographic and nano-molding techniques [94]. Gecko inspired wafer-scale nanofabrillar structures were produced by Kustandi *et al.* using the combination of colloidal nanolithography, deep silicon etching, and nanomolding methods [95]. Geim *et al.* produced microscale polyimide flexible plastic pillars using e-beam lithography and oxygen-plasma dry etching [96]. These gecko-inspired hierarchical structures exhibited high adhesion properties. Kim *et al.* used replica-molding and e-beam exposure methods to fabricate high aspect-ratio polyurethane acrylate nanohairs [97]. According to frictional-adhesion test results, these biomimetic structures showed good adhesion strength even after more than 100 cycles of attachment and detachment.

1.3 Conclusion

In this chapter, surface structures of various organisms have been presented, which inspire researchers to design high-technology materials. The underlying surface morphology and functions of these mechanisms have been summarized with biomimetic examples. It is obvious that the hierarchical surface nano/microstructures of both organisms and materials determine functional properties such as superhydrophobicity, anti-reflective, and so on. Some organisms have more than one property; for instance, gecko foot skin structures exhibit not only high adhesion but also superhydrophobicity. Nature has already developed a wide range of organisms and, therefore, just by copying their structures without any further tests, they can be used for the desired practical applications. This makes the biomimicry approach a time-saving process. We believe that when considered in the greatness of nature, undiscovered functional organism surface structures must wait for mankind to mimic and use them in many fields of technology.

References

1 Thompson, B.S. (1999) Environmentally-sensitive design: Leonardo WAS right!. *Mater. Des.*, **20**(1), 23–30.

2 Schmitt, O.H. (1969) Some interesting and useful biomimetic transforms. Third International Biophysics Congress, vol. **1069**, p. 197.

3 Benyus, J.M. (1990) *Biomimicry: Iinnovation Inspired by Nature*, William Morrow and Company, New York.

4 Bhushan, B. (2012) *Biomimetics: Bioinspired Hierarchical-Structured Surfaces for Green Science and Technology*, Springer Science & Business Media, Columbus, OH.

5 Koch, K., Bhushan, B., Jung, Y.C., and Barthlott, W. (2009) Fabrication of artificial Lotus leaves and significance of hierarchical structure for superhydrophobicity and low adhesion. *Soft Matter*, **5**(7), 1386–1393.

6 Marmur, A. (2004) The lotus effect: superhydrophobicity and metastability. *Langmuir*, **20**(9), 3517–3519.

7 Latthe, S.S., Terashima, C., Nakata, K., and Fujishima, A. (2014) Superhydrophobic surfaces developed by mimicking hierarchical surface morphology of lotus leaf. *Molecules*, **19**(4), 4256–4283.

8 Shirtcliffe, J.N., McHale, G., and Newton, M.I. (2009) Learning from superhydrophobic plants: the use of hydrophilic areas on superhydrophobic surfaces for droplet control part of the "Langmuir 25th year: wetting and superhydrophobicity" special issue. *Langmuir*, **25**(24), 14121–14128.

9 Barthlott, W. and Neinhuis, C. (1997) Purity of the sacred lotus, or escape from contamination in biological surfaces. *Planta*, **202**(1), 1–8.

10 Holloway, P.J. (1994) in *Air Pollution and the Leaf Cuticle* (eds K.E. Percy, J.N. Cape, R. Jagels, and C.J. Simpson), Springer, Berlin, pp. 1–13.

11 Martin, J.T. and Juniper, B.E. (1970) *The Cuticles of Plants*, Edward Arnold, London.

12 Ma, M. and Hill, R.M. (2006) Superhydrophobic surfaces. *Curr. Opin. Colloid Interface Sci.*, **11**(4), 193–202.

13 Young, T. (1805) An essay on the cohesion of fluids. *Philos. Trans. R. Soc. London*, **95**, 65–87.

14 Bhushan, B. (2012) *Biomimetics*, (ed. B. Bhushan) Springer, Berlin, Heidelberg, pp. 49–65.

15 Wenzel, R.N. (1936) Resistance of solid surfaces to wetting by water. *Ind. Eng. Chem.*, **28**(8), 988–994.

16 Wenzel, R.N. (1949) Surface roughness and contact angle. *J. Phys. Chem.*, **53**(9), 1466–1467.

17 Gürsoy, M. and Karaman, M. (2015) Effect of substrate temperature on initiated plasma enhanced chemical vapor deposition of PHEMA thin films. *Phys. Status Solidi C*, **12**(7), 1006–1010.

18 Barshilia, H.C. and Gupta, N. (2014) Superhydrophobic polytetrafluoroethylene surfaces with leaf-like micro-protrusions through Ar+O_2 plasma etching process. *Vacuum*, **99**, 42–48.

19 Jin, M., Feng, X., Xi, J., Zhai, J., Cho, K., Feng, L., and Jiang, L. (2005) Super-hydrophobic PDMS surface with ultra-low adhesive force. *Macromol. Rapid Commun.*, **26**(22), 1805–1809.

20 Ma, M., Mao, Y., Gupta, M., Gleason, K.K., and Rutledge, G.C. (2005) Superhydrophobic fabrics produced by electrospinning and chemical vapor deposition. *Macromolecules*, **38**(23), 9742–9748.

21 Liu, Y., Chen, X., and Xin, J.H. (2006) Super-hydrophobic surfaces from a simple coating method: a bionic nanoengineering approach. *Nanotechnology*, **17**(13), 3259.

22 Grewal, H.S., Piao, S., Cho, I.J., Jhang, K.Y., and Yoon, E.S. (2016) Nanotribological and wetting performance of hierarchical patterns. *Soft Matter*, **12**(3), 859–866.

23 Sun, M., Luo, C., Xu, L., Ji, H., Ouyang, Q., Yu, D., and Chen, Y. (2005) Artificial lotus leaf by nanocasting. *Langmuir*, **21**(19), 8978–8981.

24 Cassie, A.B.D. and Baxter, S. (1944) Wettability of porous surfaces. *Trans. Faraday Soc.*, **40**, 546–551.

25 Bhushan, B. and Her, E.K. (2010) Fabrication of superhydrophobic surfaces with high and low adhesion inspired from rose petal. *Langmuir*, **26**(11), 8207–8217.

26 Bico, J., Thiele, U., and Quéré, D. (2002) Wetting of textured surfaces. *Colloids Surf., A*, **206**(1), 41–46.

27 Feng, L., Zhang, Y., Xi, J., Zhu, Y., Wang, N., Xia, F., and Jiang, L. (2008) Petal effect: a superhydrophobic state with high adhesive force. *Langmuir*, **24**(8), 4114–4119.

28 Karaman, M., Çabuk, N., Özyurt, D., and Köysüren, Ö. (2012) Self-supporting superhydrophobic thin polymer sheets that mimic the nature's petal effect. *Appl. Surf. Sci.*, **259**, 542–546.

29 Guangming, G., Juntao, W., Yong, Z., Jingang, L., Xu, J., and Lei, J. (2014) A novel fluorinated polyimide surface with petal effect produced by electrospinning. *Soft Matter*, **10**(4), 549–552.

30 Liu, Y., Li, S., Zhang, J., Wang, Y., Han, Z., and Ren, L. (2014) Fabrication of biomimetic superhydrophobic surface with controlled adhesion by electrodeposition. *Chem. Eng. J.*, **248**, 440–447.

31 Xie, L., Tang, Z., Jiang, L., Breedveld, V., and Hess, D.W. (2015) Creation of superhydrophobic wood surfaces by plasma etching and thin-film deposition. *Surf. Coat. Technol.*, **281**, 125–132.

32 Xia, D., Johnson, L.M., and López, G.P. (2012) Anisotropic wetting surfaces with one-dimensional and directional structures: fabrication approaches, wetting properties and potential applications. *Adv. Mater.*, **24**(10), 1287–1302.

33 Wang, S., Liu, K., Yao, X., and Jiang, L. (2015) Bioinspired surfaces with superwettability: new insight on theory, design, and applications. *Chem. Rev.*, **115**(16), 8230–8293.
34 Feng, L., Li, S., Li, Y., Li, H., Zhang, L., Zhai, J., Song, Y., Liu, B., Jiang, L., and Zhu, D. (2002) Super-hydrophobic surfaces: from natural to artificial. *Adv. Mater.*, **14**(24), 1857–1860.
35 Yao, J., Wang, J., Yu, Y., Yang, H., and Xu, Y. (2012) Biomimetic fabrication and characterization of an artificial rice leaf surface with anisotropic wetting. *Chin. Sci. Bull.*, **57**(20), 2631–2634.
36 Yang, X., Song, J., Xu, W., Liu, X., Lu, Y., and Wang, Y. (2013) Anisotropic sliding of multiple-level biomimetic rice-leaf surfaces on aluminium substrates. *IET Micro Nano Lett.*, **8**(11), 801–804.
37 Zhu, D., Li, X., Zhang, G., Zhang, X., Zhang, X., Wang, T., and Yang, B. (2010) Mimicking the rice leaf from ordered binary structures to anisotropic wettability. *Langmuir*, **26**(17), 14276–14283.
38 Wu, S.Z., Wu, D., Yao, J., Chen, Q.D., Wang, J.N., Niu, L.G., Fang, H.H., and Sun, H.B. (2010) One-step preparation of regular micropearl arrays for two-direction controllable anisotropic wetting. *Langmuir*, **26**(14), 12012–12016.
39 Clasen, T., Pruss-Ustun, A., Mathers, C.D., Cumming, O., Cairncross, S., and Colford, J.M. (2014) Estimating the impact of unsafe water, sanitation and hygiene on the global burden of disease: evolving and alternative methods. *Trop. Med. Int. Health*, **19**(8), 884–893.
40 World Health Organization (2014) UN-water global analysis and assessment of sanitation and drinking-water, in *Investing in Water and Sanitation: Increasing Access, Reducing Inequalities*, World Health Organization, Geneva.
41 WHO/UNICEF Joint Water Supply, Sanitation Monitoring Programme, & World Health Organization (2014) *Progress on drinking water and sanitation: 2014 Update*, World Health Organization.
42 Parker, A.R. and Lawrence, C.R. (2001) Water capture by a desert beetle. *Nature*, **414**(6859), 33–34.
43 Zhai, L., Berg, M.C., Cebeci, F.C., Kim, Y., Milwid, J.M., Rubner, M.F., and Cohen, R.E. (2006) Patterned superhydrophobic surfaces: toward a synthetic mimic of the Namib Desert beetle. *Nano Lett.*, **6**(6), 1213–1217.
44 Garrod, R.P., Harris, L.G., Schofield, W.C.E., McGettrick, J., Ward, L.J., Teare, D.O.H., and Badyal, J.P.S. (2007) Mimicking a stenocara beetle's back for microcondensation using plasmachemical patterned superhydrophobic-superhydrophilic surfaces. *Langmuir*, **23**(2), 689–693.
45 Dorrer, C. and Rühe, J. (2008) Mimicking the stenocara beetle dewetting of drops from a patterned superhydrophobic surface. *Langmuir*, **24**(12), 6154–6158.
46 Ju, J., Bai, H., Zheng, Y., Zhao, T., Fang, R., and Jiang, L. (2012) A multi-structural and multi-functional integrated fog collection system in cactus. *Nat. Commun.*, **3**, 1247.
47 Lorenceau, É. and Quéré, D. (2004) Drops on a conical wire. *J. Fluid Mech.*, **510**, 29–45.

48 Bai, H., Tian, X., Zheng, Y., Ju, J., Zhao, Y., and Jiang, L. (2010) Direction controlled driving of tiny water drops on bioinspired artificial spider silks. *Adv. Mater.*, **22**(48), 5521–5525.

49 Cao, M., Ju, J., Li, K., Dou, S., Liu, K., and Jiang, L. (2014) Facile and large-scale fabrication of a cactus-inspired continuous fog collector. *Adv. Funct. Mater.*, **24**(21), 3235–3240.

50 Andrews, H.G., Eccles, E.A., Schofield, W.C.E., and Badyal, J.P.S. (2011) Three-dimensional hierarchical structures for fog harvesting. *Langmuir*, **27**(7), 3798–3802.

51 Phillips, M.B. and Jiang, P. (2013) in *Engineered Biomimicry* (eds A. Lakhtakia and R.J. Martin-Palma), Elsevier, pp. 305–331.

52 Parker, A.R. and Townley, H.E. (2007) Biomimetics of photonic nanostructures. *Nat. Nanotechnol.*, **2**(6), 347–353.

53 Southwell, W.H. (1991) Pyramid-array surface-relief structures producing antireflection index matching on optical surfaces. *J. Opt. Soc. Am. A*, **8**(3), 549–553.

54 Kwon, Y.W., Park, J., Kim, T., Kang, S.H., Kim, H., Shin, J., Jeon, S., and Hong, S.W. (2016) Flexible near-field nanopatterning with ultrathin, conformal phase masks on non-planar substrates for biomimetic hierarchical photonic structures. *ACS Nano*, **10**(4), 4609–4617.

55 Raut, H.K., Dinachali, S.S., Loke, Y.C., Ganesan, R., Ansah-Antwi, K.K., Góra, A., Khoo, E.H., Ganesh, V.A., Saifullah, M.S., and Ramakrishna, S. (2015) Multiscale ommatidial arrays with broadband and omnidirectional antireflection and antifogging properties by sacrificial layer mediated nanoimprinting. *ACS Nano*, **9**(2), 1305–1314.

56 Oh, S.S., Choi, C.G., and Kim, Y.S. (2010) Fabrication of micro-lens arrays with moth-eye antireflective nanostructures using thermal imprinting process. *Microelectron. Eng.*, **87**(11), 2328–2331.

57 Li, T., Lou, S., Ding, J., and Fan, T. (2014) Antireflective amorphous carbon nanocone arrays inspired from compound eyes. *Bioinspir. Biomim. Nan.*, **3**(1), 29–37.

58 Xu, H., Lu, N., Shi, G., Qi, D., Yang, B., Li, H., Xu, W., and Chi, L. (2011) Biomimetic antireflective hierarchical arrays. *Langmuir*, **27**(8), 4963–4967.

59 Zhu, C. and Gu, Z.Z. (2012) in *Bioinspiration and Biomimicry in Chemistry: Reverse-Engineering Nature* (ed. G. Swiegers), John Wiley & Sons, Inc., pp. 293–322.

60 Young, A.M. (1971) Wing coloration and reflectance in Morpho butterflies as related to reproductive behavior and escape from avian predators. *Oecologia*, **7**(3), 209–222.

61 Dushkina, N. and Lakhtakia, A. (2013) in *Engineered Biomimicry, (eds* A. Lakhtakia *and* R.J. Martin-Palma), Elsevier, pp. 267-303.

62 Sun, J., Bhushan, B., and Tong, J. (2013) Structural coloration in nature. *RSC Adv.*, **3**(35), 14862–14889.

63 Parker, A.R. (2000) 515 million years of structural colour. *J. Opt. A: Pure Appl. Opt.*, **2**(6), R15–R28.

64 Saito, A. (2005) in *Structural Colors in Biological Systems* (eds S. Kinoshita and S. Yoshioka), Osaka University Press, Osaka, p. 287.

65 Kinoshita, S. (ed.) (2013) *Bionanophotonics: an Introductory Textbook*, CRC Press.
66 Jiang, T., Peng, Z., Wu, W., Shi, T., and Liao, G. (2014) Gas sensing using hierarchical micro/nanostructures of Morpho butterfly scales. *Sens. Actuators, A*, **213**, 63–69.
67 Anderson, T.F. and Richards, A.G. Jr., (1942) An electron microscope study of some structural colors of insects. *J. Appl. Phys.*, **13**(12), 748–758.
68 Watanabe, K., Hoshino, T., Kanda, K., Haruyama, Y., and Matsui, S. (2005) Brilliant blue observation from a morpho- butterfly-scale quasi-structure. *Jpn. J. Appl. Phys.*, **44**, L48–L50.
69 Saito, A., Yoshioka, S., and Kinoshita, S. (2004) Reproduction of the *Morpho* butterfly's blue: arbitration of contradicting factors, in *Proceedings of SPIE (Optical Systems Degradation, Contamination, and Stray Light: Effects, Measurements, and Control)* (eds P.T.C. Chen, J.C. Fleming, and M.G. Dittman), SPIE, Bellingham, WA, pp. 188–194.
70 Cong, H. and Cao, W. (2004) Thin film interference of colloidal thin films. *Langmuir*, **20**(19), 8049–8053.
71 Eadie, L. and Ghosh, T.K. (2011) Biomimicry in textiles: past, present and potential. An overview. *J. R. Soc. Interface*, **8**(59), 761–775.
72 Bhushan, B. (ed.) (2012) *Encyclopedia of Nanotechnology*, Springer, Dordrecht, Netherlands.
73 Bhushan, B. (2012) Bioinspired structured surfaces. *Langmuir*, **28**(3), 1698–1714.
74 Chen, P.Y., McKittrick, J., and Meyers, M.A. (2012) Biological materials: functional adaptations and bioinspired designs. *Prog. Mater Sci.*, **57**(8), 1492–1704.
75 Bechert, D.W., Bruse, M., and Hage, W. (2000) Experiments with three-dimensional riblets as an idealized model of shark skin. *Exp. Fluids*, **28**(5), 403–412.
76 Liu, K. and Jiang, L. (2012) Bio-inspired self-cleaning surfaces. *Annu. Rev. Mater. Res.*, **42**, 231–263.
77 Lee, M. (2014) in *Remarkable Natural Material Surfaces and Their Engineering Potential* (ed. M. Lee), Springer International Publishing, pp. 15–27.
78 Bhushan, B. (2009) Biomimetics: lessons from nature–an overview. *Philos. Trans. R. Soc. London, Ser. A*, **367**(1893), 1445–1486.
79 Biomimicry Institute www.biomimicryinstitute.org (accessed 3 January 2016).
80 Carman, M.L., Estes, T.G., Feinberg, A.W., Schumacher, J.F., Wilkerson, W., Wilson, L.H., Callow, M.E., Callow, J.A., and Brennan, A.B. (2006) Engineered antifouling microtopo- graphies – correlating wettability with cell attachment. *Biofouling*, **22**(1), 11–21.
81 Wen, L., Weaver, J.C., and Lauder, G.V. (2014) Biomimetic shark skin: design, fabrication and hydrodynamic function. *J. Exp. Biol.*, **217**(10), 1656–1666.
82 Han, X., Zhang, D., Li, X., and Li, Y. (2008) Bio-replicated forming of the biomimetic drag-reducing surfaces in large area based on shark skin. *Chin. Sci. Bull.*, **53**(10), 1587–1592.

83 Chen, H., Rao, F., Shang, X., Zhang, D., and Hagiwara, I. (2013) Biomimetic drag reduction study on herringbone riblets of bird feather. *J. Bionic Eng.*, **10** (3), 341–349.

84 Smith, M.A., Jones, N.M.M. II, Page, S.L., and Dirda, M.P. (1984) Pressure-sensitive tape and techniques for its removal from paper. *J. Am. Inst. Conserv.*, **23**(2), 101–113, Article 3.

85 Qu, L., Li, Y., and Dai, L. (2012) Bioinspired surfaces I: gecko-foot mimetic adhesion, in *Bioinspiration and Biomimicry in Chemistry: Reverse-Engineering Nature* (ed. G. Swiegers), John Wiley and Sons, pp. 251–291.

86 Barnes, W.J.P. (2012) in *Encyclopedia of Nanotechnology* (ed. B. Bhushan), Springer, Dordrecht, Netherlands, pp. 70–83.

87 Eisner, T. and Aneshansley, D.J. (2000) Defense by foot adhesion in a beetle (*Hemisphaerota cyanea*). *Proc. Natl. Acad. Sci. U.S.A.*, **97**(12), 6568–6573.

88 Autumn, K. and Gravish, N. (2008) Gecko adhesion: evolutionary nanotechnology. *Philos. Trans. R. Soc. London, Ser. A*, **366**(1870), 1575–1590.

89 Autumn, K. (2006) How gecko toes stick the powerful, fantastic adhesive used by geckos is made of nanoscale hairs that engage tiny forces, inspiring envy among human imitators. *Am. Sci.*, **94**(2), 124–132.

90 Lee, M. (2014) in *Remarkable Natural Material Surfaces and Their Engineering Potential* (ed. M. Lee), Springer International Publishing, pp. 115–126.

91 Varenberg, M., Pugno, N.M., and Gorb, S.N. (2010) Spatulate structures in biological fibrillar adhesion. *Soft Matter*, **6**(14), 3269–3272.

92 Qu, L. and Dai, L. (2007) Gecko-foot-mimetic aligned single-walled carbon nanotube dry adhesives with unique electrical and thermal properties. *Adv. Mater.*, **19**(22), 3844–3849.

93 Cho, W.K. and Choi, I.S. (2008) Fabrication of hairy polymeric films inspired by geckos: wetting and high adhesion properties. *Adv. Funct. Mater.*, **18**(7), 1089–1096.

94 Davies, J., Haq, S., Hawke, T., and Sargent, J.P. (2009) A practical approach to the development of a synthetic Gecko tape. *Int. J. Adhes. Adhes.*, **29**(4), 380–390.

95 Kustandi, T.S., Samper, V.D., Yi, D.K., Ng, W.S., Neuzil, P., and Sun, W. (2007) Self-assembled nanoparticles based fabrication of Gecko foot-hair-inspired polymer nanofibers. *Adv. Funct. Mater.*, **17**(13), 2211–2218.

96 Geim, A.K., Dubonos, S.V., Grigorieva, I.V., Novoselov, K.S., Zhukov, A.A., and Shapoval, S.Y. (2003) Microfabricated adhesive mimicking gecko foot-hair. *Nat. Mater.*, **2**(7), 461–463.

97 Kim, T.I., Jeong, H.E., Suh, K.Y., and Lee, H.H. (2009) Stooped nanohairs: geometry-controllable, unidirectional, reversible, and robust gecko-like dry adhesive. *Adv. Mater.*, **21**(22), 2276–2281.

2

Chemical and Physical Modification of Surfaces

Mustafa Karaman, Mehmet Gürsoy, Mahmut Kuş, Faruk Özel, Esma Yenel, Özlem G. Şahin, and Hilal D. Kivrak

2.1 Introduction

Since the increasing demand for high-performance materials during the past century, the field of materials surface modification by various synthetic strategies has undergone an enormous expansion. Significant number of studies have been carried out regarding the development of efficient techniques for adding new functionalities to a material's surface as well as to understand the fundamental aspects of the various techniques. Surface modification processes can be categorized into two main parts: wet processes and dry processes. Wet surface modification techniques involve chemical and/or physical modification of a material's surface through the usage of various chemical agents that are either in liquid state during the processes or dissolved in a suitable solvent before their applications on the material's surface. Application of a wet chemical to a surface is easy and usually does not require special equipment. Spray coating and painting are mostly preferred for large areas, whereas dip coating, electroplating, spin coating, layer-by-layer deposition, solvent casting, or doctor blade techniques are mostly used for special purposes in small areas. Vapor deposition methods on the other hand, do not involve any wet material on a substrate surface. They have certain advantages over wet techniques. First, materials that are not compatible with the wet chemicals can be used as substrates. Besides, the transfer of a material in its vapor state is much easier than when it is in a liquid state and, therefore, very conformal and uniform coatings are possible with vapor deposition processes. Surface tension of a liquid usually prevents the transfer of liquid into the corrugated parts of the substrates, which decreases the coating conformality. Furthermore, a vapor deposition process usually occurs under vacuum or purged conditions at which only precursor or carrier molecules exist. Therefore, vapor deposited coatings are quite pure. The impurities introduced into the coatings by the solvent may lead to impure films that may not be suitable for many end uses. In the first part of this chapter, various kinds of vapor deposition processes used to deposit functional coatings onto various substrates are introduced. In the second part, some industrially important wet processes are introduced.

Surface Treatments for Biological, Chemical and Physical Applications, First Edition.
Edited by Mehmet Gürsoy and Mustafa Karaman.

2.2 Vapor Deposition Processes

Vapor deposition processes are divided into two major types:

- Physical vapor deposition (PVD)
- Chemical vapor deposition (CVD).

In PVD, vapors of materials are condensed onto the substrate, which is usually placed in a high vacuum environment. Vapors are generally produced by evaporation, sputtering, or ablation. Evaporation and subsequent condensation of the materials involve only physical changes. CVD, on the other hand, involves the creation of chemically active species in the vapor phase. The vapor phase chemical species are deposited onto the substrate surface with the help of chemical reactions that occur either in the vapor phase or on the substrate surface. Table 2.1 shows the classification of vapor deposition processes.

2.2.1 Physical Vapor Deposition

PVD uses physical processes to produce the vapor of a material, which is then condensed onto a substrate whose surface needs to be modified by a coating. In this way, a desired property can be implemented on a material's surface for a specific purpose. PVD thin films are used in a diverse range of application areas including optical, electronic, mechanical, or chemical [1–3]. Typical thicknesses of the films deposited by PVD can vary from a few nanometers to the micrometer scale. By using a multilayer coating approach, much thicker films or free-standing structures are also possible [4]. Deposition rates depend on many factors including evaporation source, chamber geometry, activation method, materials properties, and so on. PVD is now an industrially mature process, and its use is abundant especially in the production of semiconductor devices such as solar panels, metallized thin films for packaging industry, and wear resistant coatings for tool manufacture [5–7].

Compared to other deposition processes PVD offers many advantages:

- High deposition rates
- Good thickness control

Table 2.1 Classification of vapor deposition processes.

Physical vapor deposition (PVD)		**Chemical vapor deposition (CVD)**	
Basic PVD processes	Advanced PVD processes	Basic CVD processes	Advanced CVD processes
Evaporation deposition	Reactive evaporation	Atmospheric thermal CVD	Hot filament CVD
Sputter deposition	Reactive sputtering	Low-pressure CVD	Initiated CVD
Ion plating	Reactive ion plating	Plasma enhanced CVD	Photon induced CVD Oxidative CVD

- Tailored and enhanced film properties (high hardness, wear and corrosion resistance, low friction, specific optical, or electrical properties)
- Ability to use virtually any type of substrates including temperature-sensitive ones such as polymers
- Environmentally friendly than traditional wet coating processes such as electroplating and painting.

The drawbacks of the process are:

- Requires specific and costly equipment such as a vacuum chamber, pump, cooling system, and so on.
- It is generally restricted to a line-of-sight deposition.
- Some materials cannot be vaporized without change in their chemical structure.

2.2.1.1 Types of PVD Processes

Basic PVD processes fall into two main categories according to the mechanisms that are used to transfer a condensed phase into the vapor phase: evaporation and sputtering. Figure 2.1 shows the simplified schematic diagrams for the two basic PVD processes. Each process requires a vacuum chamber, and the processes are carried out under high vacuum conditions.

Evaporation The principle of evaporation which is sometimes called *vacuum evaporation* is relatively easy to understand. In this PVD mode, the temperature of the source (coating material) is raised above its boiling point under vacuum. Atoms or molecules generated after this thermal evaporation reaches to the substrate surface with little or no collision with each other during their journey between the source and substrate. The trajectory of the vaporized moieties is line-of-sight. Typical range of pressure is between 10^{-5} and 10^{-9} Torr depending on the level of impurities that can be tolerated in the as-deposited material [2]. The source material is heated with the help of basket, boat, spiral, coil, or loop

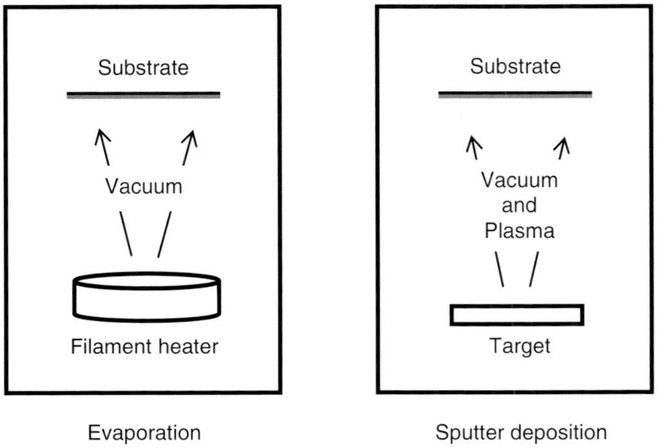

Figure 2.1 Basic PVD processing techniques: evaporation and sputtering.

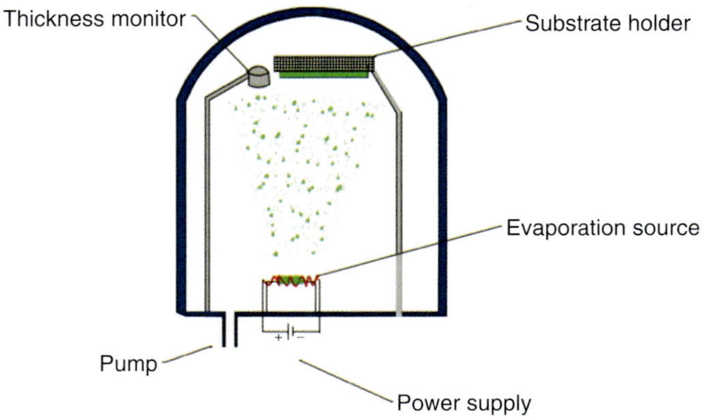

Figure 2.2 Schematic of an evaporation system.

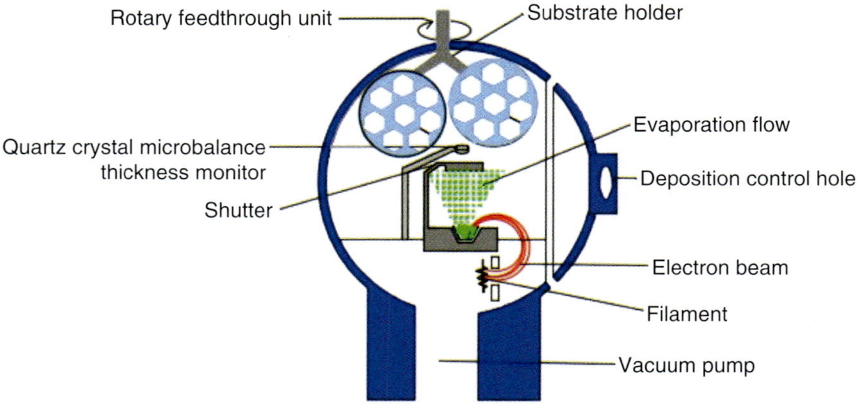

Figure 2.3 A typical e-beam evaporation system.

heaters. The heating material is usually tungsten that is resistively heated to a desired temperature using a power source. In order to protect the substrate from the radiative heat which is emitted by the hot source, the substrates are placed at a convenient distance away from the source. Vacuum evaporation is used to deposit a wide range of materials including electrically conductive coatings [8], optical coatings [9], decorative coatings, mirror coatings, barrier films on packaging materials, and corrosion resistant coatings [10] (Figure 2.2).

E-Beam Deposition E-beam deposition is a form of PVD in which the source is bombarded with an electron beam given off by a charged tungsten filament. The bombardment of the source surface by the high-energy e-beam causes atoms or molecules from the surface to be transferred into the vapor phase. A typical e-beam PVD system is composed of two main parts: an electron source that generates electrons and accelerates them as electron beam and a crucible in which the source material is placed (Figure 2.3).

Electron beams are deflected by 270° or 180° and irradiated onto the evaporation material. During this process, the deposition chamber is evacuated to pressures between 10^{-2} and 10^{-5} Pa to allow the passage of electrons from the electron source to the evaporation material. The process is highly efficient because the source is directly evaporated by the heat provided by the kinetic energy of the accelerated electrons. Metals and oxides that are hard to evaporate and substances that sublimate can be easily evaporated by the e-beam technique [11, 12]. By using several crucibles and evaporation sources, it is possible to make a multilayer film during a single processing run without breaking the vacuum [13].

Reactive Evaporation Some compounds such as refractory carbides, oxides, and nitrides have extremely high boiling points and they cannot be evaporated without a change in their chemical structure. In order to deposit such materials, a reactive evaporation process is used where at least one of the elements of the coating is introduced in the vapor phase. Usually, a metal is evaporated and the nonmetallic element is supplied in the gas phase. The reactive gases in the vapor phase are then reacted and deposited on a heated substrate surface. For example, titanium nitride films can be deposited by reactive evaporation in which titanium source is evaporated into a vacuum chamber in which a nitrogen-containing gas such as ammonia or air is present [14, 15]. Under suitable conditions titanium and nitrogen react to produce TiN which is an important coating material to harden and protect cutting and sliding surfaces for decorative purposes, and as a nontoxic exterior for medical implants.

Sputter Deposition Sputtering is the process in which a solid target is bombarded by energetic particles to eject atoms from the surface of the target. Ejected atoms are then condensed onto the substrate surface. Sputtering takes place only when the kinetic energy of the bombarding particles is high enough to break the chemical bonds that attach the surface atoms to the target. Therefore, energetic particles in the vapor phase are needed. Such energetic particles in the vapor phase are produced and accelerated by a high voltage that produces plasma. During the plasma bombardment of a material, sputtering can cause erosion of the surface and can thus be harmful. On the other hand, high-energy particles improve adhesion, film densification, and homogeneity. Besides, since the process is not thermally activated, temperature-sensitive surfaces can be coated (Figure 2.4).

The process requires high vacuum to remove all contaminant gases that could be ionized and condensed on the substrate. Usually, argon is used as the plasma gas at a pressure of around 1–75 mTorr. The pressure is a critical parameter that determines the deposition rate and film conformality [16].

Being an important and mature process of thin film formation, sputter deposition is used extensively in semiconductor and refractory coatings industries [17, 18]. Excellent coatings of metals and hard materials can be produced with good structural uniformity and adhesion on almost any kind of substrate. Several techniques used in sputtering are:

- Radio-frequency (RF) sputtering
  - Uses radio frequency above 50 kHz, insulators can be sputtered but deposition rate is usually low.

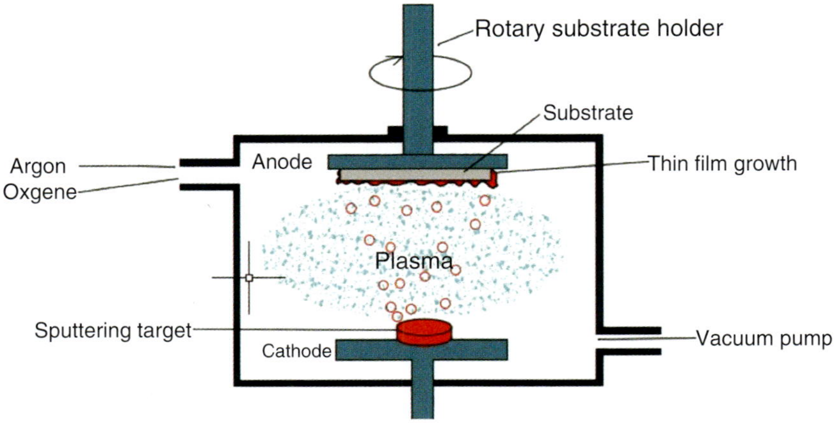

Figure 2.4 Schematic drawing of a typical sputtering system.

- DC diode sputtering
  - Uses DC current, requires an electrically conductive target, has low energy efficiency.
- Magnetron sputtering
  - Uses a magnetron to generate plasma. Deposition rates are high and process does not cause damage on the substrate surface.

Ion Plating In ion plating, the coating material is vaporized either by evaporation or sputtering of a target material, or it is directly fed as a vapor phase similar to CVD. Schematic diagrams for both vaporization techniques are given in Figure 2.5. The vapor is passed through a glow discharge where a partial ionization of the gases occurs (usually 1%). The process is also named as ion-beam-assisted deposition. The substrate is biased to a high negative potential at low pressure and it is bombarded by the high energy gas ions to sputter clean

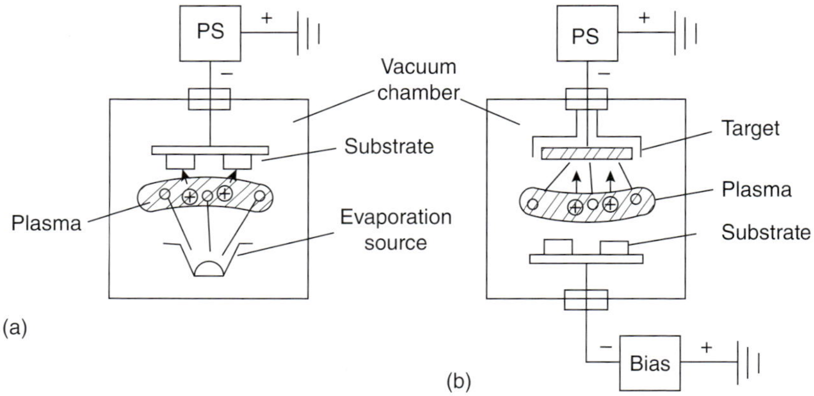

Figure 2.5 Schematic diagrams of ion plating, (a) evaporative ion plating and (b) sputter ion plating. (Randhawa 1991 [19]. Reproduced with permission of Elsevier.)

the surface. This stage is desirable for better adhesion and low-impurity content. The major drawbacks of the ion bombardment of the substrate during film deposition are heating of the substrate and decreased deposition rates because some of the deposited film is also sputtered off. When the working gas contains active gaseous species, the process is termed as *reactive ion plating*. Ion plating is commonly applied for metal films, oxides, nitrides, carbides, carbonitrides, alloy films, and multicomponent materials [19].

2.2.2 Chemical Vapor Deposition

CVD is a process of reacting gaseous species near or on a substrate surface to form a solid deposit. Manufacturing of thin film coatings, fibers and powders are possible with this versatile process. CVD is now an essential part in the manufacture of many high-tech components in the microelectronics, optical, optoelectronic, and aerospace industries, as well as in the manufacture of wear resistance parts and in the coating of tools and bearings [2, 20, 21]. CVD application is continuously increasing and the CVD market is expected to reach more than $10 billion by the end of 2020 [22]. Table 2.2 shows the various products produced by CVD and their application areas.

The earliest applications of CVD date back to the beginning of the eighteenth century, when CVD was used for refining or purification of metals, deposition of coatings for wear and corrosion, and in the production of self-standing components [20]. Although such CVD applications are still industrially important, there are many emerging applications now, especially after the requirements

Table 2.2 CVD products, production methods, and applications.

Method	Product	Application
Metallo-organic CVD	Titanium nitride	Diffusion barrier layers in integrated circuits
Plasma enhanced CVD	Diamond-like carbon	Wear resistance coatings
Low pressure CVD	Titanium carbide, titanium nitride	Tool coatings
Metallo-organic CVD	Iridium	High temperature corrosion resistance
Hot filamet CVD	Teflon	Hydrophobic finishes
Initiated CVD	Epoxides	Functionalizable coatings
LPCVD, APCVD, PECVD	Carbon nanotubes, graphene	Nanofillers in composites, high thermal and electrical conductivities
Metallo-organic CVD	Copper	Advanced semiconductor components
Atmospheric pressure CVD	Optical coatings	Glass industry
LPCVD, APCVD, PECVD	Boron carbide, boron nitride	Hard coatings
APCVD	Boron fibers	Composite additive in aerospace industry

for high-performance coatings in semiconductor and aerospace industries. As a result, many of the research efforts on CVD are now focused on coatings applications.

CVD is atomic and molecular in nature; that is, the materials being deposited are atoms and molecules. It has got many common features with PVD such as phase change (evaporation, condensation, sublimation, etc.), diffusion, heat and mass transfer, and so on. But as opposed to PVD, CVD involves chemical reactions during the depositions. The chemical bonds of the precursor molecules are broken and the resultant atoms or molecules recombine to produce solid materials. CVD is generally preferred when a desired coating cannot be implemented by just PVD because of the thermodynamics limitations [23]. Sometimes CVD is used in conjunction with PVD to produce materials that is otherwise not possible with just a single strategy. Summarized below are some advantages and limitations of the CVD processes as compared to PVD or solution-based techniques.

Advantages of CVD

- In most CVD processes, coatings are chemically bonded to the substrate surface. Therefore, there exists a strong adhesion between the substrate surface and the coating.
- Deposition rate is high and very thick coatings can be deposited in short times. Therefore, CVD is suitable for high-throughput industrial productions.
- CVD processes generally occur under moderate vacuum conditions. In many cases, atmospheric pressure CVD is also possible. Therefore, CVD does not require high or ultra-high vacuum hardware. The investment, operation, and maintenance costs of CVD are low when compared with PVD.
- CVD is a conformal process; that is, materials with complex geometries can be coated with high uniformity. However, PVD is restricted to a line-of-sight deposition.
- CVD is flexible such that coating composition, deposition rate, film conformality, and some other important parameters can be easily tuned through the selection of correct process parameters. For example, although a high deposition rate is preferred in most cases, a slow deposition rate may also be desired especially for very thin coatings and for accurate thickness control. By changing the precursor flowrates, substrate temperatures, pressure, amount, and type of energy supplied to the system can film deposition rates can be effectively controlled.

Limitations of CVD

- Not every substrate can withstand the high temperatures that are required for most of the CVD processes.
- The deposition temperatures can be lowered with the help of plasma, but in that case plasma-related damages may occur on the substrate surface.
- Chemical precursors as well as the by-products of CVD are usually volatile and toxic chemicals. Therefore, special safety, equipment, and exhaust management systems are needed.

Table 2.3 Comparison of CVD with some important solution phase methods.

Method	SAMs	Grafting	Spin coating	CVD
Use of solvent	Yes	Yes	Yes	No
Long-term stability	No	Yes	Yes	Yes
Conformality	High	High	Low	High
Real-time thickness control	No	No	No	Yes
Substrate independent	No	No	Yes	Yes
Deposition rate	Low	Low	High	High
Single-stage coating process	No	No	No	Yes

Table 2.3 gives the comparison of CVD with three important solution phase deposition methods namely spin coating, grafting, and self-assembly monolayers (SAMs). As is evident from the table, CVD has many advantages over solution phase methods. The most important advantage of CVD is its solvent-free nature; therefore, materials that are not compatible with the wet chemicals can be used as substrates in CVD [24]. Liquid-phase techniques for thin film synthesis, such as sol–gel, spin-coating, or dip-coating generally require the use of solvents that may have undesired effects on the substrates, such as the degradation of the underlying layers or alteration of the mechanical properties of the delicate substrates. Besides, many of the recent applications require conformal coatings over porous or patterned surfaces. CVD can achieve excellent coating uniformity without changing the substrate geometry [25, 26], which is not possible with solution phase techniques because of challenges arising from de-wetting and surface tension. Furthermore, CVD processes occur under vacuum or purged conditions where only precursor or carrier molecules exist. Therefore, CVD coatings are quite pure. The impurities introduced into the coatings by the solvent may lead to the formation of impure films that may not be suitable for many end uses.

2.2.2.1 CVD Reactors

Most generally, CVD reactors fall into two categories: hot wall reactors and cold wall reactors. Schematic representation of each reactor type is given in Figure 2.6. In hot wall CVD, the reactor is surrounded by heating elements. The substrates are placed into the reactor, the reactor is heated to the desired temperature, and then the precursor gases are introduced. Sometimes, pre-heating is accompanied with an inert purge gas flow or under vacuum in order to remove the impurities that may arise in the system after outgassing of some condensed materials on the substrate surface as a result of heating. The deposition temperatures are usually very high, in certain cases up to 1800 °C, limited only by the materials used for the construction of chamber and heating elements [27]. The systems run often under vacuum, on the order of a few hundred millitorrs to tens of torrs. Such thermal CVD systems that run under reduced pressures are called low pressure chemical vapor deposition (LPCVD) systems. A large number of substrates can

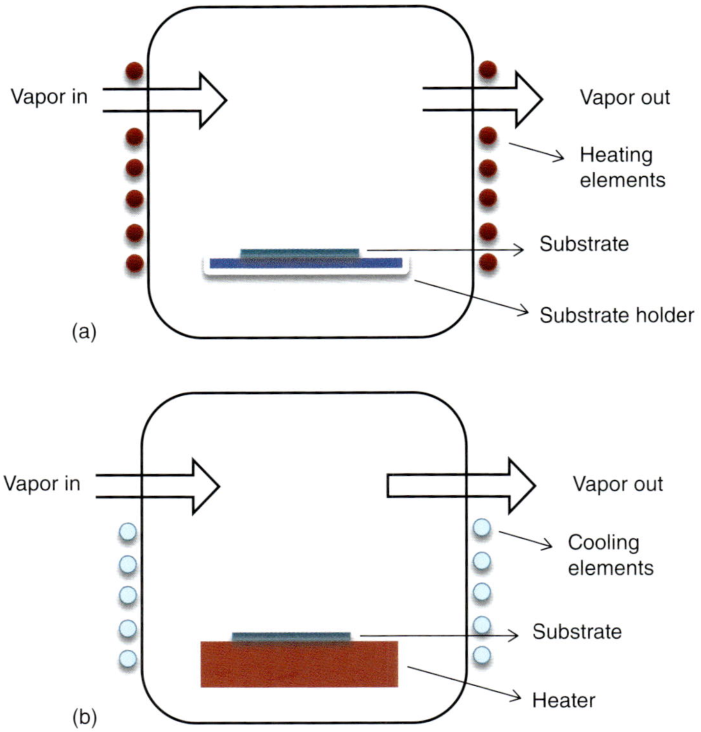

Figure 2.6 (a) A hot wall CVD reactor and (b) a cold wall CVD reactor.

be placed in a hot wall system with the help of shelves or substrate holders. After the system reaches the thermal equilibrium, the uniform substrate temperatures are achieved and thus uniform coating thickness is possible on many substrates at once. The major drawback, on the other hand, is that not only the substrate but also the whole reactor is heated and, therefore, the chamber walls are also coated, requiring frequent cleaning.

The other important class of CVD reactors is the cold wall reactors. In such systems only the substrates are heated. The heat dissipated from the substrate heaters may also reach the chamber walls with a heat transfer process (radiation, conduction, convection, or their combination) that may cause a temperature rise in the chamber walls. In order to prevent temperature rise in the walls, the walls usually contain circulation tubes or jackets through which a cooling liquid is passed during the depositions [20, 23]. Most compound semiconductor CVD processes use reactors of this type. The most apparent advantage of using a cold wall CVD system is the reduced depositions of materials on the chamber walls. Therefore, less cleaning is needed as compared with the hot wall reactors. The primary disadvantage is large temperature variations at the substrate surface, which may lead to nonuniformities of the thickness and structure of the as-deposited films. Besides, a cold wall CVD system is not suitable for large-scale depositions.

CVD reactions that are carried out in a cold wall system usually require higher pressures, from several hundred torrs to atmospheric pressure. The advantage

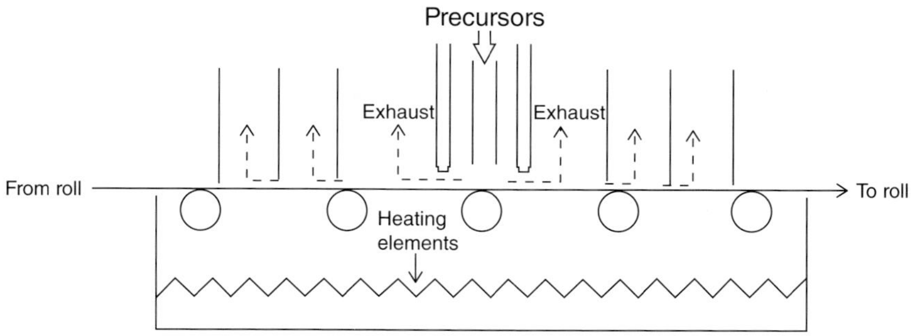

Figure 2.7 Schematic diagram of roll-to-roll atmospheric pressure CVD system.

of an atmospheric system is that the process is suitable for the roll-to-roll manufacture to produce high throughput products. The substrates are either placed on a moving belt or the belt can be directly the substrate itself; that is, a textile fabric or polymeric sheet. Figure 2.7 shows such a system, where the substrate is moved continuously over a heater and the precursor vapors are injected over the substrates. Such a system is essentially open to atmosphere and it avoids costly vacuum equipment. The major drawbacks are relatively higher gas consumptions, thickness nonuniformities, and low efficiency of precursor use.

2.2.2.2 Basic Principles of CVD: Thermodynamics, Chemistry, Heat, and Mass Transfer

In a classical CVD process, chemical reactions occur on or near a heated surface, resulting in the deposition of a thin film on that surface. Chemical reactions result in solid deposits and vapor phase by-products together with the unreacted precursors are exhausted out of the reactor. There are typically four kinds of reactions occurring in a CVD system [2];

- Thermal decomposition of a gaseous precursor.

$$AB(g) \rightarrow A(s) + B(g)$$

- Chemical reduction of a gaseous precursor by a reducing agent

$$AB(g) + {}^{1}\!/_{2}H_2(g) \rightarrow A(s) + HB(g)$$

- Exchange reactions

$$AB(g) + C(g) \rightarrow AC(s) + B(g)$$

- Disproportionation reactions

$$2AB_2(g) \rightarrow A(s) + AB_4(g)$$

In most cases, CVD is a complex process in which several types of reactions occur simultaneously. For example, for the formation of boron carbide from precursor's boron trichloride, methane, and hydrogen gases the mechanism shown in Table 2.4 was proposed [28]. This complex mechanism involves molecular

Table 2.4 Proposed mechanism for the CVD production of boron carbide [28].

$BCl_3(g) + s \rightarrow BCl_3s$
$H_2(g) + 2\,s \rightarrow 2Hs$
$CH_4(g) + 2\,s \rightarrow CH_3s + Hs$
$BCl_3s + CH_3s \rightarrow BCs + 3HCl(g) + s$
$BCs + BCl_3s + H_2(g) \rightarrow B_2Cs + 2HCl(g) + Cls$
$B_2Cs + BCl_3s + H_2(g) \rightarrow B_3Cs + 2HCl(g) + Cls$
$B_3Cs + BCl_3s + H_2(g) \rightarrow B_4C + 2HCl(g) + Cls$
$Hs + Cls \rightarrow HCl(g) + 2s$
$BCl_3(g) + H_2(g) \rightarrow BHCl_2(g) + HCl(g)$
$BCl_3(g) + CH_4(g) + H_2(g) \rightarrow B_4C(s) + BHCl_2(g) + HCl(g)$

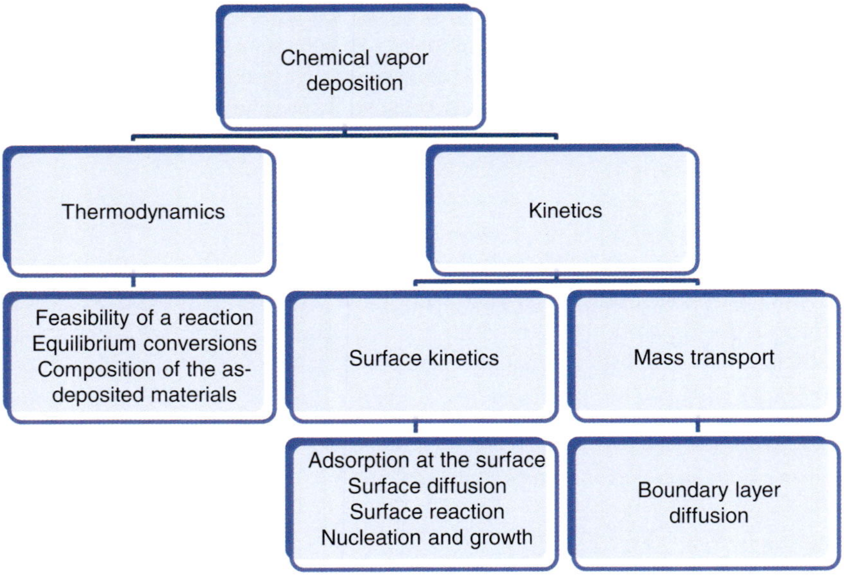

Figure 2.8 Fundamental aspects of CVD [23].

adsorption of boron trichloride and dissociative adsorption of methane and hydrogen on the substrate surface. The adsorbed intermediates, together with the gas phase hydrogen, are reacted in successive series reactions to produce finally a stoichiometric B_4C phase.

The reactions that yield solid deposits can be activated mostly by three different ways; thermal activation, plasma activation, and photon activation. Overall, CVD is a complex process and in order to fully understand the phenomena occurring during CVD requires knowledge from several disciplines [20]. Basically, a CVD reaction is governed by two fundamental subjects: thermodynamics and kinetics. While the former gives information on the feasibility of a CVD reaction, the latter deals with the speed at which the reactions occur (Figure 2.8).

Thermodynamics As with any other type of chemical reactions, CVD reactions are governed by the laws of thermodynamics, which is a valuable tool to predict the direction of a reaction. There is infinite possibility of arrangements of chemical bonds between the elements to form molecules in CVD. The purpose of a thermodynamic analysis to CVD reactions is to determine the extent to which the products are favored given the specific conditions, such as temperature and pressure. Through thermodynamic calculations we can determine how far a chemical reaction can proceed. Hence, if the thermodynamic analysis tells us that a given reaction will not precede to a significant degree, we do not need to consider it further. This information is very valuable for CVD because otherwise we would need to prove that by conducting costly and time-consuming experiments. Besides, through equilibrium calculations that are based on thermodynamic rules, one can determine the composition of a condensed phase [20, 23]. The fundamental question of a thermodynamics analysis is that: "What is the effect of temperature, pressure, and composition on the equilibrium conversion in a chemically reactive system?" [29]. To answer that fundamental question Gibbs free energy (G) analysis of a system is used. A chemical reaction is possible only if the thermodynamics is favorable; that is, the free energy change of the reaction (ΔG) is negative. The chemical equilibrium condition for a reaction at constant temperature and pressure is that G be a minimum [30]:

$$\left(\frac{\partial G}{\partial X}\right)_{T,P} = 0$$

The intermediate value of the reaction extent X for which the total Gibbs energy of the system is a minimum is denoted X^*, which is the equilibrium extent of reaction for a system (Figure 2.9).

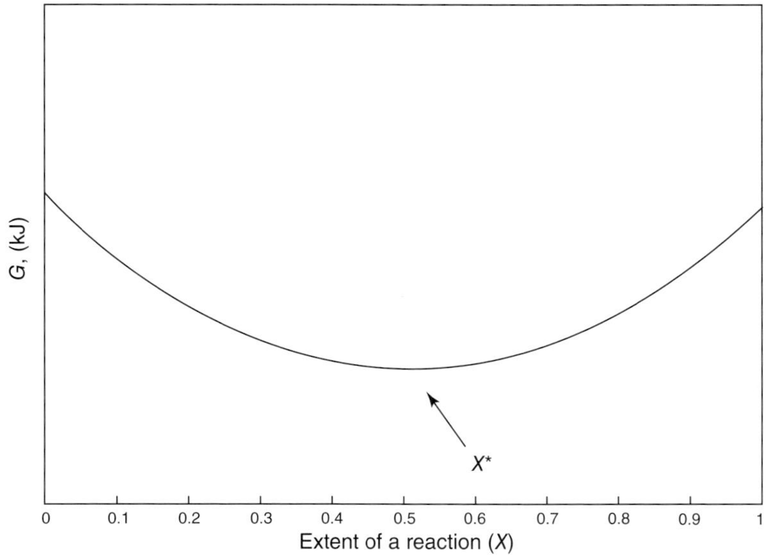

Figure 2.9 Change of Gibb's free energy during a chemical reaction.

In order to calculate the free energy change during a chemical reaction, one needs to find the standard free energies of formation (ΔG_f^0) for each component that takes part in the reaction. The free energy of formation of a component is a function of temperature and pressure.

$$\Delta G_{rxn}^0 = \Sigma(V_i \times \Delta G_f^0)$$

where V_i is the stoichiometric coefficients of reactants and products.

The free energy change for a reaction at equilibrium is zero, hence:

$\Delta G_{rxn}^0 = -RT \ln K_a$ where K_a is the equilibrium constant in terms of activities (a_i).

$$K_a(T) = \Pi_{i=1}^n a_i^{V_i}$$

This equation can be used to predict the chemical equilibrium state. It is possible to calculate the equilibrium constant provided that the free energy of formation values for products and reactants are known. Some computer programs can be used to find the set of conditions that satisfy the basic rules of thermodynamics, namely minimization of Gibbs free energy at equilibrium. Solgasmix [31] and Fact [32] are good examples for such computer-based programs that are widely used in equilibrium calculations in CVD systems. In order to run such programs, the required thermophysical properties should be entered accurately. Such thermodynamic properties can be found in many references, including Janaf tables [33].

Kinetics The purpose of using thermodynamic analysis for chemical reactions is to determine the extent to which products are favored given specified elemental composition, temperature, and pressure. However, thermodynamics is not concerned about how fast a reaction will proceed. To answer that question one must consider the reaction kinetics. The sequence of events occurring during a classical CVD process is given below:

- Reactant gases are fed to the reactor by forced flow
- Gas diffusion occurs through the boundary layer
- Gases come in contact with the substrate surface
- Deposition reactions take place on the surface of the substrate
- By-products in the vapor phase are diffused away from the substrate surface.

These stages occur in a stepwise manner, and the slowest step determines the deposition rate. Process parameters determine the rate-limiting step. The most important rate-limiting steps in the CVD process are mass transport control and surface kinetics control [34]. Deposition mechanism is also dependent on the rate-limiting step that is the main factor that affects the deposition rates as well as the structure of the as-deposited material. For example, the mechanism for which the rate-limiting step is surface kinetics, usually results in more uniform films.

If a CVD reaction is controlled by surface kinetics, the deposition rate is highly affected by the flowrates of the reactant gases. This usually happens when both temperature and pressure are low (Figure 2.10a). Low pressure results in a thin boundary layer, through which mass transfer by diffusion is fast. Besides, at low

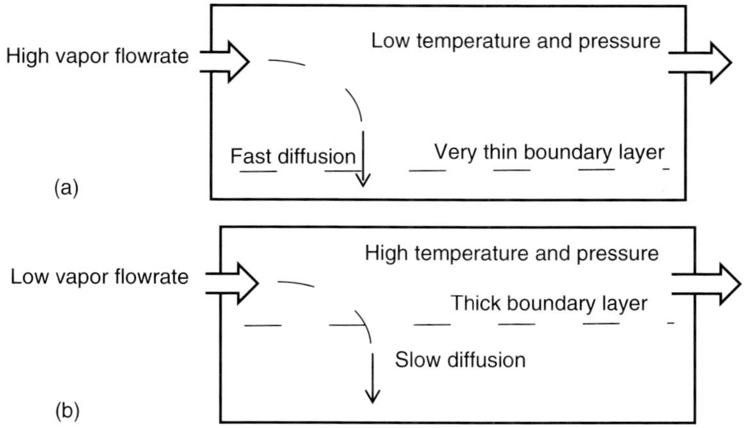

Figure 2.10 Rate-limiting steps in CVD. (a) Surface kinetics control and (b) mass-transport control.

temperature, the rate of surface reaction is low. Hence, the slowest step is the surface reaction. When pressure and temperature are higher, the rate-controlling factor is the diffusion rate of reactants and products through the thicker boundary layer (Figure 2.10b). The surface reactions occur more rapidly because the temperature is high and any molecule that interacts with the surface is consumed readily. The diffusion then becomes the rate-controlling factor. Therefore, it is possible to tune the rate-limiting step by changing the temperature, pressure, and vapor flowrate.

2.2.2.3 Various Types of CVD

Depending on the reactor type (hot-wall or cold wall reactors), pressure (low, atmospheric, or high pressures), precursor type (metallo-organic, polymer, etc.), and the type of the energy input (thermal, plasma, ion beam, UV assisted, etc.) CVD processes are categorized into different types. For a particular type of CVD process, the deposition mechanism is quite different from the other types; hence, the deposition rates, structure, and morphological properties of the as-deposited materials are also very different. Typical temperature range for classical thermal CVD is 150–1800 °C. However, with the help of enhanced CVD processes, which involve the use of plasmas, ions, photons, lasers, hot filaments, or special precursor, deposition reactions can be carried out at much lower temperatures. Among the many types of CVD processes the following are of major interest for both industrial and research practices.

Conventional (Thermal) CVD Conventional CVD technique is referred as just *CVD* in many cases. In this technique, the gaseous precursors are fed to a chamber that is held at atmospheric pressure and activated thermally in the vicinity of a hot substrate. The process is usually carried out in a tubular furnace in which a ceramic or quartz reactor chamber is placed. This type of a system has been extensively used to produce carbon nanotubes [35, 36]. During CVD, metal catalyst particles, most commonly nickel, cobalt, and iron are placed in the

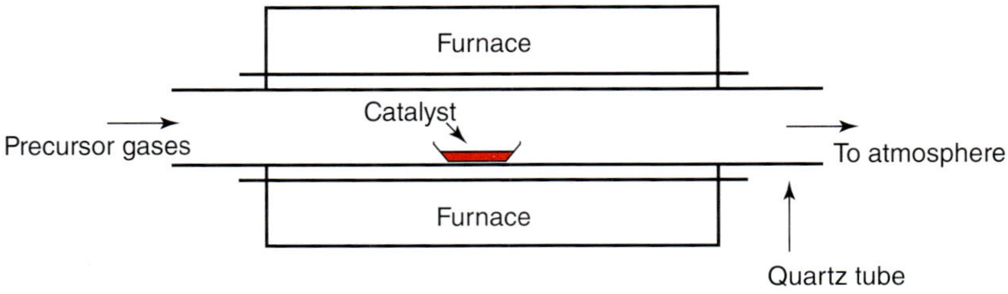

Figure 2.11 A conventional CVD system for the production of carbon nanotubes.

system. The diameters of the nanotubes that are to be grown are dependent on the size of the metal catalyst particles. The deposition temperature is approximately 600 °C. A carbon-containing gas (such as methane or acetylene) and a process gas (such as hydrogen or nitrogen) are fed into the reactor. During CVD, the carbon-containing gas is dissociated at the catalyst surface, where the nanotubes are formed (Figure 2.11).

Low-Pressure CVD As its name implies, LPCVD is applied under vacuum conditions. This technology is widely used in semiconductor processing because of some critical advantages over classical CVD techniques. First, vacuum conditions provide a very clean environment that is almost free from impurities which may cause problems for high-quality semiconductor devices. Second, by operating at a reduced pressure, the diffusivity of the gaseous species is very high, thus the species reach the substrate surface without a significant resistance. For example, decreasing the pressure from one atmosphere to 1 Torr increases the diffusibility by a factor of 1000, which means that the mass transfer cannot be the rate-limiting step. That results in a surface kinetics controlled deposition, which provides very uniform coatings. Besides, the mean free path is significantly increased at low pressures, thus a very large number of materials can be coated uniformly in a single run.

In recent years, LPCVD was successfully used to prepare graphene or graphene oxide films over a metallic catalyst surface [37, 38]. For the synthesis of graphene layers with lower impurities and residual oxygen, the base pressure of the reactor must be lowered. For example, the controlled growth of large-grain single-crystalline graphene with grain size up to 100 μm has been reported [39]. Depending on the vacuum level, controlled growth of graphene flowers with four to six lobes were achieved. LPCVD is also widely used to deposit Si, Si_3N_4, and SiO_2 thin films over heated substrates [40–42].

Plasma Enhanced CVD As stated before, there are many types of chemical reactions occurring in a CVD system. In order to start a chemical reaction, classical CVD depends on thermal energy to supply the required activation energy. In plasma-enhanced chemical vapor deposition (PECVD), also known as *plasma-assisted chemical vapor deposition* (PACVD) the reactions are activated by cold plasma. Therefore, deposition temperatures are substantially lower as compared with the classical CVD.

Plasma is one of the four fundamental states of matter, the others being solid, liquid, and gas. Plasma can be created by heating a gas above a certain temperature or subjecting it to a strong electromagnetic field. Under the influence of heat or an electromagnetic field, gas molecules are dissociated and eventually ionized. Plasma consists of positively charged ions, negatively charged electrons, and neutral atoms [43]. Properties of the resulting plasma are quite different from those of the initial gas. Ionizing a gas using an electric discharge is the most preferred way of creating plasma because the thermally created plasmas usually require extremely high temperatures, up to 5000 K. First use of PECVD dates back to the 1960s, when silicon oxide and nitride thin films were deposited for semiconductor applications [44]. Today, the process is being used for an extensive range of applications.

For PECVD, two types of plasma are mostly used; glow-discharge plasma and arc-discharge plasma. Glow-discharge plasma is generated by a high-frequency electric field at reduced pressures. The most used frequencies for CVD are 2.45 GHz micro-wave and, 13.56 MHz radio frequencies. Once the gas is ionized, the high-energy electrons collide with the gas molecules, which results in dissociation and generation of reactive species in the gas phase. Typical equipment for a glow-discharge PECVD system consists of microwave (MW)- or RF-generators, a vacuum reactor, and conductive electrodes. Figure 2.12 shows a schematic representation of an RF PECVD system that is used to deposit polymeric coatings for low surface energy finishes over various substrates [45].

Arc plasma operates at lower frequency, at which both positively and negatively charged particles acquire energy and their temperatures are almost equally increased. The response of particles is slow, which allows for thermal equilibrium, so such plasmas are often called *isothermal*. Compared with the glow discharge

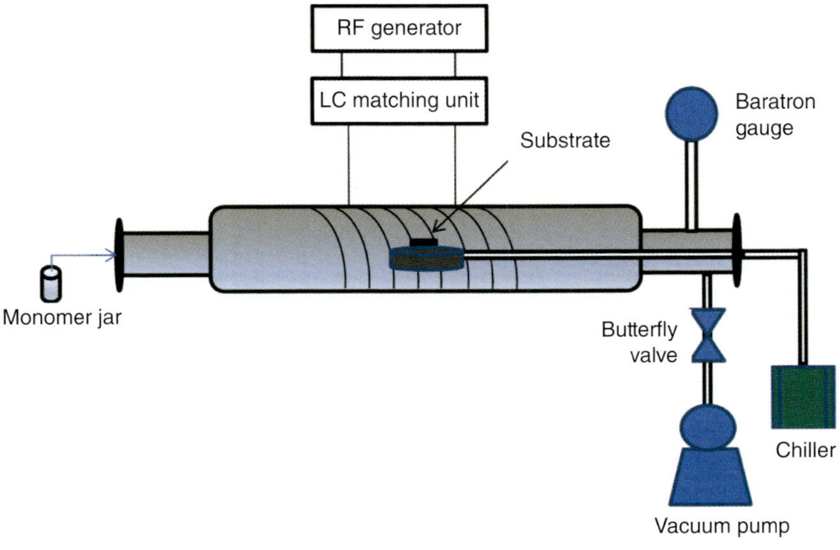

Figure 2.12 An RF PECVD system for polymer deposition. (Karaman 2015 [45]. Reproduced with permission of Wiley.)

plasma, arc plasma is carried out at relatively high pressures – up to atmospheric pressure; therefore, the mean free part is very low. As a result, collisions of species in plasma are very frequent and particles heat quickly. Arc plasma is extremely hot and requires a lot of energy. Therefore, it is not preferred for most of the practical applications.

One of the most important advantages of using glow-discharge PECVD is its ability to lower the deposition temperatures. Films that can be deposited at substantially high temperatures (100–1300 °C) using conventional CVD can be deposited by PECVD at much lower temperatures (200–500 °C) [46, 47]. Since the process is usually carried out under vacuum, the deposition mechanism is controlled by surface kinetics, which improves film uniformity over complex substrates.

Hot-Wire CVD Hot-wire chemical vapor deposition (HWCVD) is a type of CVD in which the activation of the gas phase reactions is achieved via heated filaments. The filament or filament array (tungsten) is placed close to the substrate and it is heated to very high temperatures. HWCVD is an established CVD technique to deposit diamond films [48]. Very conformal films can be deposited by this low-pressure method on rigid or flexible substrates. The technique is plasma-free and the substrate temperature is very low; hence plasma or temperature-related damages on the substrate surface are minimized [49].

2.2.2.4 Chemical Vapor Deposition of Polymeric Thin Films

Polymer thin films are desired for many practical applications because of their flexible chemical and mechanical properties. A large number of polymers that have specific functional groups can be deposited by CVD on almost any kind of substrate surface. Polymer deposition by CVD is usually carried out on an unheated substrate and, therefore, temperature-sensitive materials can also be coated. Especially in the last decades, polymer thin film deposition using CVD has become very popular. Among many other polymer synthesis methods, CVD has unique advantages and has been used for many applications such as in protective coatings, barrier layers, corrosion resistant films, optical devices, microelectronics, and biomedicals. The required energy to start the polymerization reactions is supplied in various ways, such as plasma, heated wire, or laser. The reactions are usually carried out in cold wall reactors.

Plasma Polymerization (PECVD) Plasma polymerization is the formation of polymeric materials under the influence of a partially ionized gas. Plasma polymerization is a specific type of plasma chemistry and involves reactions between plasma species, between plasma and surface species, and between surface species [50]. As a result of such reactions, materials in the form of a thin film or as a powder can be deposited if a suitable vapor is fed to the reactor. The precursors for plasma polymerization are usually organic or organometallic monomers, but some reactants that do not have conventionally polymerizable groups such as methane or ethane may also be used [51]. Today, plasma polymerization is an important process in the formation of entirely new kinds of functional films. The properties of as-deposited materials are considerably different from the conventionally synthesized polymers.

PECVD of inorganic thin films usually involves deposition on a heated substrate, which decreases the deposition rate and allows for extended contact between the plasma and surface. During the long contact between plasma and surface, one of the roles of plasma is to remove organic content and hydrogen in the film, leaving mostly the inorganic materials behind. The deposition of TiO_2 or SiO_2 are examples; in both, an organometallic precursor is fed to the reactor and in the end the organic part and hydrogen of the precursors are eliminated to leave only metal oxides [52]. However, in an organic synthesis, the interaction of plasma with the reactants and surface should be at a minimum in order to preserve the chemical structure of the organic molecules [53].

If a desired polymeric coating is desired to be placed on a substrate surface, the following conventional procedure is followed:

1) Synthesis of monomer
2) Polymerization of the monomer
3) Dissolving the polymer in a suitable solvent
4) Application of the coating solution on the substrate surface
5) Removal of the solvent
6) Curing of the coating.

This stepwise procedure is time consuming and the solvent can be trapped in the as-deposited polymer. Besides, this lengthy procedure increases the risk of contamination from other impurity sources. Plasma polymerization, on the other hand, is carried out in just one step that starts from a simple gas or monomer [54]. Chemical reactions that occur under plasma conditions are very complex and such reactions are not observed in any other types of chemically reactive systems. Therefore, these complex reactions usually result in unique polymers that cannot be produced by conventional techniques. Plasma polymers are often highly branched and cross-linked, which are desired properties for many end applications, such as diffusion barriers and passivation coatings.

Atom transfer radical polymerization (ATRP) [55] and spin coating [56] are widely used wet techniques to obtain polymeric thin films. Plasma polymerization offers many advantages over such wet processes, due to its all-dry and conformal nature [57]. However, it is not straightforward to obtain a deposit of pure material in plasma polymerization. In many cases, functional group retention is low due to extensive monomer fragmentation and substrate ion bombardment [58, 59]. Carrying out depositions at low plasma power or pulsing the plasma may increase the functional group retention to some extent, but such conditions usually result in low deposition rates [60]. Recently, a new technique called initiated plasma-enhanced chemical vapor deposition (iPECVD) has been developed as an alternative way to improve the functional group retention. The use of the initiator together with the monomer vapor allowed plasma depositions to be carried out at low power inputs [61, 62]. As an example, Gürsoy *et al.* applied iPECVD in a planar system (Figure 2.13) to deposit functional poly(hydroxyethyl methacrylate) (PHEMA) thin films at power values as low as

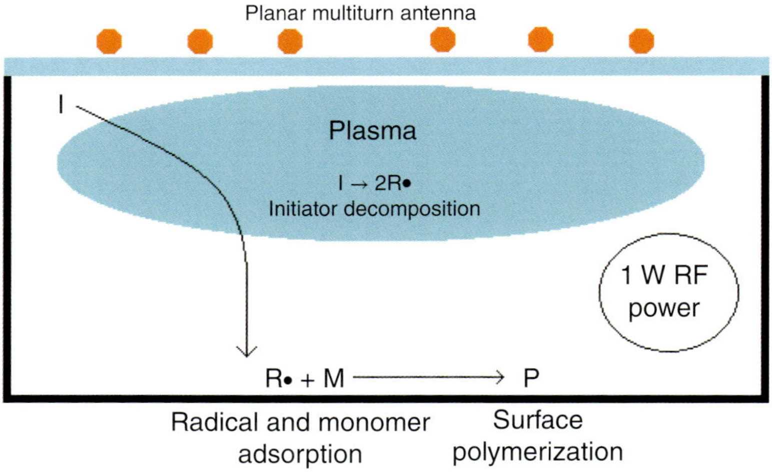

Figure 2.13 An iPECVD system to deposit PHEMA thin films at 1 W plasma power. (Gürsoy 2016 [63]. Reproduced with permission of Wiley.)

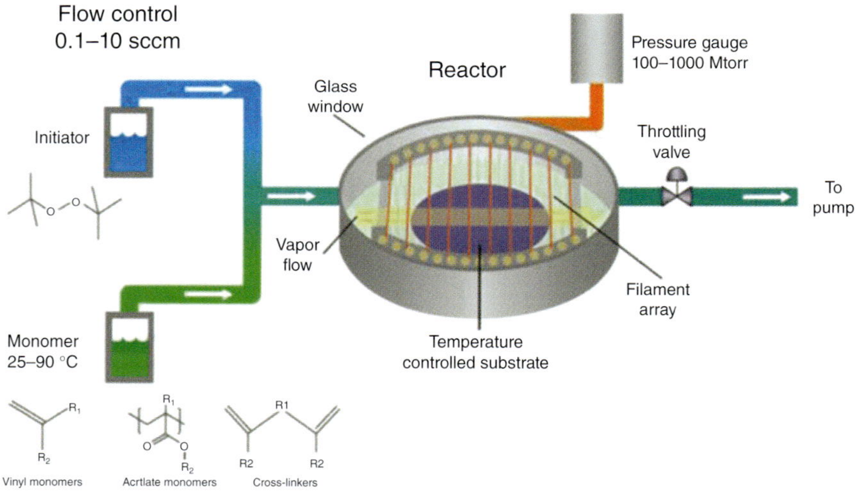

Figure 2.14 Schematic diagram of iCVD process. (Reeja-Jayan 2014 [64]. Reproduced with permission of Wiley.)

1 W [63]. With the help of initiator, structural group retention was almost 100% while keeping the deposition rates very high.

Initiated CVD Initiated CVD (iCVD) is a low-energy process to produce polymeric thin films from monomer and initiator species. Monomers that contain polymerizable vinyl bonds have been successfully used in iCVD processes [64]. The iCVD technique falls within the broader class of hot-wire CVD process. In both iCVD and HWCVD techniques, the energy required for the reaction is provided by the heated filaments (Figure 2.14). The usage of initiator in iCVD is the major distinction between iCVD and HWCVD [65]. Complex geometry substrates can be coated by low-pressure iCVD conformally, without solvent damages on the substrate surface, which are observed in conventional wet processes [66]. The use of initiator species allows polymerization to be carried out at low filament temperatures. Therefore, the substrate to be coated remains free from the high temperatures, plasma, or light sources, which can alter the chemical and/or physical nature of the substrates as well as the polymer being deposited [67]. Hence, very high retention of the functional groups of the monomers is observed [68, 69]. Besides, flowing two or more monomers simultaneously into the iCVD reactor creates copolymer or terpolymer thin films [70]. PECVD process is known for producing highly cross-linked polymeric films. In iCVD, on the other hand, the degree of cross-linking can be controlled easily by employing divinyl or trivinyl monomeric species [71].

The mechanism of polymerization in iCVD is considered to be the analog of the mechanism of solution-phase free radical polymerization. In iCVD, chain growth monomers (acrylates, methacrylates, styrenes, etc.) are vaporized first, and delivered into a cold wall vacuum chamber in the vapor phase. The initiating species (e.g., organic peroxides) are fed simultaneously with the monomer vapors. iCVD reactors contain cooling stages on which the substrates are placed. Once the monomer vapors are fed into the chamber, the monomer molecules preferentially accumulate on the cold substrate surface and eventually they are adsorbed. Initiator molecules, on the other hand, have high vapor pressures, and their adsorption on any surface is not quite possible. The rate of polymerization depends on the amount of adsorbed monomer onto the substrate surface. The concentration of monomer on the substrate surface depends on various parameters; substrate temperature, reactor pressure, and monomer/initiator flow ratio. For example, maintaining the substrate at low temperature leads to high deposition rates because at low substrate temperatures the adsorption of monomer is promoted. However, care must be taken to ensure that the substrate temperature is not too low, which might cause condensation. Usually, the dimensionless ratio between monomer partial pressure (P_{m}) and saturation pressure of the monomer at the substrate temperature (P_{sat}) is defined to relate the surface concentration of monomer at a given temperature [72]. The preferred range of $P_{\mathrm{m}}/P_{\mathrm{sat}}$ ratio is 0.3–0.7 for most iCVD processes. Operating within this range promotes uniform film growth without leading to undesirable liquid-phase condensation. As with any other type of CVD processes there is a trade-off between the film uniformity and the deposition rates, and in iCVD, $P_{\mathrm{m}}/P_{\mathrm{sat}}$ value is a valuable tool to tune the deposition rates and uniformity.

Control over the deposition rate, structure, and conformality of iCVD polymers requires a clear understanding of the complicated reaction mechanisms. Figure 2.15 depicts an iCVD deposition mechanism in a classical vacuum reactor (Figure 2.15). Monomer (M) and initiator (I_2) are fed simultaneously into the reactor. In iCVD, filament temperature is kept between 180 and 350 °C, which is not high enough to activate a monomeric species. Therefore, without an initiator, it is not possible to deposit a reasonable amount of material on a substrate surface. The initiator species, on the other hand, reacts on or near the hot filaments to form free-radical initiating species, which diffuse to the substrate surface and

(1) Initiation

$I_2 \rightarrow 2I^*$

$I^* + M(\text{ad}) \rightarrow M_1^*(\text{ad})$

(2) Propagation

$M^*(\text{ad}) + M_n^*(\text{ad}) \rightarrow M_{(n+1)}^*(\text{ad})$

(3) Termination

$M_n^*(\text{ad}) + M_m^*(\text{ad}) \rightarrow M_{n+m}(\text{ad})$

(4) Primary radical termination

$M_n^*(\text{ad}) + I^* \rightarrow M_n(\text{ad})$

Figure 2.15 The plausible reaction mechanism during iCVD polymerization (M: monomer, I: Initiator).

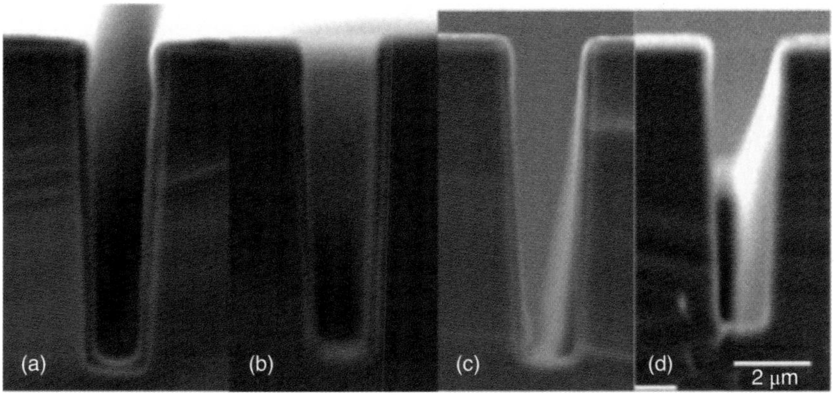

Figure 2.16 Scanning electron microscopy (SEM) images of iCVD deposited poly(cyclohexyl methacrylate) on microtrenched silicon wafer. (a) $P_m/P_{sat} = 0.11$, (b) $P_m/P_{sat} = 0.19$, (c) $P_m/P_{sat} = 0.36$, and (d) $P_m/P_{sat} = 0.72$. (Baxamusa 2008 [73]. Reproduced with permission of Wiley.)

react with the adsorbed monomer to initiate polymerization. The polymerization proceeds in a manner analogous to solution phase bulk polymerization. The film will continue to grow as monomer and initiator are fed to the reactor.

In iCVD polymerization, monomers arrive at the substrate surface through vapor phase diffusion. Unlike wet processes, iCVD does not suffer from surface tension effects, which usually lead to nonuniform film thicknesses. Under appropriate deposition conditions, iCVD films are highly conformal (Figure 2.16). P_m/P_{sat} ratio was found to be the most important factor that determines film conformality [73]. For a given substrate geometry, a low P_m/P_{sat} ratio results in perfect conformality. As P_m/P_{sat} ratios increase, the as-deposited material tends to become nonuniform and less conformal (Figure 2.16d).

The rate-limiting step for iCVD depends on the processing conditions. For most iCVD polymers, the deposition rate is inversely related to the substrate temperature. Hence, there is apparently negative activation energy, indicating that the surface adsorption of monomer is the rate-limiting step. In that adsorption-limited regime, higher substrate temperature decreases the rate of adsorption on the surface, leading to lower deposition rates. In that case, film uniformity and conformality are greatly improved. Apart from the substrate temperature, filament temperature also strongly affects the deposition rates, since the initiating radicals are formed near the heated filaments. At low filament temperatures, deposition rates increase with increasing filament temperature according to the Arrhenius equation. At high filament temperatures, the dependence between filament temperature and deposition rate is weaker, indicating that the rate-limiting step is not the initiator decomposition step anymore; instead, the mass transport of initiating radicals to the surface becomes the rate-limiting step.

Like many other polymer deposition processes, iCVD is carried out under high to moderate vacuum conditions. At such reduced pressures, most monomers are easily evaporated and fed to the reactor without the need of a carrier gas flow. Some other monomers that have relatively low vapor pressures at room temperature are heated to suitable temperatures to increase their vapor pressures. The flowrate of a monomer (F_M) into a vacuum chamber can be calculated using the following equation [74]:

$$F_M = C_s(P_{sat} - P_d)$$

where, C_s is the tubing or valve conductance, P_{sat} is the saturation vapor pressure of the monomer, and P_d is the downstream (reactor) pressure. It is obvious from the equation that in order to have a sufficient flowrate, the difference between P_{sat} and P_d must be high. This can be achieved in two ways; either by increasing the temperature of the jar in which the liquid monomer is placed, or by decreasing the reactor pressure. However, both should be done carefully because there are some limiting situations. For example, heating the monomer beyond a certain temperature may initiate self-polymerization of the monomer inside the jar, while decreasing the reactor pressure too much decreases the P_m/P_{sat} ratio too much, at which the deposition might not be physically possible. In some situations where the monomer has too low vapor pressures, pumping the monomer jar directly cannot provide a high monomer flow. In such situations, a small carrier flow through a bubbler can be the solution. A typical bubbler arrangement is shown in Figure 2.17. In that arrangement, the carrier is assumed to be saturated completely by the monomer vapor and exits the bubbler at a higher pressure, P_o where:

$$P_o = P_i + P_r$$

The flow of monomer is given by Hersee and Ballingall [74]:

$$F_M = F_c P_r/(P_o - P_r)$$

This equation gives an accurate description of precursor flow for output pressures in the range $P_o > 3P_r$. For lower pressures the following modified equation is used [74]:

$$F_M = F_c P_r((P_d - P_r + (F_c + F_r)/C_s))$$

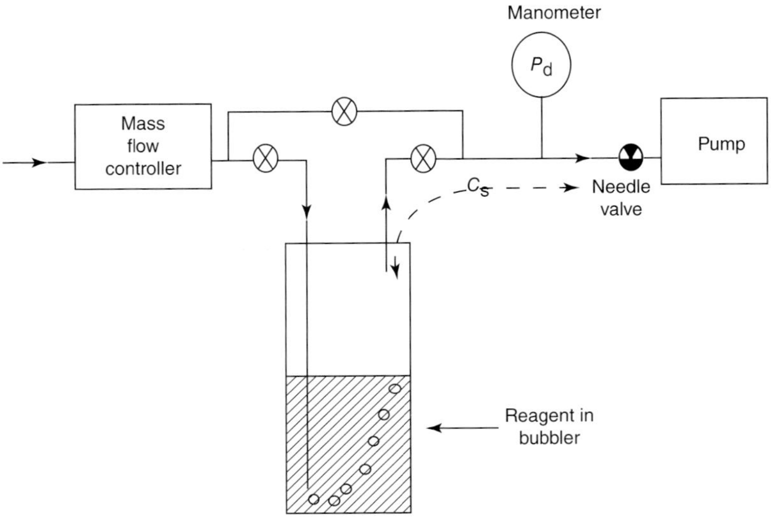

Figure 2.17 A typical bubbler arrangement. (Hersee 1990 [74]. Reproduced with permission of American Vacuum Society.)

2.2.3 Atomic Layer Deposition (ALD)

Atomic layer deposition (ALD) can be considered a subclass of CVD process. In ALD, a film is deposited on a substrate by exposing its surface to alternate gaseous species. In contrast to CVD, ALD precursors are never fed simultaneously into the reactor, but they are introduced as a series of sequential pulses. In each of these pulses the precursor molecules react with the surface in a self-limiting way, so that the reaction terminates once all the reactive sites on the surface are consumed [75]. A schematic representation of ALD which use that self-limiting mechanism is shown in Figure 2.18.

Al_2O_3 was one of the first and important materials deposited by ALD. Al_2O_3 ALD is usually carried out using trimethylammonium (TMA) and water as the precursors. The Al_2O_3 ALD deposition occurs during alternating exposures to

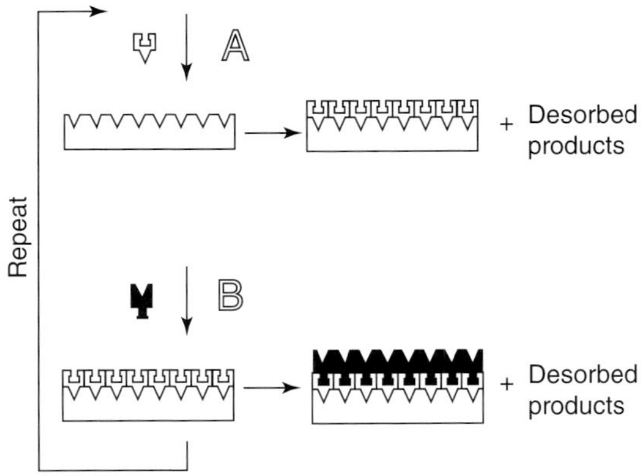

Figure 2.18 Schematic representation of ALD. (George 1996 [76]. Reproduced with permission of American Chemical Society.)

TMA and H_2O. As a result very strong Al—O bonds are formed [76]. Typical deposition rates are around 1 A per cycle.

The as-deposited materials after ALD are smooth and extremely conformal. The overall reaction for Al_2O_3 ALD is:

$$2Al(CH_3)_3 + 3H_2O\,Al_2O_3 + 3CH_4$$

Normally one complete ALD growth cycle consists of four steps: exposure of the first precursor, purge of the reaction chamber, exposure of the second precursor, a further purge of the reaction chamber. The number of growth cycles can be repeated as many times as required in order to obtain the required film thickness. The overall deposition rate depends on the time required to complete each individual cycles.

As stated before, ALD is closely related to CVD. In a binary precursor CVD system, the reactants A and B are fed simultaneously into the reactor and a film is formed continuously on the substrate surface. In ALD, the surface is exposed to the A and B reactants individually and the film is produced in a stepwise manner. Usually a CVD process that is based on binary precursors is first found and then the same reactants re applied sequentially to deposit the same film through ALD. TiO_2 and ZnO are some important materials that are based on binary precursors and they can be deposited by both CVD and ALD. Such depositions are activated thermally by heating the substrate. The other common thermal ALD systems are binary metal oxides and nitrides, such as ZrO_2, HfO_2, Ta_2O_5, TiN, W_2N, and TaN. Like CVD, some ALD processes are activated with the help of plasma. Plasma enhanced ALD is utilized mostly to deposit single-element ALD materials. Single element metals or semiconductors are very difficult to deposit using thermal ALD. H_2 plasma is used to generate hydrogen radicals that reduce the precursors. Figure 2.19 shows a schematic diagram for plasma enhanced ALD of metals. Titanium element, for example, can be deposited by ALD, using $TiCl_4$ and hydrogen radicals from H_2 plasma. $TiCl_4$ is first exposed to the surface, and

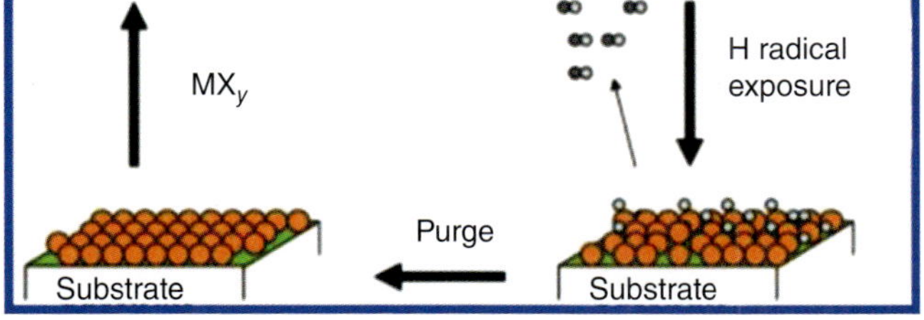

Figure 2.19 Schematic diagram of a typical plasma enhanced ALD. (George 2010 [75]. Reproduced with permission of American Chemical Society.)

then the hydrogen radicals reduce the Ti atoms and remove the chlorine from the surface. Although the deposition rates are very low during Ti ALD, the films are extremely conformal and pin-hole free [77].

2.3 Wet Coating Techniques

Liquid phase coating methods depend on the formation of thin films of a variety of materials by deposition from the liquid phase. The formation of these films from liquid phases is based mainly on chemical reactions (electrochemical and chemical deposition processes). Among many types of solvent-based processes the following are of major interest for both industrial and research practices.

2.3.1 Sol–Gel Coating

Sol–gel technique is a facile method to synthesize high-purity ceramics in liquid phase. The method also allows preparing functional and high-quality coatings on different substrates. The terminology of sol–gel is derived from the chemical process of Solution and Gelation. Sol can be defined to be colloidal particles or polymers in a liquid phase. The continuous reactions in sol lead a formation of a gel consisting of a three-dimensional solid network surrounded liquid phase. Because the demanded final products are strongly influenced by the direction of chemical reactions, understanding the undergoing reactions during the synthesis is a key parameter to control the process. The basis of the process is hydrolysis and condensation of organometallic precursors, mainly metalalkoxides. The first step starts with the hydrolysis of metalalkoxide. The direction of hydrolysis, of course, depends on some parameters such as water content and pH, which will be discussed later. The second step is the condensation of hydrolyzed products with itself as well as with unreacted precursors [78]. Figure 2.20 shows the summary of sol–gel reactions.

The most commonly used metal alkoxides are Tetraethoxysilane (TEOS), Titanium tetraisopropoxide (TTIP), Tetramethoxysilan (TMOS), Zr(IV) propoxide, and so on. Figure 2.21 shows commonly used metalalkoxides used in sol–gel processes.

The first condensed product formed in the solution is called as *sol*. The colloids continue to form three-dimensional solids, in another mean polymeric structure, leading to gelation, which is called as *gel*. Further condensation reactions result in a dense gel and in agglomeration of polymers. All these processes can be interrupted at any step giving different sol–gel products. For instance, the gel and the sol forms are mostly proper for the applications in coating technology, while the formation of dense gel is a need to obtain dense and uniform particles [79]. Figure 2.22 shows the relationship between products and applied routes of sol–gel process.

As mentioned before, the final product of a sol–gel process is strongly influenced by parameters such as pH, the type of precursors, water content, and drying conditions.

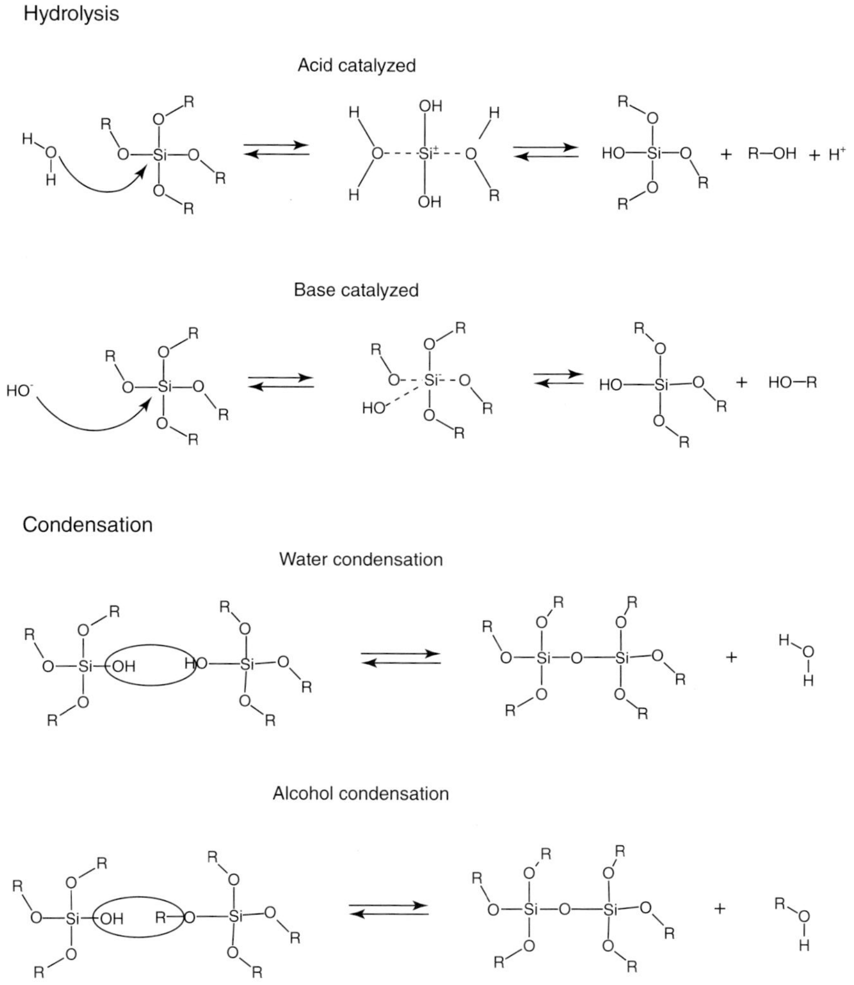

Figure 2.20 Chemical reactions of sol–gel process.

2.3.1.1 Effect of pH

As is well known, hydrolysis reactions strongly depend on the pH value. It is also valid for a sol–gel process that directly affects the structure of the final product. In acidic media, as more alkoxy groups are hydrolyzed, the reaction rate decreases leading to the formation of a linear structure. Thus, the final product gives sphagetti-like structures as given in Figure 2.23a. On the other hand, basic media is vice versa. The reaction rate increases by the hydrolysis of metal alkoxides which leads to the formation of a branched structure. The structure of the final product may be described as appearing to be broccoli-like (see Figure 2.23b).

2.3.1.2 Water Content

The amount of water is one of the most critical parameters for a sol–gel process. Completion of hydrolysis requires four steps of hydrolysis reaction as can

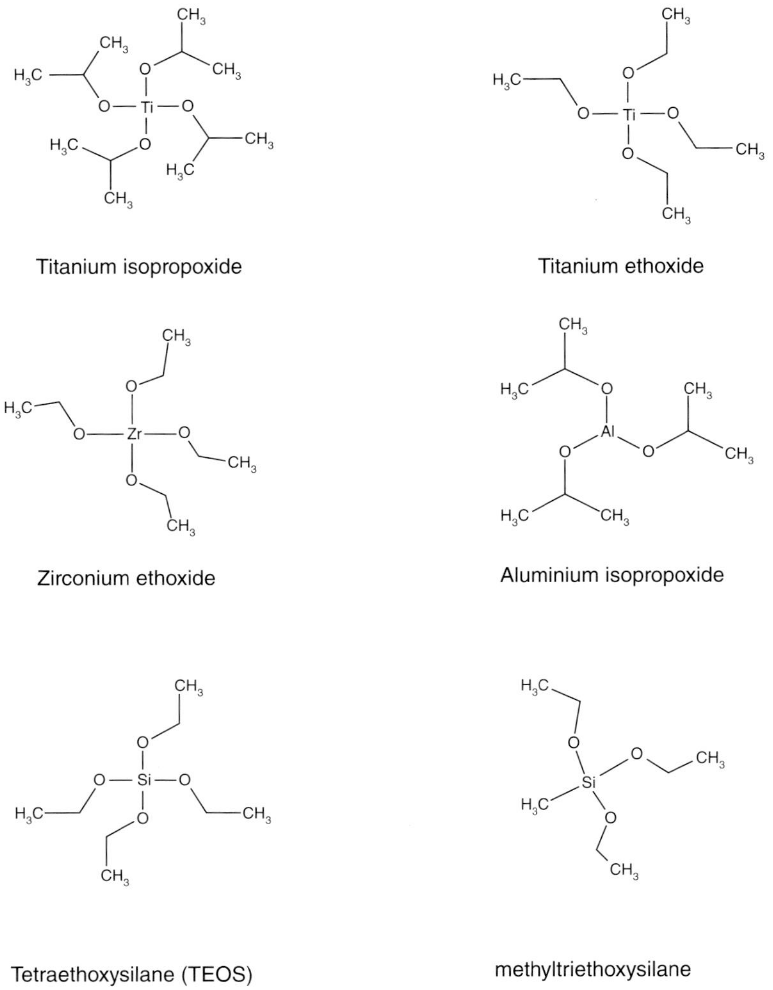

Figure 2.21 The structures of Metalalkoxides for sol–gel synthesis.

be understood from the structure of $M(OR)_4$. Water is a reactant in hydrolysis while it is a product in condensation reactions. Small amounts of water lead to a decrease in the rate of hydrolysis due to the low concentration of the reactant (water). On the other hand, large amounts of water also decrease the rate due to the dilution of the reactant ($M(OR)_4$). Therefore, the amount of water must be adjusted considering the expected products and reaction rates.

2.3.1.3 The Types of Precursors

The sol–gel process consisting of both hydrolysis and condensation reactions is clearly influenced by metal alkoxide precursors depending on their chemical structures. We may first consider the alkoxide groups for the same central metal atoms. The more electronegative alkoxide groups make the central metal atoms more electrophilic and thus facilitate the nucleophilic attack through to the metal

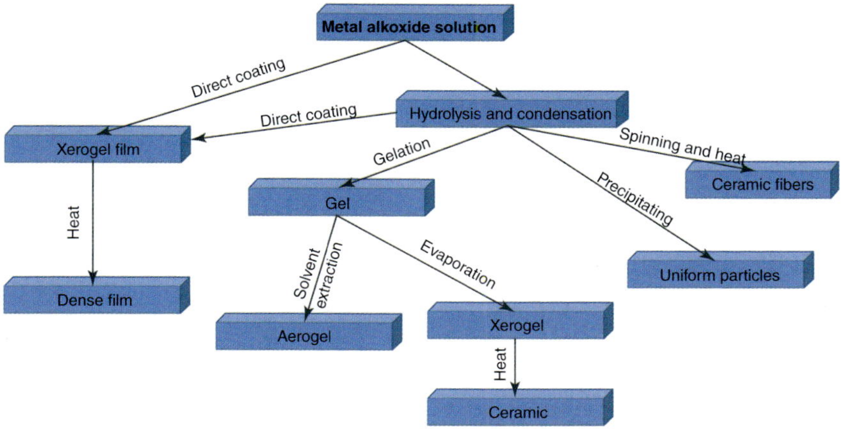

Figure 2.22 The relationship between applied route and products.

Figure 2.23 The structural shapes of acid and base catalyzed sol–gel reactions.

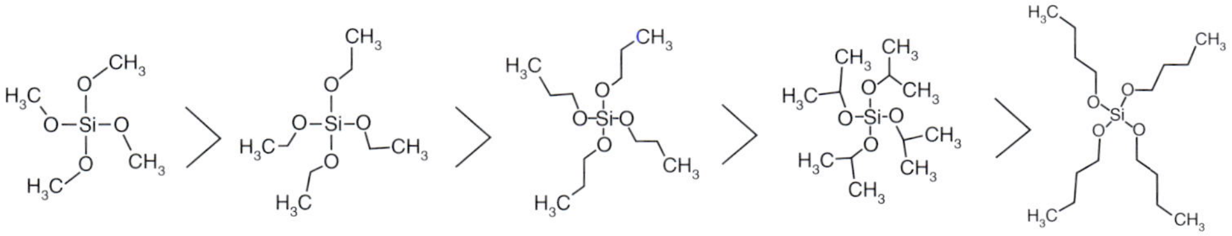

Figure 2.24 The reactivity of Si based alkoxides depending on alkoxide chains.

atoms by water. On the other hand, side chains of alkoxides create a steric effect. The branched and longer side chains decrease the hydrolysis rate. The following sequence maybe proposed based on this approach (Figure 2.24).

The central atom also plays a role in reaction rate. More electronegative and larger central atoms are more attractive for nucleophilic attacks. The proposed reactivity for different central atoms for the same alkoxide chains are shown in Figure 2.25.

2.3.1.4 Temperature, Drying, and Aging

The final products of sol–gel synthesis are also influenced by drying conditions. The drying process involves the evaporation of alcohol, water, and other volatile material. Evaporation of those materials results in a wrinkled structure of the gel.

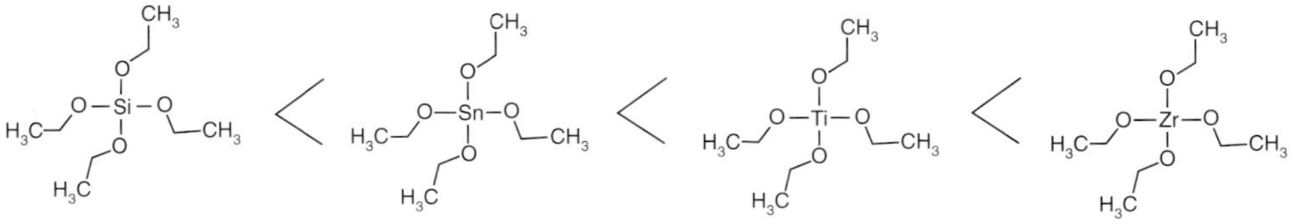

Figure 2.25 The reactivity of same alkoxides depending on central atom.

Thus, random cracks and deformations are usually formed in the gel matrix. To prevent such events, some additives that lead to relaxation during the drying process can be used in synthesis. Another issue is thermal treatment for completing the sol–gel process. Since hydrolysis and condensation reactions are reversible and are still going on, heat treatment must be applied to terminate the chemical reactions and obtain the final products. Heat treatment between 100 and 500 °C is generally used depending on the precursors to form and complete M—O—M bonds. The most critical parameter for heat treatment is the phase transition temperature of metal oxides. Therefore, the maximum temperature must be adjusted by considering the phase transition points besides the demanded, such as amorphous or which crystal, structure.

2.3.1.5 Sol–Gel Coatings

Sol–gel techniques provide many advantages such as low temperature processing, tunable thickness, high optical quality, and purity, in coating technology. Since sol–gel is a liquid-based process, different coating techniques may be suggested for sol–gel based solutions. Among those coating techniques, spin coating, dip coating, spray coating, roll-to-roll coating, and screen printing may be listed. Coating techniques are preferred depending on the application area of coatings. For instance, to prepare thin and homogenous films, spin coating maybe suggested while for thicker films dip coating is preferred. Roll-to-roll coating is a useful route for large area liquid based coatings.

Sol–Gel based coatings are widely used in the preparation of corrosion-protective coatings in metal industry and in functional coatings for electronic and surface-related technological applications. The stable metal oxides such as SiO_2, ZrO_2, Al_2O_3, and TiO_2, are useful materials to guard against corrosion of metals. Besides these protective properties, photoactive metal oxides such as TiO_2 behave as photocatalysts for cleaning the surfaces. On the other hand, semiconductors such as ZnO and TiO_2 can be prepared based on the sol–gel process and is advantageous for electronic applications. This process is widely used in thin film electronic device technologies. Organic–inorganic hybrid sol–gel products are also interesting materials for functional surfaces. Super hydrophilic or hydrophobic surfaces, functional surfaces for biotechnological applications are other application fields of sol–gel processes [80].

2.3.2 Electrospinning

Electrospinning is a fiber fabrication technique that uses a high-voltage electric field to form solid fibers from a polymeric fluid stream delivered through a millimeter-scale nozzle. Electrospinning participates of properties of both electrospraying and conventional solution dry spinning of fibers [81]. This process suggests proper capabilities for producing novel natural nanofibers and fabrics with controllable pore structure [82].

The process does not necessitate the use of coagulation chemistry or high temperatures to produce solid threads from solution. This makes the process especially appropriate for the production of fibers using large and complex molecules.

Electrostatic attraction of a liquid was observed for the first time by William Gilbert in 1600. John Francis Cooley took out the first electrospinning patent in 1900 [83]. In 1934, Antonin Formhals patented a technique that used a voltage of 57 kV for electrospinning cellulose acetate using acetone and monomethyl ether of ethylene glycol as solvents. Bernard Vonnegut and Raymon Newbauer designed a simple equipment for electrical atomization and fabricated streams of highly electrified uniform droplets of approximately 0.1 mm diameter in 1952 [84]. Since 1995 there have been many theoretical developments in the mechanisms of the electrospinning process [83].

There are constitutively three components to bring about the process: a high voltage supplier, a capillary tube with a pipette or needle of small diameter and a metal collecting screen. Electrospinning is carried out at room temperature in atmospheric conditions. The polymer solvent solution goes out from the injector and forms a droplet. The injector is charged, which transfers its electrical energy to the droplet. At a significant voltage, the droplet will change shape forming what is known as the *Taylor cone*. Before arriving at the collecting screen, the solution jet evaporates or solidifies, and is picked up as a leashed web of small fibers [84] (Figure 2.26).

Some of the most important parameters in the electrospinning process are given below.

Molecular weight: Molecular weight of the polymer effects the morphologies of electrospun fiber. Lowering the molecular weight of the polymer shows a tendency to form beads rather than smooth fiber. By increasing the molecular weight, a smooth fiber can be attained [85].

Solution properties: Solution viscosity is critical in specifying the fiber morphology. Continuous and smooth fibers cannot be acquired in very low viscosity. High viscosity brings about difficulties in ejection of jets from solution. In the electrospinning process viscosity, polymer concentration, and polymeric molecular weight are related to each other [85].

Electric potential: Applied voltage is a considerable factor in the electrospinning process. Charged jets ejected from Taylor Cone can occur if the applied voltage is higher than the threshold voltage [84, 85].

Flow rate: The flow rate of the polymer from the syringe is an important factor for electrospinning due to the affect of the jet velocity and the material transfer rate. On the condition that the flow rate is very high, bead fibers with thick diameter will appear rather than the smooth fiber with thin diameter owing to the short drying time and low stretching forces [84].

Concentration: In cases where the concentration is suitable, smooth nanofibers can be attained. If the concentration is very low, polymeric microparticles will

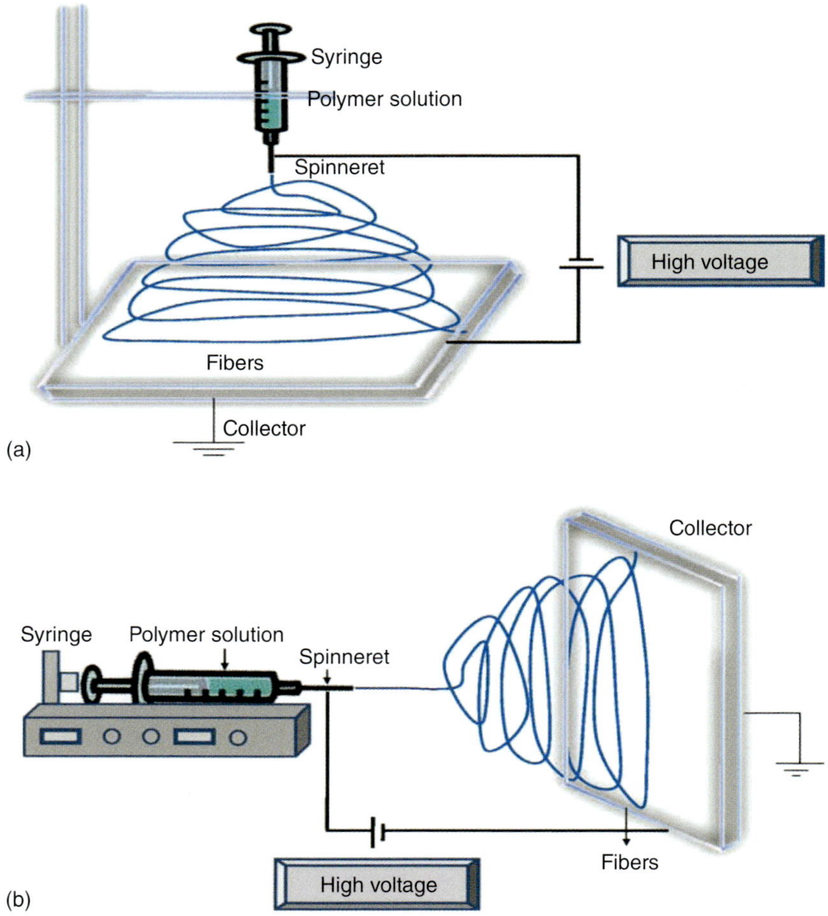

Figure 2.26 Schematic diagram of set up of electrospinning apparatus (a) typical vertical set up and (b) horizontal setup of electrospinning apparatus. (Reprinted from [84], copyright (2010), with permission from Elsevier.)

be formed. If the concentration is very high, spiral curve-shaped microribbons will be formed [85].

Distance between the capillary and collection screen: It has been proven that the distance between the collection screen and the capillary can also affect fiber morphology [85].

Ambient parameters: Ambient parameters such as humidity and temperature can affect the fiber diameters and morphologies [85].

Needle gauge: By decreasing the needle diameter, the increase in the polydispersity of the nanofiber diameters can be monitored [86].

The laboratory setup for electrospinning process is composed of a hypodermic syringe needle connected to a high-voltage (5–50 kV) direct current power supply, a syringe pump, and an earthed collector [87].

Electrospinning is suitable for scaling up using the following techniques:

- Bubble electrospinning
- Rotating roller electrospinning
- Multiplying the needles
- Wire electrospinning
- Spiral coil electrospinning
- Rotary cone electrospinning
- Ball electrospinning
- Hollow tube electrospinning
- High speed electrospinning
- Plate edge electrospinning
- Bowl electrospinning.

Alteration of the characteristics of the system can provide to create fibers with unique structures and features. Emulsion electrospinning, coaxial electrospinning, and melt electrospinning are some examples.

2.3.2.1 Emulsion Electrospinning

Emulsion electrospinning is usually a more unwieldy process as compared to coaxial spinning because of the greater number of variables, which must be explained in creating the emulsion. A water phase and an immiscible solvent phase are mixed in the presence of an emulsifying agent to form the emulsion. Throughout the electrospinning process the emulsion droplets within the fluid are stretched and progressively closed down leading to their coalescence. Provided that the volume fraction of the inner fluid is adequately high, a continuous inner core can be formed [88].

2.3.2.2 Coaxial Electrospinning

Coaxial electrospinning method uses a multiple solution feed system that enables the injection of one solution into another at the tip of the spinneret. The cover fluid is believed to behave as a carrier that draws in the inner fluid at the Taylor Cone of the electrospinning jet. On the condition that the solutions are immiscible, a core shell structure is generally observed. On the other hand, miscible solutions can result either in porosity or a fiber with apparent phases because of phase separation during solidification of the fiber [89].

2.3.2.3 Melt Electrospinning

Another method used for nanofiber is the melt electrospinning method and the working principles of this method are as follows:

1) Polymers are melted when exiting from the die by hot air
2) Melt polymers are thinned by the cold air
3) Finally, thinned polymer is collected above the collector.

Generally, diameters of obtained fibers are between 0.5 and 10 μm. Melt electrospinning method is suitable for more fiber production but the strength of the fibers produced by this method is low.

A wide range of materials can be produced by the electrospinning method; hence as-produced nanofibers have extensive usage in many areas including composites, filtration, biomedical, textile, medical, and catalysts.

Nanofibers have extensive usage in composite industry because of their suitable properties such as being small and light weight. Composites nanofibers obtained via electrospinning have properties such as high elastic modulus and strength/weight rate. Composites nanofibers have superior properties, since nanofibers exhibit better mechanical properties than microfibers.

In biomedical engineering, electrospinning submits a lot of advantages as a nanofiber producer, which gain favor to researchers in drug delivery, bone tissue engineering, heart tissue engineering, and so on [90]. Electrospinning also makes it possible to produce protective clothes that provide prevention against chemical and biological dangers. Breathing poisonous gases and contact of skin with some chemicals is a vital problem for troops in war periods. Electrospinning technology offers solutions to overcome this problem through production of nonwoven fabrics. Electrospun nanofibers can also be used as catalyst carriers in the chemical industry, whereby the extremely large surface could provide vast amount of active sites, which enhance catalytic activity [91].

2.3.3 Electrolytic Anodization

Anodizing is an electrochemical process to form oxide coatings for increasing the performance of the surface. The first commercial anodizing process was patented in 1923 by Bengough and Stuart. The most common anodized metal is aluminum [92, 93]; however, several metals are suitable for anodizing such as magnesium, zinc, titanium, niobium, and tantalum. The most important applications of electrolytic anodization are corrosion protection, electrical insulation, and decorative coatings.

In the electrolytic anodization process, the anodized metal at the anode is coated by applying a direct current through an electrolytic solution (Figure 2.27). The electric current releases hydrogen at the cathode and oxygen at the surface of the anode.

There are many different types of anodizing processes (e.g., using sulfuric acid, chromic acid, phosphoric acid) that produces oxide layers with different properties. The process involves pre-treatment, anodizing, coloring, and sealing.

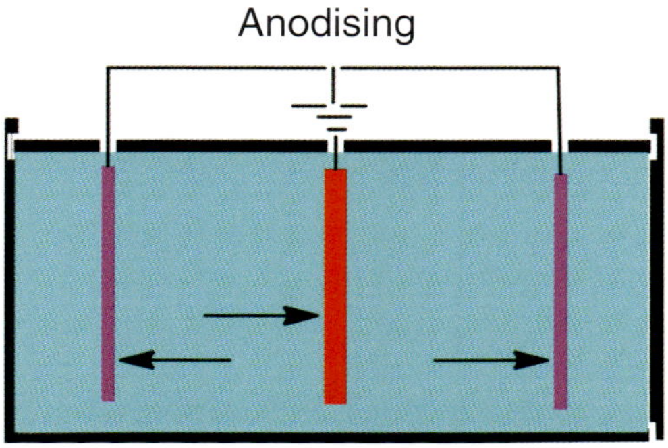

Figure 2.27 Scheme of electrolytic anodization.

Pre-treatment: Surfaces has been etched with NaOH to remove surface imperfections.
Anodizing: The electric current has been applied for the formation of oxide layer with controlled thickness.
Coloring: Coloring has been achieved by one of the coloring methods (electrolytic coloring, integral coloring, organic dyeing and interference coloring).
Sealing: Pores in the oxide film have been sealed to close the pores in the surface making it resistant to staining, abrasion, crazing, and color degradation.

The advantages of anodizing includes: thin, durable, and hard coating, excellent corrosion protection, fade resistant, cheap, and simple process. The disadvantages of anodizing surfaces are limited color choices have surface defects and low thermal conductivity.

2.3.4 Electroplating

Electroplating is an electrochemical process of coating a metal onto other metallic/plastic surface to prevent corrosion or for decorative purposes. Electroplating was first discovered by Luigi Brugnatelli in 1805 through using the electrodeposition process for the electroplating of gold. The most common electroplating metal is nickel [94]; however, there are also specific types of electroplating metals such as copper [95], silver [96], and chromium. Electroplating is generally used in industries (automobile, airplanes, electronics, jewelry, and toy) for mechanical and engineering properties, appearance, protection, and some special properties desired for the product.

The electroplating cell consists of a positive electrode (anode), a negative electrode (cathode), and an electrolyte solution (containing the metal ions) through which electric current flows and the plated part is generally the cathode (Figure 2.28).

There are different types of electroplating such as mass plating, rack plating, continuous plating, and line plating. The electroplating sequence involves cleaning, rinsing, stripping, and plating.

Cleaning: Surface must be free of contaminants for bonding to take place.
Rinsing: Sandblasting of parts to etch surface and improve adherence of plating metal.

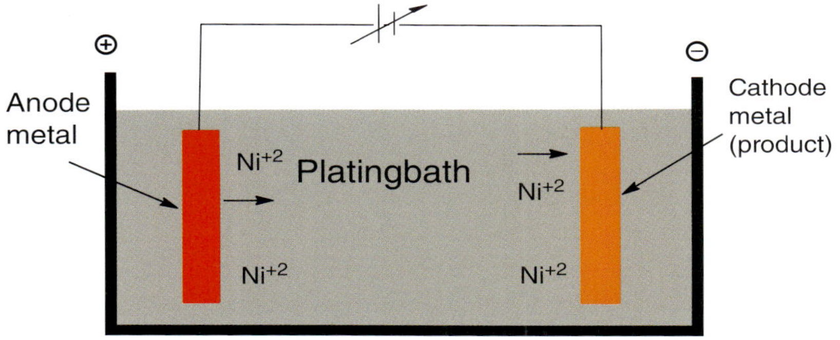

Figure 2.28 Scheme of electroplating.

Stripping: To remove chemical residue/unsatisfactory deposits before or after plating.

Plating: Process that uses electrical current to drive the chemical reaction by reducing metal ions and enable control of the plating process.

There are a number of advantages of this process: more durability, longer lasting, more resistance to damage, and so on. The disadvantages of the process are high cost, time consuming, nonuniform plating, and environmental pollution.

2.3.5 Electroless Plating

Electroless plating is a deposition method based on the chemical reduction of metal (copper, nickel, silver, gold, or palladium) ions in an aqueous solution and deposition of metal without using electrical energy [97, 98]. The reducing agent actually converts the metal ions (M^+) to the metal (M) that gets plated over a catalytic surface as shown in Figure 2.29. It was discovered by A. Brenner and G. E. Riddel in 1944. Electroless plating is mainly used in aerospace, automotive, military, electronic applications oil/gas production, and food processing industries.

Electroless plating is the most commonly used technique for metal deposition on surfaces due to its advantages, such as corrosion resistance and high hardness, low cost, simple and uniform coating on complex shapes, and capability to deposit on either conductive or nonconductive parts [99].

The electroless plating consists of degreasing, etching, seeding, electroless plating steps.

Degreasing: Cleaning process that uses alkaline/acid solutions containing surfactants to remove oils and organic chemicals.

Etching: Chromic acid/sulfuric acid are used for chemical etching to roughen the sample surface for adhesion enhancement. The roughening actually creates a bonding site for further plating of samples.

Seeding: Surface seeding with a catalyst involves treating the surface with a catalytically active metal.

The advantages of electroless plating are exceptional corrosion resistance and high hardness and simple and uniform deposition process on complex shapes. The main disadvantages of electroless plating are aging of the solution and reaction by products affecting the deposition [99].

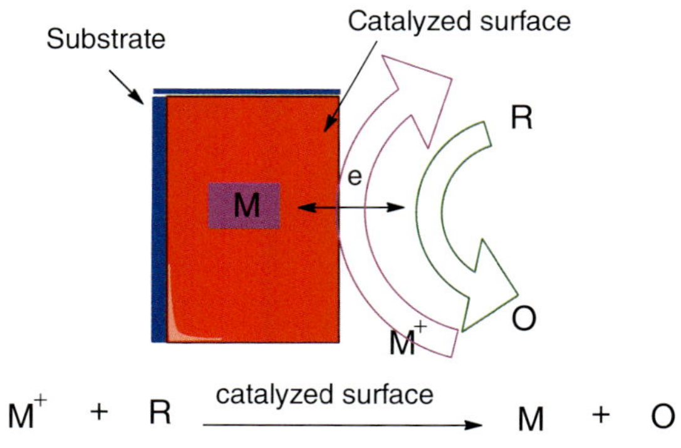

Figure 2.29 Scheme of electroless plating.

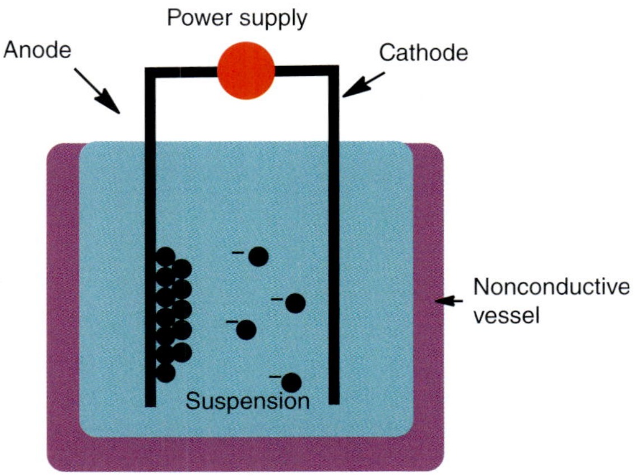

Figure 2.30 Scheme of electrophoretic deposition.

2.3.6 Electrophoretic Deposition

Electrophoretic deposition (EPD) is a colloidal processing technique for the deposition of materials from charged nanoparticles in suspension by the application of an electric field. The first patent for the use of EPD was awarded in 1917 to Davey and General Electric. Several applications including coatings, nanoscale assembly, micropatterned thin films, near shape ceramics and glasses, solid oxide fuel cells, laminated or graded materials, hybrid materials, and so on, have been described [100].

EPD is a two-step process. First, the charged particles suspended in a liquid medium move toward the oppositely charged electrode under the effect of an electric field. Then, the particles deposit on the electrode to form a film (Figure 2.30).

In early studies, EPD was used mainly in traditional ceramic coating processing, but nowadays, the field of applications and materials is broad and include a wide range of nanomaterials, polymers, and composites [101]. The process conditions (concentration of particles in solution, applied electric field, time) affect the thickness of the film. EPD has many advantages such as low cost, simple equipment, applicability to a wide range of materials, uniform coating thickness, relatively high speed of coating, high purity, flexibility, and efficiency. The main disadvantage of EPD, compared with other colloidal processes, is that it cannot use water as the liquid medium, because the application of a voltage to water causes the evolution of hydrogen and oxygen gases at the electrodes, which could adversely affect the quality of the deposits formed.

2.3.7 Dip Coating

Dip coating is the oldest and most commonly used process among the various wet chemical thin film deposition methods [102]. The first commercial applied coating process was patented in 1939 by Jenaer Glaswerk Schott & Gen.

The dip coating process may be separated into five stages (Figure 2.31):

Dipping: Dipping of the substrate into the coating solution.

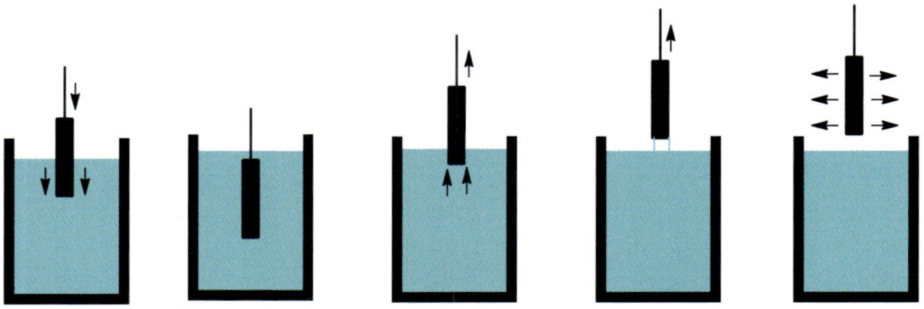

Figure 2.31 Scheme of dip coating.

Immersion: the substrate is immersed in the solution at a constant speed to allow for the coating material to apply itself to the substrate.
Withdrawing: The thin layer deposits itself on the substrate while it is pulled up. The withdrawing is carried out at a constant speed. The speed determines the thickness of the coating.
Drainage: Excess liquid will drain from the surface.
Evaporation: The solvent evaporates from the liquid, forming the thin layer.

The advantages of dip coating includes: simple, corrosion resistance, alternative colors, wide range of thickness, and adaptable to all shapes and sized objects.

Dip coating has some disadvantages, such as: light parts tend to float and fall from the conveyor and film thickness can be different from top to bottom.

References

1 Selvakumar, N. and Barshilia, H.C. (2012) Review of physical vapor deposited (PVD) spectrally selective coatings for mid- and high-temperature solar thermal applications. *Sol. Energy Mater. Sol. Cells*, **98**, 1–23.
2 Bunshah, R. (2000) *Handbook of Hard Coatings*, Noyes Publications, New Jersey.
3 Helmersson, U., Lattemann, M., Bohlmark, J., Ehiasarian, A.P., and Gudmundsson, J.T. (2006) Ionized physical vapor deposition (IPVD): a review of technology and applications. *Thin Solid Films*, **513**, 1–24.
4 Sproul, W.D. (1994) Multilayer, multicomponent, and multiphase physical vapor deposition coatings for enhanced performance. *J. Vac. Sci. Technol., A*, **12**, 1595.
5 Lange, J. and Wyser, Y. (2003) Recent innovations in barrier technologies for plastic packaging—a review. *Packag. Technol. Sci.*, **16**(4), 149–158.
6 Nakamura, K., Inagawa, K., Tsuruoka, K., and Komiya, S. (1977) Applications of wear-resistant thick films formed by physical vapor deposition processes. *Thin Solid Films*, **40**, 155–167.
7 Laudise, R.A., Kloc, C., Simpkins, P.G., and Siegrist, T. (1998) Physical vapor growth of organic semiconductors. *J. Cryst. Growth*, **187**, 449–454.

8 Mattox, D.M. (2010) *Handbook of Physical Vapor Deposition (PVD) Processing*, 2nd edn, Elsevier, Oxford.
9 Barreau, N., Marsillac, S., Bernède, J.C., Nasrallah, T.B., and Belgacem, S. (2001) Optical properties of wide band gap indium sulphide thin films obtained by physical vapor deposition. *Phys. Status Solidi A*, **184**(1), 179–186.
10 Liua, C., Leyland, A., Bia, Q., and Matthewsa, A. (2001) Corrosion resistance of multi-layered plasma-assisted physical vapour deposition TiN and CrN coatings. *Surf. Coat. Technol.*, **141**, 164–173.
11 Ohmi, S., Takeda, M., Ishiwara, H., and Iwai, H. (2004) Electrical characteristics for Lu2 O 3 thin films fabricated by E-beam deposition method. *J. Electrochem. Soc.*, **151**(4), G279–G283.
12 Mukaia, T., Suresha, S., Kitab, K., Sasakib, H., Kobayashi, N., Higashic, K., and Inoued, A. (2003) Nanostructured Al–Fe alloys produced by e-beam deposition: static and dynamic tensile properties. *Acta Mater.*, **51**(14), 4197–4208.
13 Perry, M.D., Boyd, R.D., Britten, J.A., Decker, D., Shore, B.W., Shannon, C., and Shults, E. (1995) High-efficiency multilayer dielectric diffraction gratings. *Opt. Lett.*, **20**(8), 940–942.
14 Suri, A.K., Nimmagadda, R., and Bunshah, R.F. (1980) Synthesis of titanium nitrides by activated reactive evaporation. *Thin Solid Films*, **72**(3), 529–533.
15 Peebles, D.E. and Pope, L.E. (1989) Reactive evaporation of thin titanium nitride films in ultrahigh vacuum and their friction and wear behavior as a function of contact stress. *Thin Solid Films*, **173**(1), 19–37.
16 Assunção, V., Fortunato, E., Marques, A., Águas, A., Ferreira, I., Costa, M.E.V., and Martins, R. (2003) Influence of the deposition pressure on the properties of transparent and conductive ZnO:Ga thin-film produced by r.f. sputtering at room temperature. *Thin Solid Films*, **427**, 401–405.
17 Venables, J.A. (2000) *Introduction to Surface and Thin Film Processes*, Cambridge University Press, Cambridge.
18 Rossnagel, S.M. (1999) Sputter deposition for semiconductor manufacturing. *IBM J. Res. Dev.*
19 Randhawa, H. (1991) Review of plasma-assisted deposition processes. *Thin Solid Films*, **196**, 329–349.
20 Pierson, H.O. (1999) *Handbook of Chemical Vapor Deposition: Principles, Technology, and Applications*, 2nd edn, Noyes Publications, New York.
21 Sherman, A. (1987) *Chemical Vapor Deposition for Microelectronics: Principles, Technology, and Applications*, Noyes Publications, New York.
22 MarketsandMarkets. (2015) Vapor Deposition Market by Technology (Chemical Vapor Deposition, Physical Vapor Deposition), by End-user Industry (Microelectronics, Cutting tools, Industrial & Energy, Medical, Decorative Coating) – Global Trends & Forecast to 2019.
23 Park, J.H. and Sudarshan, T.S. (2001) *Chemical Vapor Deposition*, Surface Engineering, vol. **2**, ASM International, Ohaio.
24 Ma, M., Mao, Y., Gupta, M., Gleason, K.K., and Rutledge, G.C. (2005) Superhydrophobic fabrics produced by electrospinning and chemical vapor deposition. *Macromolecules*, **38**(23), 9742–9748.

25 Hausmann, D., Becker, J., Wang, S., and Gordon, R.G. (2002) Rapid vapor deposition of highly conformal silica nanolaminates. *Science*, **298**(5592), 402–406.

26 Alf, M.E., Asatekin, A., Barr, M.C., Baxamusa, S.H., Chelawat, H., Ince, G.Ö., Petruczok, C.D., Sreenivasan, R., Tenhaeff, W.E., Trujillo, N.J., Vaddiraju, S., Xu, J., and Gleason, K.K. (2009) Chemical vapor deposition of conformal, functional, and responsive polymer films. *Adv. Mater.*, **22** (18), 1993–2027.

27 Kordina, O., Hallin, C., Ellison, A., Bakin, A.S., Ivanov, I.G., Henry, A., Yakimova, R., Touminen, M., Vehanen, A., and Janzén, E. (1996) High temperature chemical vapor deposition of SiC. *Appl. Phys. Lett.*, **69**(10), 1456.

28 Karaman, M., Sezgi, N.A., Doğu, T., and Özbelge, H.Ö. (2009) Mechanism studies on CVD of boron carbide from a gas mixture of BCL3, CH4, and H2 in a dual impinging-jet reactor. *AlChE J.*, **55**(3), 701–709.

29 Koretsky, M. (2013) *Engineering and Chemical Thermodynamics*, 2nd edn, John Wiley & Sons, Inc., Hoboken, NJ.

30 Çengel, Y.A. and Boles, M.A. (2011) *Thermodynamics: An Engineering Approach*, 7th edn, McGraw-Hill, New York.

31 Besmann, T.M. (1977) *SOLGASMIX-PV, A Computer Program to Calculate Equilibrium Relationship in Complex Chemical Systems*, Oak Ridge National Laboratory, Oak Ridge, TN.

32 Bale, C.W., Pelton, A.D., and Thompson, W.T. (1996) *FACT 2.1*, Ecole Polytechnique de Montreal/Royal Military College, Canada, Kingston, Ontario.

33 Chase, M.W. (1985) JANAF thermochemical tables. *Am. Chem. Soc. Am. Inst. Phys.*, **13**, 1.

34 Dobkin, D.M. and Zuraw, M.K. (2003) *Principles of Chemical Vapor Deposition*, Springer, Dordrecht, Netherlands.

35 Kong, J., Cassell, A.M., and Dai, H. (1998) Chemical vapor deposition of methane for single-walled carbon nanotubes. *Chem. Phys. Lett.*, **292**, 567–574.

36 Choi, G.S., Cho, Y.S., Hong, S.Y., Park, J.B., Son, K.H., and Kim, D.J. (2002) Carbon nanotubes synthesized by Ni-assisted atmospheric pressure thermal chemical vapor deposition. *J. Appl. Phys.*, **91**, 3847.

37 Liu, W., Li, H., Xu, C., Khatami, Y., and Banerjee, K. (2011) Synthesis of high-quality monolayer and bilayer graphene on copper using chemical vapor deposition. *Carbon*, **49**(13), 4122–4130.

38 Lee, S., Lee, K., and Zhong, Z. (2010) Wafer scale homogeneous bilayer graphene films by chemical vapor deposition. *Nano Lett.*, **10**(11), 4702–4707.

39 Zhang, Y., Zhang, L., Kim, P., Ge, M., Li, Z., and Zhou, C. (2012) Vapor trapping growth of single-crystalline graphene flowers: synthesis, morphology, and electronic properties. *Nano Lett.*, **12**(6), 2810–2816.

40 Watanabe, K., Tanigaki, T., and Wakayama, S. (1981) The properties of LPCVD SiO_2 film deposited by SiH_2Cl_2 and N_2O mixtures. *J. Electrochem. Soc.*, **128**(12), 2630–2635.

41 Martin, F., Bertin, F., Sprey, H., and Granneman, E. (1991) LPCVD Si3N4 growth retardation on silicon native oxide compared with in situ HF vapour-deglazed silicon substrates. *Semicond. Sci. Technol.*, **6**(11), 1100–1103.

42 Mastrangelo, C.H., Tai, Y.C., and Muller, R.S. (1990) Thermophysical properties of low-residual stress, Silicon-rich, LPCVD silicon nitride films. *Sens. Actuators, A*, **23**, 856–860.

43 Roth, J.R. (1995) *Industrial Plasma Engineering: Principles*, vol. **1**, Institute of Physics Publishing, Bristol and Philadelphia, PA.

44 Joyce, R.J., Sterling, H.F., and Alexander, J.H. (1968) Silicon oxide and nitride films deposited by an r.f. glow-discharge. *Thin Solid Films*, **1**(6), 481–494.

45 Karaman, M. and Yenice, E. (2015) Plasma enhanced chemical vapor deposition of poly(2,2,3,4,4,4-hexafluorobutyl acrylate). *Thin Films*, **21**, 188–195.

46 Lee, I.S., Kim, J.W., Youn, C.J., Park, S.K., and Hahn, Y.B. (1996) Preparation and characterization of TiO_2 thin films by PECVD on Si substrate. *Korean J. Chem. Eng.*, **13**(5), 473–477.

47 Deshmukh, S.C. and Aydil, E.S. (1994) Low-temperature plasma enhanced chemical vapor deposition of SiO_2. *Appl. Phys. Lett.*, **65**, 3185.

48 Moustakas, T.D. (1989) The role of the tungsten filament in the growth of polycrystalline diamond films by filament-assisted CVD of hydrocarbons. *Solid State Ionics*, **32–33**, 861–868.

49 Schropp, R.E. (2009) Hot wire chemical vapor deposition: recent progress, present state of the art and competitive opportunities. *ECS Trans.*, **25**(8), 3–14.

50 d'Agostino, R. (1990) *Plasma Deposition, Treatment, and Etching of Polymers*, Academic Press, Inc., San Diego, CA.

51 Yasuda, H. and Hirotsu, T. (1978) Critical evaluation of conditions of plasma polymerization. *J. Polym. Sci., Polym. Chem. Ed.*, **16**(4), 743–759.

52 Yang, W. and Wolden, C.A. (2006) Plasma-enhanced chemical vapor deposition of TiO_2 thin films for dielectric applications. *Thin Solid Films*, **515**(4), 1708–1713.

53 Casserly, T.B. and Gleason, K.K. (2006) Effect of substrate temperature on the plasma polymerization of poly(methyl methacrylate). *Chem. Vap. Deposition*, **12**(1), 59–66.

54 Yasuda, Y. (1985) *Plasma Polymerization*, Academic Press Inc., Orlando, FL.

55 Stratakis, E., Mateescu, A., Barberoglou, M., Vamvakaki, M., Fotakis, C., and Anastasiadis, S.H. (2010) From superhydrophobicity and water repellency to superhydrophilicity: smart polymer-functionalized surfaces. *Chem. Commun.*, **46**, 4136.

56 Feng, M., Morales, A.B., Beugeling, T., Bantjes, A., Van der Werf, K., Gosselink, G., De Grooth, B., and Greve, J. (1996) Adsorption of high density lipoproteins (HDL) on solid surfaces. *J. Colloid Interface Sci.*, **177**, 364.

57 Lau, K.K.S., Bico, J., Teo, K.B.K., Chhowalla, M., Amaratunga, G.A.J., Milne, W.I., McKinley, G.H., and Gleason, K.K. (2003) Superhydrophobic carbon nanotube forests. *Nano Lett.*, **3**, 1701.

58 Vasudev, M.C., Anderson, K.D., Bunning, T.J., Tsukruk, V.V., and Naik, R.R. (2013) Exploration of plasma-enhanced chemical vapor deposition as a method for thin-film fabrication with biological applications. *ACS Appl. Mater. Interfaces*, **5**, 3983.

59 Bodas, D.S., Desai, S.M., and Gangal, S.A. (2005) Deposition of plasma-polymerized hydroxyethyl methacrylate (HEMA) on silicon in presence of argon plasma. *Appl. Surf. Sci.*, **245**, 186.

60 Wu, Q.G. and Gleason, K.K. (2003) Plasma-enhanced CVD of organosilicate glass (OSG) films deposited from octamethyltrisiloxane, bis(trimethylsiloxy) methylsilane, and 1,1,3,3-tetramethyldisiloxane. *Plasmas Polym.*, **8**, 31.

61 Pfluger, C.A., Carrier, R.L., Sun, B., Ziemer, K.S., and Burkey, D.D. (2009) Cross-linking and degradation properties of plasma enhanced chemical vapor deposited poly(2-hydroxyethyl methacrylate). *Macromol. Rapid Commun.*, **30**, 126.

62 Coclite, A.M. and Gleason, K.K. (2012) Initiated PECVD of organosilicon coatings: a new strategy to enhance monomer structure retention. *Plasma Processes Polym.*, **9**, 425.

63 Gürsoy, M., Uçar, T., Tosun, Z., and Karaman, M. (2016) Initiation of 2-hydroxyethyl methacrylate polymerization by *tert*-butyl peroxide in a planar PECVD system. *Plasma Processes Polym.*, **13**(4), 438–446.

64 Reeja-Jayan, B., Kovacik, P., Yang, R., Sojoudi, H., Ugur, A., Kim, D.H., Petruczok, C.D., Wang, X., Liu, A., and Gleason, K.K. (2014) A route towards sustainability through engineered polymeric interfaces. *Adv. Mater. Interfaces*, **1**, 1400117.

65 Martin, T.P., Lau, K.K.S., Chan, K., Mao, Y., Gupta, M., O'Shaughnessy, W.S., and Gleason, K.K. (2007) Initiated chemical vapor deposition (iCVD) of polymeric nanocoatings. *Surf. Coat. Technol.*, **201**(22–23), 9400–9405.

66 Karaman, M. and Çabuk, N. (2012) Initiated chemical vapor deposition of pH responsive poly(2-diisopropylamino)ethyl methacrylate thin films. *Thin Solid Films*, **520**(21), 6484–6488.

67 Sarıipek, F. and Karaman, M. (2014) Initiated CVD of tertiary amine-containing glycidyl methacrylate copolymer thin films for low temperature aqueous chemical functionalization. *Chem. Vap. Deposition*, **20**, 373–379.

68 Karaman, M., Kooi, S., and Gleason, K.K. (2008) Vapor deposition of hybrid organic–inorganic dielectric bragg mirrors having rapid and reversibly tunable optical reflectance. *Chem. Mater.*, **20**(6), 2262–2267.

69 Mao, Y. and Gleason, K.K. (2004) Hot filament chemical vapor deposition of poly(glycidyl methacrylate) thin films using *tert*-butyl peroxide as an initiator. *Langmuir*, **20**, 2484–2488.

70 Sarıipek, F., Akın, İ., Arslan, G., and Karaman, M. (2016) Vapor deposition of blanket polymeric nanocoatings on porous substrates for the facile preparation of thin film composite membranes. *Vacuum*, **128**, 17–26.

71 Lau, K.K.S. and Gleason, K.K. (2008) Initiated chemical vapor deposition (iCVD) of copolymer thin films. *Thin Solid Films*, **516**(5), 678–680.

72 Alf, M.E., Asatekin, A., Barr, M.C., Baxamusa, S.H., Chelawat, H., Ince, G.O., Petruczok, C.D., Sreenivasan, R., Tenhaeff, W.E., Trujillo, N.J.,

Vaddiraju, S., Xu, J., and Gleason, K.K. (2010) Chemical vapor deposition of conformal, functional, and responsive polymer films. *Adv. Mater.*, **22**, 1993–2027.
73 Baxamusa, S.H. and Gleason, K.K. (2008) Thin polymer films with high step coverage in microtrenches by initiated CVD. *Chem. Vap. Deposition*, **14**, 313–318.
74 Hersee, S.D. and Ballingall, J.M. (1990) The operation of metalorganic bubblers at reduced pressure. *J. Vac. Sci. Technol., A*, **8**(12), 800–804.
75 George, S.M. (2010) Atomic layer deposition: an overview. *Chem. Rev.*, **110**, 111–131.
76 George, S.M., Ott, A.W., and Klaus, J.W. (1996) Surface chemistry for atomic layer growth. *J. Phys. Chem.*, **100**, 13121–13131.
77 Rossnagel, S.M., Sherman, A., and Turner, F. (2000) Plasma-enhanced atomic layer deposition of Ta and Ti for interconnect diffusion barriers. *J. Vac. Sci. Technol., B*, **18**, 2016–2020.
78 Brinker, C.J. and Scherer, G.W. (eds) (2013) *Sol–Gel Science: The Physics and Chemistry of Sol–Gel Processing*, Academic Press Inc., San Diego, CA.
79 Milea, C.A., Bogatu, C., and Duta, A. (2011) The influence of parameters in silica sol–gel process. *Bull. Transilvania Univ. Brasov*, **4**(53), 59–66.
80 Wang, D. and Bierwagen, G.P. (2009) Sol–gel coatings on metals for corrosion protection. *Prog. Org. Coat.*, **64**(4), 327–338.
81 Ziabicki, A. (1976) *Fundamentals of Fiber Formation*, John Wiley & Sons, Ltd, London.
82 Zussman, E., Theron, A., and Yarin, A.L. (2003) Formation of nanofiber crossbars in electrospinning. *Appl. Phys. Lett.*, **82**, 973–975.
83 Tucker, N., Stanger, J.J., Staiger, M.P., Razzaq, H., and Hofman, K. (2012) The history of the science and technology of electrospinning from 1600 to 1995. *J. Eng. Fibers Fabr.*, **7**(3), 63–73.
84 Bhardwaj, N. and Kundu, S.C. (2010) Electrospinning: a fascinating fiber fabrication technique. *Biotechnol. Adv.*, **28**, 325–347.
85 Li, Z. and Wang, C. (2013) in *One-Dimensional Nanostructures* (eds Z. Li and C. Wang), Springer, Berlin, Heidelberg, pp. 15–28.
86 Macossay, J., Marruffo, A., Rincon, R., Eubanks, T., and Kuang, A. (2007) Effect of needle diameter on nanofiber diameter and thermal properties of electrospun poly(methyl methacrylate). *Polym. Adv. Technol.*, **18**, 180–183.
87 Merritt, S.R., Exner, A.A., Lee, Z., and Recum, H.A. (2012) Electrospinning and imaging. *Adv. Eng. Mater.*, **14**(5), 266–278.
88 Xu, X., Zhuang, X., Chen, X., Wang, X., Yang, L., and Jing, X. (2006) Preparation of core-sheath composite nanofibers by emulsion electrospinning. *Macromol. Rapid Commun.*, **27**(19), 1637–1642.
89 Bazilevsky, A.V., Yarin, A.L., and Megaridis, C.M. (2007) Co-electrospinning of core–shell fibers using a single-nozzle technique. *Langmuir*, **23**(5), 2311–2314.
90 Kanani, A.G. and Bahrami, S.H. (2010) Review on electrospun nanofibers scaffold and biomedical applications, trends biomater. *Artif. Organs*, **24**(2), 93–115.

91 Fang, J., Wang, X., and Lin, T. (2011) *Nanofibers – Production, Properties and Functional Applications* (ed., Lin Tong), InTech–Open Access Publisher, Rijeka, pp. 287–326.

92 John, S., Balasubramanian, V., and Shenoi, B.A. (1985) Studies on anodizing of aluminium in alkaline electrolyte using alternating current. *Surf. Technol.*, **26**(3), 207–216.

93 Tsangaraki-Kaplanoglou, I., Theohari, S., Dimogerontakis, T., Wang, Y.M., Kuo, H.H.H., and Kia, S. (2006) Effect of alloy types on the anodizing process of aluminum. *Surf. Coat. Technol.*, **200**(8), 2634–2641.

94 Hili, K., Fan, D., Guzenko, V.A., and Ekinci, Y. (2015) Nickel electroplating for high-resolution nanostructures. *Microelectron. Eng.*, **141**, 122–128.

95 Zhang, Y., Wang, H., Sun, Y., Wu, K., Wang, H., Cheng, P., and Ding, G. (2016) Copper electroplating technique for efficient manufacturing of low-cost silicon interposers. *Microelectron. Eng.*, **150**, 39–42.

96 Ren, F.Z., Yin, L.T., Wang, S.S., Volinsky, A.A., and Tian, B.H. (2013) Cyanide-free silver electroplating process in thiosulfate bath and microstructure analysis of Ag coatings. *Trans. Nonferrous Met. Soc. China*, **23**(12), 3822–3828.

97 Muench, F., Felix, E.M., Rauber, M., Schaefer, S., Antoni, M., Kunz, U., Kleebe, H.J., Trautmann, C., and Ensinger, W. (2016) Electrodeposition and electroless plating of hierarchical metal superstructures composed of 1D nano-and microscale building blocks. *Electrochim. Acta*, **202**, 47–54.

98 Baudrand, D. and Bengston, J. (1995) Electroless plating processes: developing technologies for electroless nickel, palladium, and gold. *Met. Finish.*, **93** (9), 55–57.

99 Shacham-Diamand, Y., Osaka, T., Okinaka, Y., Sugiyama, A., and Dubin, V. (2015) 30years of electroless plating for semiconductor and polymer micro-systems. *Microelectron. Eng.*, **132**, 35–45.

100 Besra, L. and Liu, M. (2007) A review on fundamentals and applications of electrophoretic deposition (EPD). *Prog. Mater. Sci.*, **52**(1), 1–61.

101 Corni, I., Ryan, M.P., and Boccaccini, A.R. (2008) Electrophoretic deposition: from traditional ceramics to nanotechnology. *J. Eur. Ceram. Soc.*, **28** (7), 1353–1367.

102 Muresan, L., Popovici, E.J., Tomsa, A.R., Silaghi-Dumitrescu, L., Barbu-Tudoran, L., and Indrea, E. (2008) Preparation by dip coating method and characterisation of WO3 thin films. *J. Optoelectron. Adv. Mater.*, **10**(9), 2261–2264.

3

Surface Characterization Techniques

Gökhan Erdoğan, Günnur Güler, Tuğba Kiliç, Duygu O. Kiliç, Beyhan Erdoğan, Zahide Tosun, Hilal D. Kivrak, Uğur Türkan, Fatih Özcan, Mehmet Gürsoy, and Mustafa Karaman

3.1 Introduction

The chemical, physical, and morphological features of surfaces play a very crucial role in material properties including wettability, optical, adhesive, mechanical, and so on. Therefore, diagnosis of surface structures is very important for materials science. The properties of natural and fabricated material surfaces can be determined using characterization techniques. In this chapter, some of the most common characterization methods are given.

3.2 Surface Characterization Methods

3.2.1 X-ray Spectroscopy Techniques

The discovery of X-rays in 1845 by Wilhelm Conrad Rontgen gave rise to significant scientific advancement that benefited a variety of fields by making the invisible to the visible. In 1901, Wilhelm Rontgen received the Nobel Prize in physics for the invention of X-rays. The discovery of X-ray led to significant improvement in the diagnosis for medical applications. In the years subsequent to this discovery, X-rays were used for military applications. Initially, the harmful effects of radiation were not comprehended. Later, it was observed that those who were exposed to X-ray radiation encountered severe burns and skin damages. Thomas Edison's assistant who worked extensively with X-rays, died of skin cancer in 1904. Then, the risk of X-ray radiation was fully understood and this led to the use of protection shields. Today, X-ray technology is being widely employed in numerous fields such as medicine, material analysis, and devices [1].

X-rays are electromagnetic energy waves acting like light rays. X-rays have wavelengths in the angstrom range. Furthermore, X-rays are highly energetic and are employed not only to penetrate solids but also to probe their internal structure. As a result of the atom excitation via high energy photons or electrons, an electron is removed from the atom, leading to an electron hole at the inner orbit. These electron holes are filled by outer electrons. As a result

Surface Treatments for Biological, Chemical and Physical Applications, First Edition.
Edited by Mehmet Gürsoy and Mustafa Karaman.

of this phenomenon, X-rays are produced. The energy released by the excited atom is specific for an element and the intensity related to the amount of the element.

3.2.1.1 X-rays Florescent Spectroscopy

X-ray florescent spectroscopy technique is a powerful and sensitive technique as it allows fast and accurate qualitative as well as quantitative determination of elements. In this technique, the sample is irradiated by the X-rays coming from an X-ray source. X-rays absorbed by the sample are sufficiently energetic to remove electrons located in the inner K and L shell. There are several transitions that yield a fluorescent photon with a characteristic energy equal to the difference in energy of the initial and final orbital as (i) L → K transition is called K_α, (ii) M → K transition is called K_β, and (iii) M → L transition is called L_α [2].

As a result of electron removal, the unstable ions are formed. They become more stable and electron transfer occurs from the outer shells to the inner shells leading to X-ray production. The schematic representation of electron removal from the outer shells to the inner shells is demonstrated in Figure 3.1. This X-ray is called *secondary X-ray*, which is different from the basic X-ray obtained from the X-ray source. The radiation released during the electron transfer from the outer layer to the inner layer is called *fluorescence radiation*. The energy of X-ray fluorescence depends on the energy difference of the electrons transferred between the orbitals.

The intensity of the fluorescent radiation depends on the concentration and the amount of the sample. The fluorescent radiation wavelength can be calculated from Planck's law, given in Eq. (3.1):

$$\lambda = \frac{hc}{E} \tag{3.1}$$

λ is the wavelength
h is the Planck's constant, 6.626×10^{-34} J s
c is the speed of the light, 3.00×10^{8} m s^{-1}
E is the energy.

X-ray florescent spectroscopy (XRF) is one of the most commonly used methods. The samples are not destroyed during the analysis and this method could be

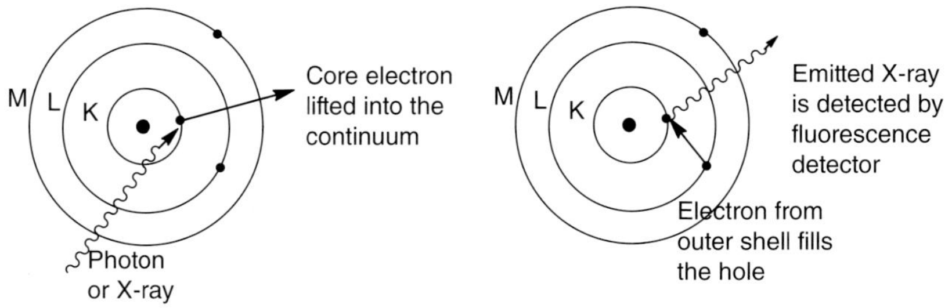

Figure 3.1 The illustration of the electron removal and the X-ray fluorescence radiation.

applied to solids, liquids, and gases. Samples of solid powder are pressed into pellets, liquid specimens are introduced into the cell made from plastic or metal, and gas samples are analyzed at high-pressure vessels. Orbital of transition between the outer and inner layers is between the K and L layers and, conventionally, is often referred to as the transition K_α. The samples are usually in the solid form for XRF applications. Hovewer, the analysis of the solutions can be performed by using appropriate equipments. The solid samples could be transferred on a carriri via pressing a tablet. For the nonhomogenous samples, samples could be melted or dissolved. XRF spectrum is a graphical representation of X-ray intensity peaks as a function of energy peaks. In the spectrum, one can note that the peak energy identifies the element and the peak intensity is indicative of the elemental concentration [3].

3.2.1.2 X-ray Diffraction Technique

X-ray diffraction (XRD) is an analytical technique primarily used for phase identification of a crystalline material. In addition, XRD is employed for monitoring the phase transformation as well as for determining the particle size. It has three basic units: an X-ray tube, a sample holder, and an X-ray detector.

X-rays are produced in a cathode ray tube by heating a filament to produce electrons. The electrons are accelerated through a target by applying a voltage and bombarding the target material with electrons following which a characteristic X-ray spectrum is obtained. In the spectrum there are several components such as K_α and K_β. The commonly used target materials are as follows: Cu, Fe, Mo, Cr. Most researchers prefer the copper target due to its availability with CuK_α radiation (1.5418 Å) [4]. Consider two parallel planes of atoms A–A and B–B in order to understand the diffraction in Figure 3.2. These two planes have the same miller indices h, k, and l. The incoming beam with an angle of θ is monochromatic and coherent. Two rays in this beam, labeled 1 and 2, are scattered by atoms P and Q. Constructive interference of the scattered rays 1 and 2 occurs also at an angle to the planes, if the path length difference between 1-P-1 and 2-Q-2 is equal to a whole number, n, of wavelengths. That is, the condition for diffraction is $n\lambda = SQ + QT$ or $n\lambda = 2d \sin \theta$.

$$n\lambda = 2d \sin \theta \tag{3.2}$$

λ is the wavelength of the X-ray beam
d is the distance between the two lattice
θ is the angle between the incoming X-rays and the normal to the reflecting lattice plane
n is the integer called the *order of reflection*.

The XRD analysis can be done in two different geometries. The first method is generally known as *powder method*, and can be defined as a symmetric method since the system always detects the scattered X-rays at a diffraction angle that is equal to the angle of X-rays incident of the surface. Both the sample and the detector move step by step during the measurement. While the samples rotate at an angle (θ) the detector moves two times this angle (2θ), which is also confirmed in Bragg's law. This method, therefore, always detects the diffracted X-rays

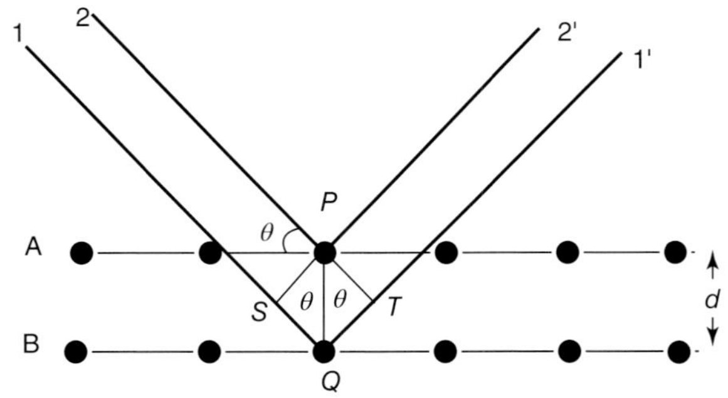

Figure 3.2 X-rays scattered by atoms in an ordered lattice.

from grains that are only oriented parallel to the surfaces and not the others. One disadvantage of this geometry is that the effective depth probed by the incident beam always changes during the scan due to the change in the angle of the incident beam. This property of the method might cause some misinterpretation if it is not taken notice on examining, for example, a material having a layered structure. This introduces the concept of fixing the angle of the incident beam and scanning the scattered beam; for that kind of analysis the following method will be considered.

The second method is used to determine the layered structures. In this system, the incident X-ray beam is fixed to a predetermined value on the sample and detector scans 2θ degrees. Contrary to the powder method, this method facilitates diffraction from the planes not parallel to the surface, never the parallel ones. Normally, this method uses very small angles of incidence on the surface providing information from quite thin layers and is ideally suitable for examination of layer thicknesses, which is on the order of 0.1 μm. The main power behind this method, therefore, involves providing the information layer by layer by just changing the angle of X-ray beam incident on the sample surface. Due to extremely low incident angles the scattered beam mainly lacks enough intensity and parallelism, which might lead to missing or wrong information. Using a Göbel mirror and parallel beam optics may solve this problem. Further information can be found in the literature.

Diffraction patterns are mainly used to identify the crystallographic phases in material characterization. Diffraction peaks are only observed when the sample possesses sufficient long-range order, which is an important limitation. Diffraction lines from perfect crystals are very narrow. The Scherrer equation and XRD spectra was used to calculate the dimensions of the materials. A typical XRD spectra is given in Figure 3.3. Scherrer's equation as follows:

$$\langle L \rangle = \frac{K\lambda}{\beta \cos \theta} \tag{3.3}$$

β is the integral breadth of peak at 2θ values ($\beta = \theta_1 - \theta_2$)
λ is the wavelength of the X-ray beam ($\lambda_{\text{Cu K}_\alpha} = 1.54\,\text{nm}$)
θ is one half of the deviation of the diffracted beam and the normal reflecting line
L is the interplanar spacing for a plane.

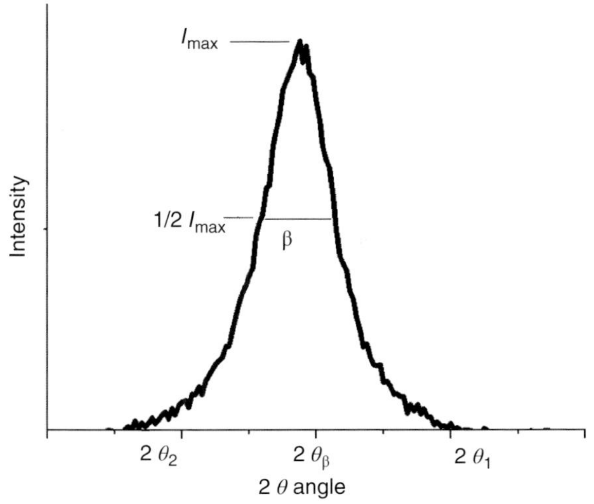

Figure 3.3 X-ray diffraction pattern.

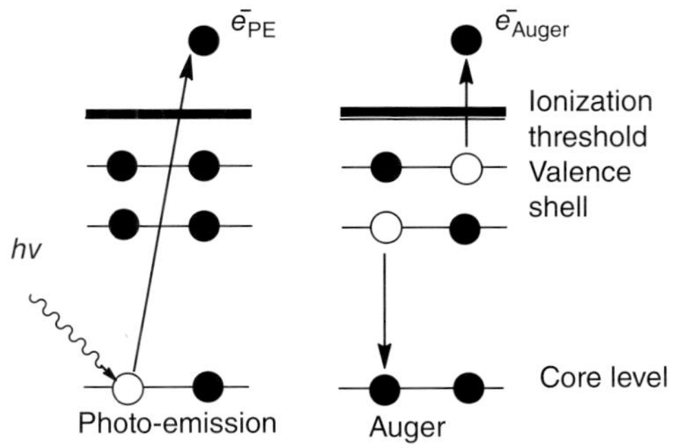

Figure 3.4 Auger processes and X-ray process.

3.2.1.3 X-ray Photoelectron Spectroscopy

X-ray photoelectron spectroscopy (XPS) provides information on the elemental composition and the oxidation state of the elements. Photoelectric effect is crucial for the XPS theory. Mg K_α (1253.6 eV) and Al K_α (1486.3 eV) are the most commonly used X-ray sources. In XPS, one measures the intensity of photoelectrons as a function of their kinetic energy.

The sample is irradiated with a monochromatic X-ray beam. An ejected photoelectron leaves behind a core hole in the atom. The hole created in the shell gives rise to a photoelectron. This hole if filled by an electronic transition from the other shell. The energy released is associated with the X-ray photon (see Figure 3.4). Each atom has core level electrons. The binding energy of each core level electron is specific. The measured kinetic energy of a core level photoelectron peak can be related directly to its characteristic binding energy. The energy of core level electrons depends on the chemical state of the atom. Binding energies on the XPS spectrum are not only element specific but also contain chemical information. As a rule of thumb, the binding energy increases with the oxidation state or with the electro negativity of its neighbors [4, 5].

Photoelectron peaks are labeled according to the electron quantum numbers. An electron has an orbital momentum l (0, 1, 2, 3… Indicated as s, p, d, f…) and a spin momentum. The total momentum of this electron could be predicted as $j = l + s$. Each level has two levels as the spin may be either up ($s = +1/2$) or down ($s = -1/2$). The energy difference is called the *spin orbit splitting*.

3.2.2 Surface Characterization with FTIR Spectroscopy

Fourier transform infrared (FTIR) spectroscopy is a kind of molecular vibrational spectroscopy that provides information about the chemical composition, bonding properties, and atomic and molecular structure (and environment) of materials at the molecular scale. FTIR spectroscopy has been extensively used as a surface characterization, analytical, or biophysical technique for the study of a variety of materials including chemical and biological species as well as drugs, thin films, coatings, synthetic polymers, natural biopolymers (DNA, protein, etc.), biomaterials, nanoparticles, and biomedical samples (dental implants etc.) [6–12]. FTIR technique enables to measure all type of materials (solids, liquids, and gases).

Infrared (IR) spectroscopy is based on the interaction of IR radiation with an oscillating electric dipole moment of the molecules. The IR region covers the range of 0.7–500 μm (14 000–20 cm^{-1}) and is divided into three parts as being near-IR (0.78–2.5 μm), mid-IR (2.5–50 μm), and far-IR (50–500 μm) radiation. The mid-IR range (4000–400 cm^{-1}) is the most useful region of the electromagnetic spectrum for the analysis of biological macromolecules, chemicals, and polymeric materials.

Atoms and molecules interact with electromagnetic radiation in a wide variety of ways (absorbing and/or emitting). Absorption of electromagnetic radiation stimulates different types of motion in atoms and/or molecules. For example, absorption of the IR radiation by the sample excites the interatomic vibrations (or normal modes). These normal modes are described by stretching (symmetric and antisymmetric), bending, rocking vibrations, and so on. If the frequency of IR light coincides with the frequency of molecular vibration and if stretching of bonds between the atoms creates a dipole moment change along with the vibration (IR-active), then the incident IR beam is absorbed by a molecular vibration of the sample, generating an IR spectrum (Figure 3.5). The pattern of absorption is called *spectrum (sing.) or spectra (pl.)*. The amount of IR light (or intensity)

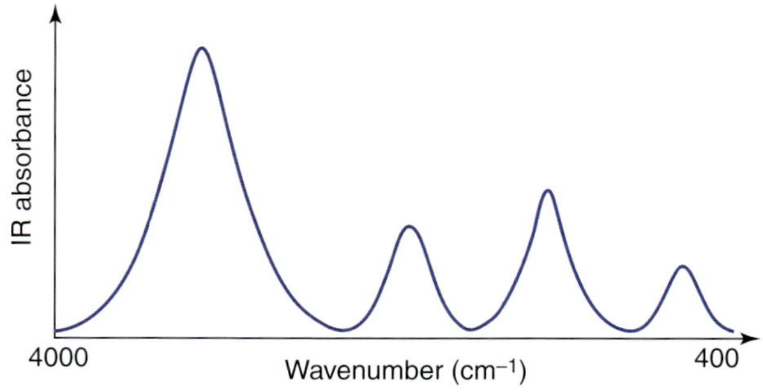

Figure 3.5 A schematic representation of an FTIR spectrum. (Image Courtesy of Günnur Güler.)

absorbed by the material is plotted as a function of wavenumber. Wavenumber is the number of waves in a space and is depicted by the reciprocal of wavelength. Wavenumber (cm^{-1}) is used as the unit of energy in a spectrum since it is proportional to the frequency and energy due to

$$\text{wavenumber} = \frac{1}{\lambda} = \frac{\vartheta}{c} \tag{3.4}$$

$$\text{energy} = h\vartheta \tag{3.5}$$

where h is the Planck constant, ϑ is the frequency, and c is the speed of light. Vibrational modes that do not generate a dipole moment change in the molecule are called *IR-inactive* (not observed in the IR spectrum) such as symmetric stretching modes of CO_2 and homonuclear diatomic molecules (H_2, N_2, and O_2) while heteronuclear diatomic molecules (CO and HCl) are IR active having induced dipole moment due to the change in the atomic positions (dipole moment = charge × distance).

FTIR technique is used to determine the presence of the functional groups (i.e., OH, NH, CH, CO, and PO) in materials. The IR radiation absorbed by the vibrations of functional groups gives rise to the characteristic IR signals at specific wavenumbers (or frequencies). These signals are unique for a given compound and the absorbed intensity is associated with the quantity of the absorber. Atoms within a molecule oscillate with a frequency of

$$\vartheta_{vib} = \frac{1}{2\pi}\sqrt{\frac{k}{\mu}} \tag{3.6}$$

where μ is the reduced mass of the atoms and k is the force constant of the bond between molecules or atoms. Different vibrational modes are detected at different frequencies in the spectrum because a characteristic frequency depends on two important parameters. These are the masses of atoms and the bond length between atoms. This makes IR spectroscopy very sensitive to any changes in the molecular environment. Frequency decreases with increasing atomic mass while it increases with increasing bond energy (k). Therefore, an IR spectrum provides a fingerprint pattern of molecular structure of the material, which is composed of thousands of atoms.

3.2.2.1 FTIR Spectrometers

An FTIR spectrometer is an instrument that displays the IR spectrum of a sample. In the 1940s, the dispersive IR spectrometers were developed with a grating monochromator used as an analytical technique for the characterization of compounds. However, modern FTIR spectrometers have been used for advanced researches since the late 1970s, which use a Michelson interferometer to record an IR spectrum.

A Michelson interferometer is the most important component of the spectrometer [13, 14]. It harbors a source, beam splitter, two plane mirrors (one fixed and one movable), a laser, and a detector. The collimated IR beam, which passes through the interferometer, is split into two beams via the beam splitter. After recombination, a constructive or destructive interference pattern is created and is directed to the output. The output IR beam passes through the sample

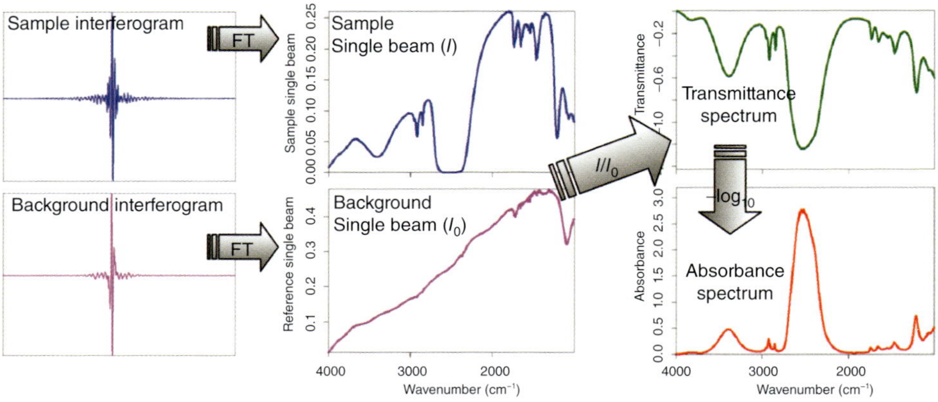

Figure 3.6 The process of formation of a Fourier transform (FT) infrared spectrum drawn by OPUS software [15]. (Image Courtesy of Günnur Güler.)

where some energy is absorbed and some is transmitted. The transmitted light is focused toward the detector. The optical path length of a beam varies with a movable mirror. The detector signal (output) is collected as an interferogram (Figure 3.6). This interferogram is transformed to the frequency domain by a computer using the mathematical operation called Fourier-transformation at discrete points to display the IR single beam spectrum. To produce *absorbance or transmittance spectrum*, the *sample single beam spectrum* is collected taking the reference as *background single beam spectrum* (Figure 3.6). This reference spectrum is recorded in the absence of a sample; for example, the air spectrum. The final spectrum (i.e., transmittance spectrum) can be converted to one another (i.e., absorbance spectrum) by calculating the negative logarithm ($A = -\log_{10} T$) of the data points. The absorbed IR radiation (A) by a sample can be directly calculated by *Beer-Lambert law:*

$$A = -\log_{10}\left(\frac{I}{I_0}\right) \tag{3.7}$$

where (I) and (I_0) are the intensities of the single beam spectrum of the sample and reference, respectively.

For FTIR spectrometers, photon detectors (semiconductor detectors such as mercury-cadmium-tellurium or MCT) and thermal detectors (deuterated triglycine sulfate or DTGS) are available to measure the intensity of the incident beam. MCT detectors operate at liquid nitrogen temperature of about 77 K while DTGS detectors work at room temperature.

3.2.2.2 Methods and Sampling Techniques

An IR spectrum can be recorded by FTIR spectrometers combined with attenuated total reflection unit-Fourier transform infrared (ATR-FTIR) or in the transmission mode. Both are absorption techniques. FTIR can identify even small concentrations of materials (typically 0.1–1 mM).

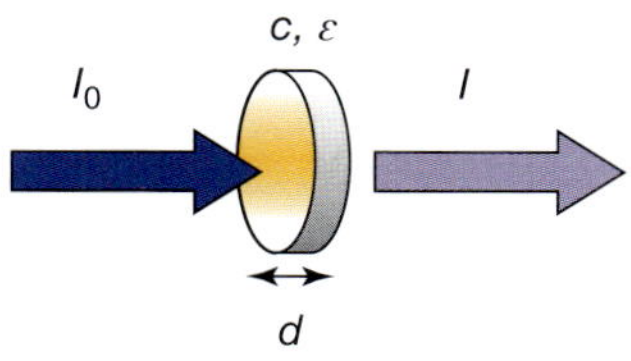

Figure 3.7 A schematic representation of an absorption measurement with FTIR in the transmission mode [15]. (Image Courtesy of Günnur Güler.)

FTIR in transmission mode: In transmission measurements, the IR beam is passed through a cuvette before it reaches the detector (Figure 3.7)

$$A = \varepsilon c d = -\log_{10}\left(\frac{I}{I_0}\right) \tag{3.8}$$

The absorbance is calculated from the Beer-Lambert's law, where c is the concentration of the absorbing molecules, ε is the molar extinction coefficient (or molar absorption index), and d is the path length of the measuring cuvette. Thus, this technique is suitable for quantitative analysis. Importantly, the FTIR cuvettes (or optical materials) should be transparent in the mid-IR spectral range so that only IR signals arising from the measuring sample are observed in the spectrum. Such optical materials can be ZnSe, AgCl, KRS-5, KBr, CsBr, CsI, KCl, NaCl, CaF_2, and BaF_2 that have different stability against aqueous solution and different refractive indices. A suitable path length has to be selected for liquid samples. For example, CaF_2 is the most widely used cuvette to study aqueous solutions of materials due to its insolubility in water and low refractive index.

In comparison to aqueous solutions, analyzing of solid samples is rather complex for a transmission measurement. A certain amount of solid sample must be diluted with KBr that is transparent in the IR range. Subsequently, the mixture of KBr and the solid sample is pressed to make "KBr-pellet." Homogenization of this pellet is crucial to obtain a high quality and reproducible spectra. Therefore, ATR-FTIR spectroscopy is mostly preferred to get rid of preparing of KBr-pellets.

ATR-FTIR spectroscopy: One of the most frequently applied experimental methods for surface characterization is the ATR-FTIR spectroscopy. It is an absorption technique that soft materials (biological and chemical species, polymers, gels, pastes, semi-dry films, 2D crystals, and proteoliposomes of protein–membrane complexes) as well as hard materials (thin films, powders, metals, etc.) can be easily analyzed without particular sample preparation.

The IR beam is directed to a trapezoidal-shaped IR-transparent crystal with a high refractive index, n (Figure 3.8). The crystal material is mostly capped with zinc selenide ($n = 2.42$), diamond ($n = 2.4$), or germanium ($n = 4$), where the sample material is placed (Figures 3.8 and 3.9). The incident light is reflected at the interface between two materials. which is made up of different refractive indexes. The reflected IR beam is attenuated and protrudes a few micrometers (0.5–5 μm) into the sample medium ($n_{sample} = {\sim}1.5$). IR radiation is partially absorbed by the sample deposited on the surface of crystal plate. Therefore, ATR-FTIR is a kind of surface sensitive technique and sometimes also stands for evanescent wave spectroscopy. It is easy to use because the material is just placed on the top of the ATR unit. Solid materials such as thin films, coatings, powders, metals, papers, rubber, fibers, leather-like samples, and polymers can pressed with a pressure tip

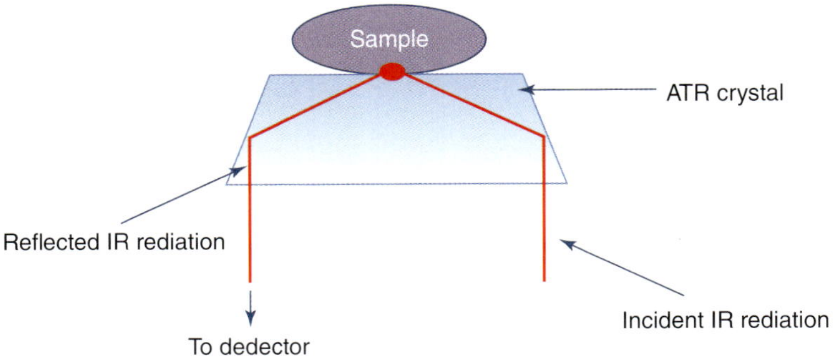

Figure 3.8 A schematic representation of an ATR unit. IR beam undergoes internal one-single reflection in the ATR crystal of high refractive index. The sample is in contact with the crystal. (Image Courtesy of Günnur Güler.)

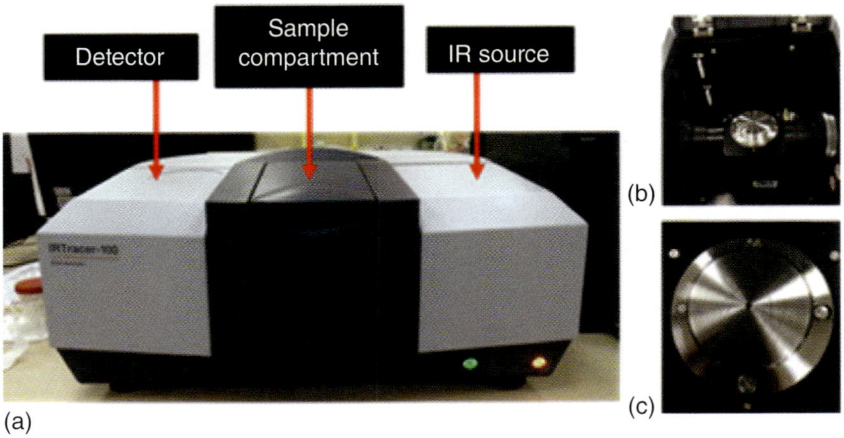

Figure 3.9 The FTIR spectrometer (a) equipped with an ATR unit and crystal plate (b). Image at (c) shows the sampling area made up of diamond. (Image Courtesy of Günnur Güler.)

of a high-pressure clamp to provide an intimate physical contact with the crystal plate. This technique requires only a tiny amount of sample but should be in good contact with the sampling area to produce a strong and reproducible IR signal. Therefore, it is well suited particularly for the surface characterization of the materials.

3.2.2.3 Advantages and Disadvantages of FTIR Spectroscopy

In comparison to other techniques, FTIR spectroscopy has many advantages. FTIR spectrometers provide high spectral accuracy with high signal-to-noise ratio, high resolution, and sensitivity to weak peaks. The technique is also economic, nondestructive, and fast. It is easy to use and one spectrum is measured very quickly within seconds or a couple of minutes. It does not change the structure of the samples due to small photon energy (0.05–0.5 eV). FTIR requires a small amount of sample; particularly, a few microliters (or one drop) of liquids and one pinch of solid material are enough for the FTIR measurements.

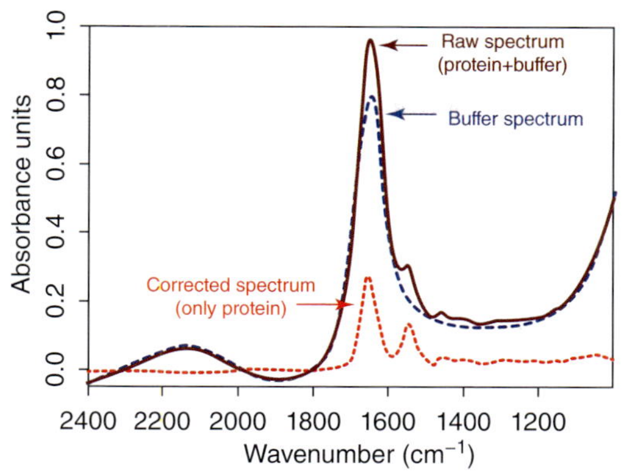

Figure 3.10 The FTIR absorbance spectra of the osmosensor protein BetP in *aqueous solutions* before (raw spectrum) and after (corrected spectrum) H_2O buffer subtraction [15]. (Image Courtesy of Günnur Güler.)

However, FTIR spectroscopy has some limitations. Particularly, studying with aqueous solutions has certain difficulties because water has an intense absorbance due to O—H vibrational signals that overlap strongly with the signals of other functional groups. This problem can be overcome by taking the water as a background before measurement of the real sample or the spectrum of the water/buffer solution can be mathematically subtracted/corrected from the real sample spectrum (see Figure 3.10) so that we can get rid of the water absorption in the spectrum. The other disadvantage is that FTIR provides only average information from one type of functional group. For example, the C=O vibrational modes of two different types of materials are observed at the same frequency range with a minimal frequency shifting depending on the local and global environment of the C=O molecules.

3.2.2.4 Applications of FTIR Spectroscopy

IR spectroscopy is one of the classical methods used as a characterization technique to explore chemical bonds in surface atoms in many research areas, including organic and inorganic chemistry, solid state physics, material science, biotechnology, biomedical engineering, biophysics, biochemistry, and pharmacy.

FTIR is a time-saving and nondestructive technique that requires low set up and running cost. Importantly, it provides label-free analysis of materials. This makes FTIR spectroscopy a valuable tool for the characterization of structural and physicochemical properties of materials such as dynamics and flexibility, stability, purity, ionization (protonation or deprotonation), thermodynamics, binding, kinetics of a reaction, structure determination, identification of chemical and molecular composition, surface modifications, and structural inhomogeneity. FTIR is also sensitive to single bonding changes, thus, FTIR technique can determine with great confidence the bond length, bond strength, H-bonding properties, conformational freedom, charge distributions, redox state, and electrostatic interactions in a material.

Surface characterization studies using FTIR spectroscopy for chemical analysis have been performed on a variety of nanoparticles that show potential for drug delivery systems [6, 7]. For instance, zein–pectin capsule was developed for target delivery [6]. Here, the ATR technique was used to analyze the characteristic chemical interactions between pectin (a carbohydrate) and zein (a protein) which are the natural biopolymers. Formation of the zein – pectin complex via hydrogen bonding was confirmed in the IR spectra.

FTIR spectroscopy has been primarily used for surface chemistry of nanostructures and nanoparticles as well [16, 17]. For example, different types of nanostructures (nanobelts, nanobricks, and nanosheets) of polypyrrole (PPy), which are very promising polymers for the electrochemical supercapacitors, were fabricated [16]. Polymerization of pyrrole was achieved electrochemically. Subsequently, they were analyzed with FTIR spectroscopy to confirm the formation of conducting PPy thin films. Observation of characteristic IR signals absorbing at $1194\,cm^{-1}$ (C—N stretching vibrations) and $922\,cm^{-1}$ (C=C in-plane bending vibrations of the pyrrole ring) reflect the doping state of PPy thin film. ATR-FTIR spectroscopy is a powerful tool for the surface characterization of polymeric materials. It provides surface-specific information since the penetration depth of the IR beam can protrude deeply into the surface of a material. For example, polymers of HEMA (2-hydroxyethyl methacrylate) are used in many biomedical applications like drug delivery and biosensors. Thin films of poly-HEMA at various conditions have been extensively analyzed with FTIR technique [18].

FTIR is particularly well suited for the analysis of polymer processing (surface modifications, coatings, etc.). Surface modifications of commercial ultrafiltration polyethersulfone membranes achieved by a graft polymerization technique were characterized by the ATR technique to reveal the chemical changes during modification [12]. In this work, FTIR could also recognize the adsorption of the albumin protein on the surface of these membranes. It was also able to detect the conformational changes of the albumin during adsorption. Thus, the ATR technique is very useful for binding or adsorption studies [19] with biological systems such as proteins, nucleic acids lipids, and so on. Since ATR is a surface-sensitive technique, conformation analysis of adsorbed proteins, orientation of molecules/functional groups, surfactants, and drugs can be studied well.

When compared to other techniques, FTIR spectroscopy is highly effective for surface characterization. It has been commonly used also for the characterization of thin films, nanowires, and coatings in terms of molecular ordering/disordering, bonding arrangements, trace impurities as well as associated chemical imperfections. For example, SiO_2 or silica-like thin films and nanowires deposited directly on the ATR crystal have been studied for many years [20–23]. In these studies, the Si—O—Si bonding vibrations (—O—Si—O—) detected in the IR range of $1000-1250\,cm^{-1}$ are related to interface roughness and structural disordering. Particularly, a major absorption band centered around $1070\,cm^{-1}$ refers to the structural disordering. An increment of the intensity of this band was correlated with the structurally disordered amorphous SiO_2.

FTIR spectroscopy has been also used in the field of material science and solid state physics to investigate the dynamics of atoms in crystals due to phonons [24–27]. Normal modes in molecules and phonons in crystals are the detected vibrational modes of materials. IR spectra of rock minerals such as calcite ($CaCO_3$), which has a trigonal structure with two molecules per unit cell as well as dolomite ($CaMg(CO_3)_2$) which has a hexagonal structure, have been analyzed [24]. Since their structures are very similar, their IR spectra resemble, but FTIR spectroscopy can distinguish these crystals by following the spectral shifts in wavenumbers. Other crystal lattices such as $(GaAS)_n/(AlAs)_m$ [26, 27] and InAsSb/GaSb superlattices [28] have been also characterized with FTIR spectroscopy.

Characterization of the surface modifications and biochemical properties of the implants (bone and dental implants, etc.) and biomaterials (calcium hydroxyapatite, etc.) are crucial for their interactions with the surrounding tissue/cell since it affects the bioactivity and biocompatibility. Thus, FTIR spectroscopic characterization has become increasingly valuable in the biomedical field [29–31]. Titanium-based implants and Ca^{2+}-alginate hydrogel implants are some examples analyzed with FTIR spectroscopy. Dental implants coating with hydroxyapatite using two different processes (plasma spray and modified biomimetic process) was also characterized with FTIR [8].

FTIR technique is frequently used for analyzing of tissue, cell, and biological macromolecules like peptides, proteins, lipids, carbohydrates, and nucleic acids (DNA, RNA, and microRNA), and even individual amino acids. For example, molecular and biochemical characterization of cells [9] and tissues [13, 30], 2D crystals and proteoliposomes of protein–membrane complexes [10, 15], analysis of secondary structure of proteins and characterization of enzyme–substrate reactions [14, 32–34], lipid–protein interactions [10, 35], and nucleic acid–drug interactions [36–38] are some examples studied by FTIR spectroscopy. These studies provide information on biochemical and biophysical characteristics, such as determination of conformational changes and functional features of proteins, inter- and intra-molecular interactions with their environment, bonding arrangements, dynamics and flexibility of molecules, stability, protonation–deprotonation status.

3.2.3 Nuclear Magnetic Resonance Spectroscopy

Nuclear magnetic resonance spectroscopy (NMR spectroscopy), is a nondestructive analytical technique that determines the unique structure and also enables conformational analysis of a compound by exploiting the magnetic properties of certain atomic nuclei. The technique has been discovered simultaneously by two research groups Edward M. Purcell, Howard C. Torrey, Richard V. Pound from Harvard University and Felex Bloch, William W. Hansen, and Martin Packard from Stanford University upon observing the phenomenon of NMR in both solids and liquids [39]. NMR spectroscopy has been applied to a wide range of disciplines including chemical sciences, biological and biochemical research, food industry, pharmaceutical development and production, polymer industry, and even geophysics. It has also led to the development of magnetic resonance imaging (MRI), which uses the same principle as NMR to get an image of a soft tissue [40, 41].

As one of the surface characterization techniques, NMR spectroscopy has been widely used for the characterization of solid materials both in bulk and surface structure. It has also been implemented to understand the surface reactivity, oxidation state, and structure of active species as a complementary technique to other alternative methods like XRD and high-resolution electron microscopy [42].

3.2.3.1 Theory of NMR Spectroscopy

Subatomic particles included in a nuclei are considered as spinning on their own axes. However, in many atoms (like ^{12}C) that have even number of protons and neutrons, these spins are paired and the overall spin, I, of the nucleus becomes zero (so called NMR inactive). Otherwise, the atomic nucleus has an overall spin; when both the numbers of neutrons and protons are odd numbers, it will be an integer (1, 2, 3, etc.); on the other hand, if the sum of the number of protons and number of neutrons is odd, it will be a half-integer (1/2, 3/2, etc.). The overall spin generates a randomly oriented magnetic field. When an external magnetic field B_0 is applied, two spin states exist: +1/2 (lower energy level) and −1/2 (higher energy level) and the nuclear magnetic field align itself either with or oppose the external magnetic field due to absorption and re-emission of electromagnetic radio frequency radiation (roughly 4–900 MHz). When the spin of the nuclei flipped, the nuclei are said to be in resonance with B_0. The frequency at which the *resonance* occurs is called *Larmor frequency* measured in an NMR experiment and is a unique property of a nucleus since every nucleus absorbs electromagnetic radiation at different wavelengths/frequencies (Figure 3.11).

Additionally, every specific nuclei enters resonance at different energies and frequencies depending on their chemical and electronic environment. Since charged particles surround the nucleus by moving in a loop to create a magnetic field, the "experienced" magnetic field by the nucleus is slightly different from the external magnetic field applied. This phenomenon is called *nuclear shielding* and the resultant change in energy levels requires a different frequency for resonance which in return results in a new peak in an *NMR spectrum*. The basis of structural and conformational analysis of molecules relies on nuclear shielding and *chemical shift* (δ) which is a term used to describe the relative resonance frequency of a certain compound to a reference compound and represented by the formula:

$$\delta = \frac{\upsilon_{\text{sam}} - \upsilon_{\text{ref}}}{\upsilon_{\text{ref}}} \tag{3.9}$$

where υ_{sam} is the absolute resonance frequency of a certain sample while υ_{ref} is an absolute resonance frequency for a standard compound under the same external

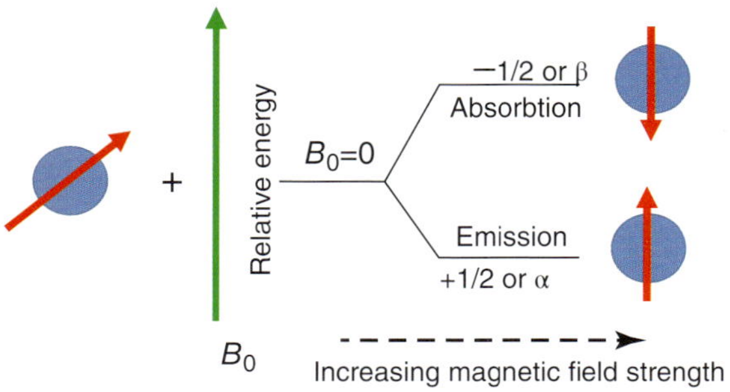

Figure 3.11 Energy levels of a nucleus with $I = ½$ after an exertion of a magnetic field B_0.

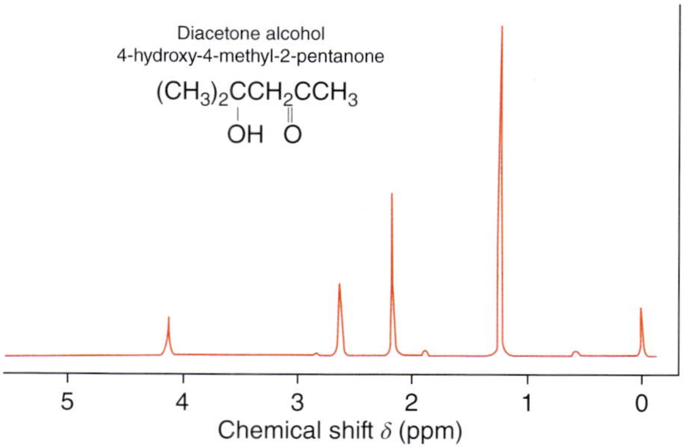

Figure 3.12 Sketch of a proton NMR spectrum included in Becker, Chapter 1.

magnetic field. As a standard compound, generally tetramethylsilane (TMS) has been used.

In an NMR spectrum that consists of a set of peaks as exemplified below, each peak corresponds to a distinct compound. The x-axis shows the frequency in parts per million and the y-axis shows the absorption peaks (Figure 3.12).

3.2.3.2 Types of NMR Spectroscopy

There are various types of NMR spectroscopy depending on:

- The type of instrumentation used
  - *Continuous wave NMR*: The absorbance of the sample under resonance is recorded as a function of either radio frequency (RF) excitation or magnetic field. In continuous-wave (CW) NMR, both RF excitation and magnetic field are continuously on.
  - *Fourier transform NMR*: The sample of interest is irradiated with a specific range of frequencies for a specific time and the decay of signal over time (FID; free induction decay) is observed till the relaxation (return back to the ground state energy level) and then converted to signal-frequency domain.
- The type of nucleus involved
  - *Carbon-13 NMR*: used for identification of carbon atoms, only ^{13}C isotope that accounts for 1% of all organic molecules since ^{12}C is not NMR active, in a substance and has a wide variety of use in organic chemistry for structure elucidation.
  - *^{1}H NMR*: Also known as *proton NMR* used to identify the structure of a compound with respect to hydrogen by providing information regarding the number of different H present in a molecule, the electronic environment of the different types of hydrogens, the number of hydrogen neighboring a hydrogen. For solution-based measurements, inert solvents that do not contain protons (deuterated, ^{2}H = Deuterium/D) like; D_2O (deuterated water), $CDCl_3$ (deuterated chloroform) are used.
- The physical state of the sample
  - *Solid state NMR*: NMR conducted in solids. The NMR spectra seems to be broadened due to dipolar interactions and chemical shift anisotropy (orientation dependence of the chemical shift). For high resolution solid state

NMR, various methods have been used: magic angle spinning, dilution, multiple pulse sequences, cross-polarization [43].
  - *Solution state NMR*: most of the NMR experiments are done in solution phase due to rapid molecular motion of solutions.
  - *Gas state NMR*: used to expand the molecular electronic properties, to yield temperature-dependent experimental data and observe intermolecular effects on molecular electronic properties [44].
- The data collection
  - *One dimensional NMR* (i.e., Carbon-13 NMR, ^{1}H NMR): only x axis has a chemical shift scale, y axis shows the intensity.
  - *Two-dimensional NMR* (i.e., homonuclear 2D, heteronuclear 2D): both the x and y axes have chemical shift scales and the data is collected in 2D as a grid-like map. The structural determination is done by looking at the spectra at the peaks in the grid and by matching them to the x and y axes. Intensities used as third axis. There are several types of 2-D NMR; COSY (correlation spectroscopy, both axes corresponds to ^{1}H NMR), HETCOR (heteronuclear correlation spectroscopy where ^{1}H NMR and ^{13}C NMR are used in x- and y-axis).

3.2.3.3 Instrumentation and Sample Handling

A typical NMR instrument basically consists of four main components: a magnet, a frequency generator, a detector, and a recorder as illustrated in Figure 3.13. The magnets used in NMR instruments are superconducting responsible for providing a homogeneous magnetic field at 60–100 MHz. The induction of the magnetic field is provided by magnetic coils when alternating current created by frequency generator flows through them. The frequency generator could work either as a continuous wave or pulsed. The detector subtracts the base frequency

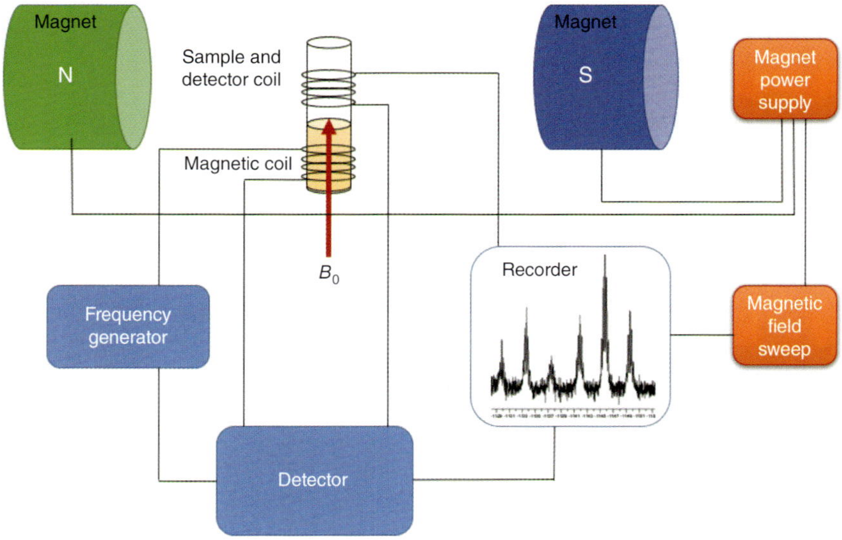

Figure 3.13 Typical sketch of an NMR instrument.

from the output frequency and recorder can be of in various types: oscilloscopes, computer, and so on.

3.2.3.4 Applications of NMR

NMR has become an indispensable tool for analytical technology and has taken a pivotal role in a wide variety of applications in many disciplines. Not only scientific disciplines but also various industries take advantage of sophisticated NMR for routine analysis and basic research. Some examples have been given below to give insight about the usage of NMR. For detailed explanations and other examples other reference books and reviews dedicated for this aim could be followed [40, 45–50].

Biochemistry: to analyze the biomacromolecules up to 100 kDa for structure determination, to identify the protein function and protein folding, to do atomic resolution structure analysis on membrane or such environments, motional properties of biomacromolecules, determination of ionization state of the enzymes, determination of interior water of proteins, and analysis of hydrogen bonding for several biomolecule interactions.
Pharmaceutical: drug screening, drug design, drug–protein drug/enzyme interactions, and drug–receptor interactions.
Chemistry: structural elucidation and conformational analysis of compounds.
Material science: polymer chemistry, physics, silicates (i.e., minerals, zeolites).
Food analysis: determination of oil content of a product (seeds, meals, and meat, etc.) and determination of humidity in various foodstuffs.

3.2.4 Electron Microscopes

An electron microscope uses electrons to form a magnified image. This magnified image is the ratio of the image size on the screen to the size of the object that is being examined under a microscope. To understand why it is essential to use electrons instead of light for high magnified images we need to know the definition of resolution and depth of field.

Resolution is the minimum distance that a microscope can resolve. Resolution of an optical microscope is limited due to the nature of visible light. The wavelength of visible light is between 400 and 700 nm. The electron energy in an electron microscope can change from 0.1 up to 300 keV. The wavelength changes with the energy and, therefore, the wavelength is not constant in a whole electron energy range

$$E = hc/\lambda \tag{3.10}$$

For 10 keV of electron energy wavelength of an electron based on Eq. (3.10) is 0.12 nm. When the wavelength of visible light is compared with the wavelength of an electron it is seen that the resolution of an electron microscope is more powerful than that of an optical microscope.

The depth of a field is related to the features in the image. It is a measure of how many of these features are in focus at the same time. Electron microscopes have higher depths of field when compared to light microscopes.

Electron microscopes not only give information about topography and morphology but also of composition and structure. These are the information that a light microscope cannot supply.

There are two types of electron microscopes, scanning electron microscope (SEM) and transmission electron microscope (TEM). We will now investigate them in detail.

3.2.4.1 Scanning Electron Microscope (SEM)

SEM is generally preferred for a wide range of applications such as metals, polymers, ceramics, and surface coatings due to its wide range of capabilities such as getting high resolution and three-dimensional images and compositional, topographical, and morphological information about any solid materials.

The first SEM idea was proposed by Knoll in 1935. It was far beyond being a SEM but Von Ardenne in 1938 built a scanning transmission electron microscope (STEM) by adding a scanning coil to a TEM. STEM mode is still a standard imaging technique in modern microscopes. In 1942, Zworykin made some improvements in SEM design. Oatley from Cambridge University built a microscope with a better resolution in 1948 and, in 1965, the first commercial SEM was built by Cambridge instruments [51].

SEM, as it is in the other types of electron microscope, uses a focused beam of electrons that are produced in the electron source at the top of the microscope. The electron source has an area and this area is dependent on the type of the source that is being used. There are three types of electron sources: thermionic sources, Schottky emission sources, and field emission sources. Sources have an effect on the resolution and the best resolution is achieved in field emission electron sources. The electron beam follows a vertical path due to electromagnetic lenses through the column of the microscope. Electromagnetic lenses focus and direct the beam down to the sample. When the beam hits the sample the electrons interact with the specimen. The interaction is directly proportional to the energy of incident beam. This is called *interaction volume* and it becomes broader when the energy is increased. The electrons do not follow a straight line in the sample as it is seen in Figure 3.14. They follow a zig–zag path instead.

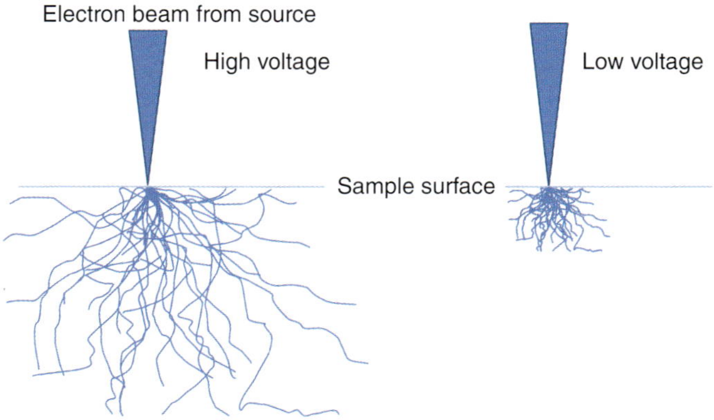

Figure 3.14 Interaction volume for high and low voltage.

Following this interaction a variety of characteristic signals are generated and every signal is detected by its own detector. Secondary electrons are detected by Secondary Electron (SE) detector, Back scattered electrons are detected by Back Scattered Electron (BSE) detector, X-rays are detected by Energy Dispersive X-ray Spectroscopy (EDS) detector and Transmitted electrons are detected by Scanning Transmission Electron Microscopy (STEM) detector as it is seen in Figure 3.15.

Characteristic signals after being detected by specific detectors are converted to a signal on the screen as an image. In Figure 3.16, an SE image of ZnO crystals is seen. This image is acquired using SE detector and it shows the topography of the ZnO crystals.

In Figure 3.17, a BSE image of gold particles on textile fibers using BSE detector is shown. This detector uses atomic contrast. A brighter BSE intensity correlates with higher average atomic number. The brighter the region of interest, the higher the atomic number and the darker the region of interest the lower the atomic number is.

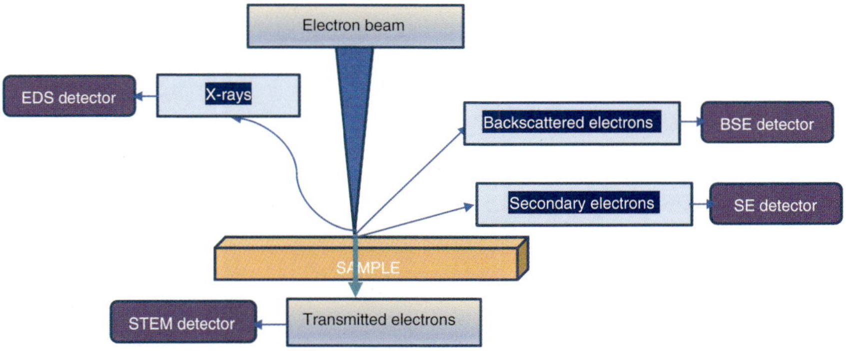

Figure 3.15 Electron beam and sample interaction.

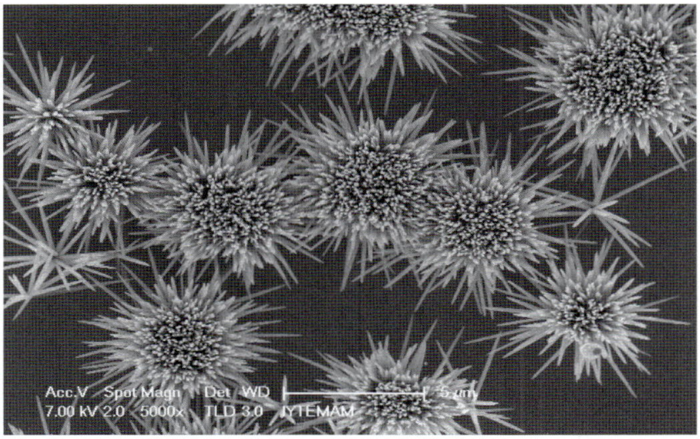

Figure 3.16 ZnO crystals.

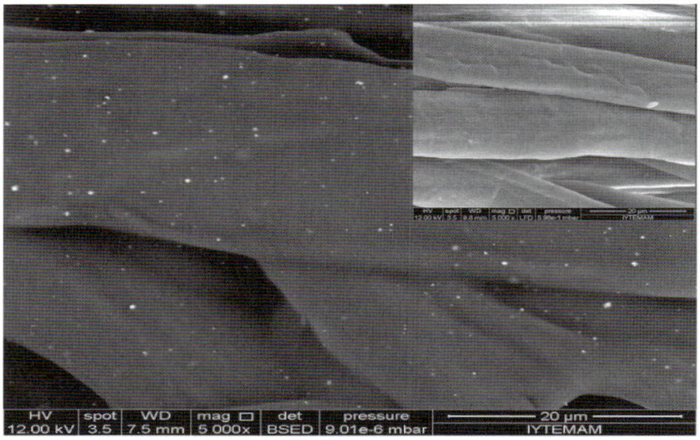

Figure 3.17 BSE image of gold particles.

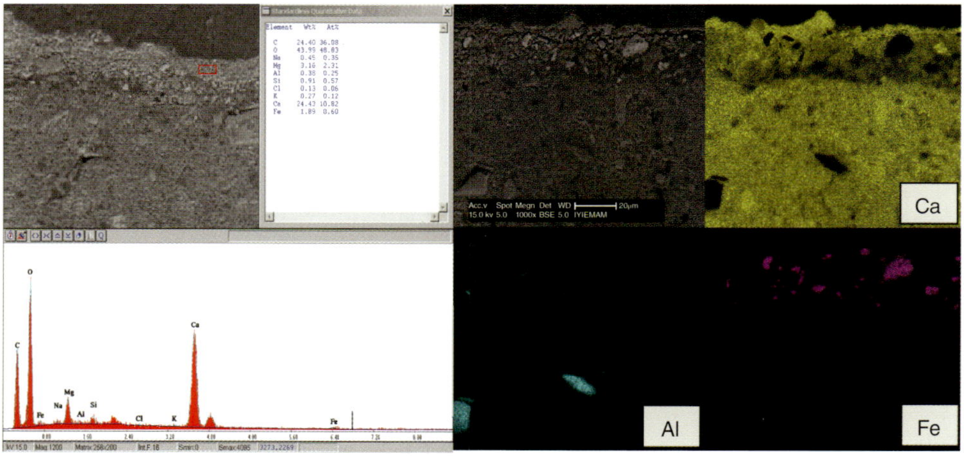

Figure 3.18 EDS image and elemental mapping of a ceramic sample.

Using X-ray imaging techniques one can get information about the composition of the sample by examining the qualitative and quantitative analysis and the elemental mapping of the sample. In Figure 3.18, a quantitative analysis of a ceramic material and its elemental mapping are seen.

Using a very thin sample, prepared by TEM sample preparation methods that will be described in Section 3.2.4.3, electrons pass through the sample and these transmitted electrons are observed using a STEM detector. In Figure 3.19a, a STEM image of gold particles on polystyrene latex is seen.

The detectors used to acquire the images discussed above are seen in Figure 3.20.

Sample Preparation for SEM Samples examined under SEM must be conductive, humid, and oil free. If the sample to be examined is not conductive, a sample coating procedure must be followed. One of the sample coating procedures for a SEM sample preparation is sputter coater. In this coating system Ar gas is used

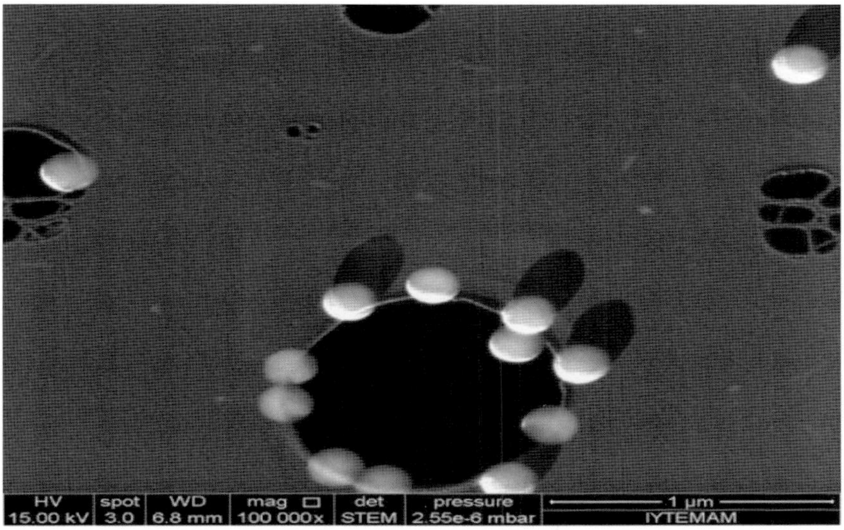

Figure 3.19 STEM image of gold particles on polystyrene latex.

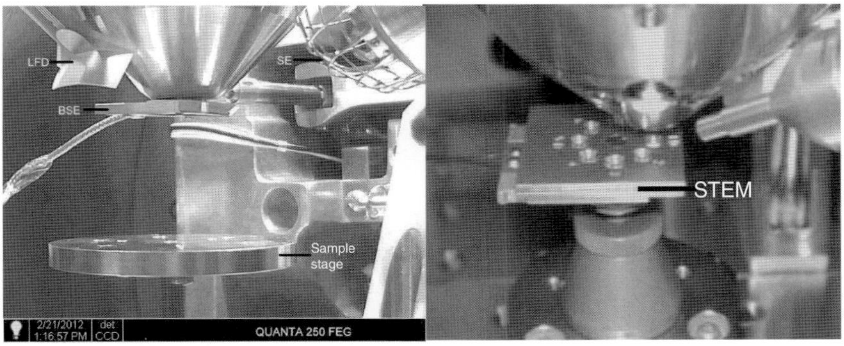

Figure 3.20 Detectors inside a SEM.

as an inert gas. Ar is ionized at high voltages. Positive Ar ions are attracted by the target and bombards it. The target is usually made of Au or Au/Pd. Target atoms are sputtered onto the sample making its surface conductive. The schematic of sputtering procedure is seen in Figure 3.21.

3.2.4.2 Environmental Scanning Electron Microscopy (ESEM)

The environmental scanning electron microscopy (ESEM) was developed in the mid-1980s, and differs from conventional SEM by changing the sample environment through a range of pressures, temperatures, and gas compositions [52]. Moist, wet, hot, dirty, oily, nonconductive samples can be examined in their natural state without modification or preparation. This technology allows imaging of samples and adds variables such as hydration, dehydration, sublimation, thermal cycling, and introduction of gas to characterize *in situ* dynamic changes such as tension, compression, deformation, crack propagation, and adhesion. Naturally hydrated samples, whose properties will change with drying, are easily character-

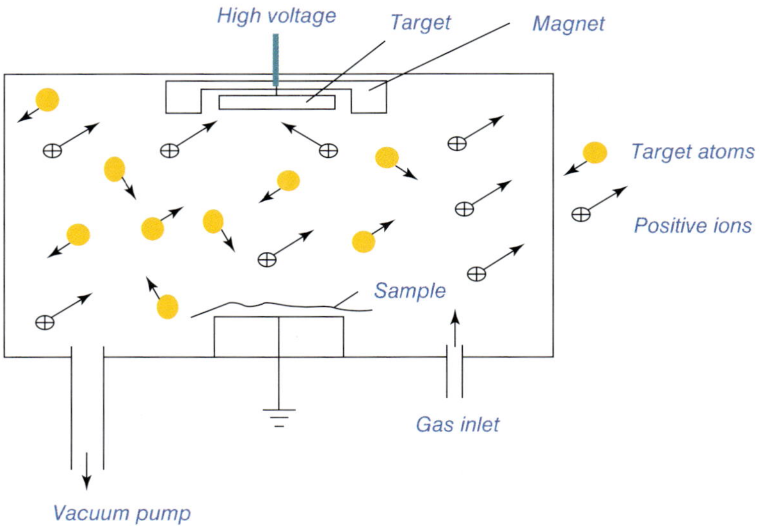

Figure 3.21 Sputter coating procedure for SEM sample preparation.

Table 3.1 The vacuum types of SEM for examined various sample type.

Vacuum types	Chamber pressure	Sample types
High vacuum	$<1.3\times10^{-2}$ Pa	Dry and oil free samples
Low vacuum	10–130 Pa (0.1–1 Torr)	Uncoated and nonconductive samples
ESEM vacuum	130–2600 Pa (1–20 Torr)	Moist, wet, hot, oily, and dirty samples (control humidity and temperature)

ized using water vapor and a temperature control stage. The temperature can be controlled between −25 and 1500 °C with special devices and heating, cooling, freezing, melting information of the samples can be obtained [51].

Many kinds of samples have been investigated by the electron microscopy technology using different vacuum system as shown in Table 3.1.

In conventional SEM, a single pressure limiting aperture (PLA) is used. However, multiple PLAs are used to separate the sample chamber from the column in the ESEM. In spite of the column is still high vacuum, the chamber pressures variable. By using this technique, the column and image quality do not damage the contaminants that come from the samples. At the ESEM mode, special devices such as environmental secondary electron detector (ESED) or gaseous secondary electron detector (GSED) are used for the same purpose [53].

When the electron beam goes to the nonconductive samples, positive ions are fastened on the sample surface as the charge accumulates. The gas ionization is used to amplify the SE signal. Thus, they effectively allay charging artifacts and, in this way, the specimens do not require to be coated with a conductive film. Therefore, using high accelerating voltage ESEM can acquire X-ray data from nonconductive samples [52] (Figure 3.22).

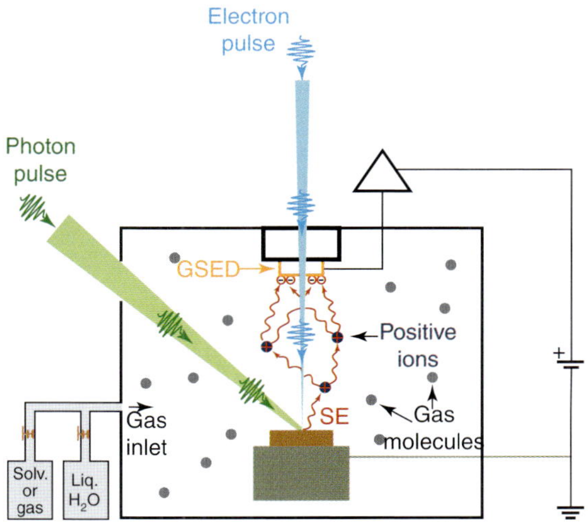

Figure 3.22 Environmental scanning electron microscope chamber with GSED detector. (Yang 2013 [53]. Reproduced with permission of Wiley.)

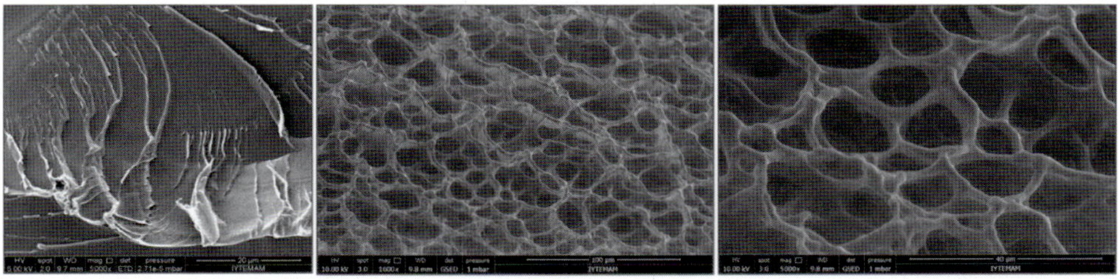

(a) (b) (c)

Figure 3.23 (a) SEM images of dried hydrogel sample with high vacuum Everhart–Thornley detector (ETD) detector. (b) SEM images of wet hydrogel sample with ESEM GSED detector. (Magnification: 1000×). (c) SEM images of wet hydrogel sample with ESEM GSED detector. (Magnification: 5000×).

At the Izmir Institute of Technology Center for Materials Research, some materials were studied by ESEM and differences between high vacuum and ESEM were showed. Hydrogels possess a highly porous structure in their wet, hydrolyzed state. Studying the pore structures of hydrogels using SEM poses challenges as the normal technique would require the use of dried samples; the material loses its pore structure when dried. Below are images comparing dry and wet samples, the latter generated with ESEM using GSED detectors (Figure 3.23).

In addition, chitosan nanoparticles have spherical shapes in wet conditions but lose this property when dried. Images from samples of dried (scanning electron microscope Everhart—Thornley detector, SEM-ETD) and wet particles are shown below. Wet samples were generated with ESEM using cooling stage and GSED detector (Figure 3.24).

3.2.4.3 Transmission Electron Microscope

Techniques using electron microscopes are very powerful because they produce high-resolution images, allowing for a wide range of applications in the fields of

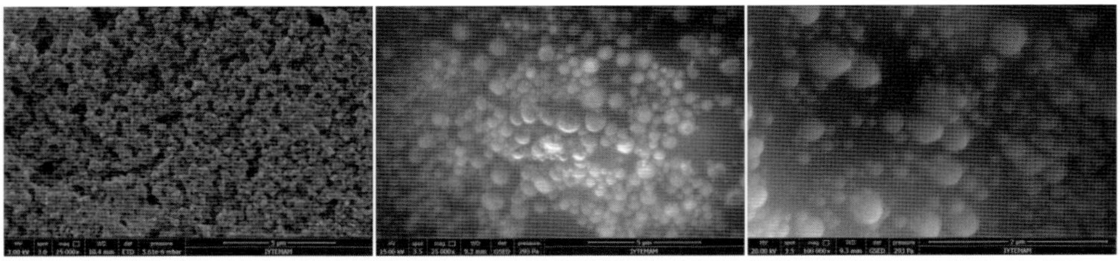

(a) (b) (c)

Figure 3.24 (a) SEM images of dried chitosan nanoparticles with high vacuum ETD detector. (b) SEM images of wet chitosan nanoparticles with ESEM GSED detector. (c) SEM images of wet chitosan nanoparticles with ESEM GSED detector with cooling stage.

education, science, and industry. With TEM, a high energy beam of electrons are used to provide morphologic, compositional, and crystallographic information on samples [54]. It is also used to collect information on characteristics such as the material's size, shape, crystallinity, composition, and elemental mapping during the interaction of the electron beam with the specimen [55].

TEM provides information to researchers on the sample at a molecular level, and is used in different fields such as nanotechnology, life sciences, medical, biological and material research, forensic analysis, gemology, and metallurgy as well as in industry and education.

The first electron microscope was developed by Knoll and Ruska at 1932. This invention earned the Nobel Prize in 1986. The first commercial TEM called The Metropolitan Vickers EM1 was built in 1936 [56].

The TEM and the light microscope operate on the same basic principles. TEM uses electrons instead of light. The wavelength of electrons is much smaller than that of light. Because of this the optimal resolution of TEM images is better than that from a light microscope. Thus, the details of the internal structure of a sample, insomuch that as small as individual atoms, can be examined. The glass lenses used in light microscopes are replaced by magnetic lenses and the projection screen is replaced by a fluorescent screen [57].

In TEM technology, a beam of high-energy electrons passes through a specimen of nanometer thickness [55]. A magnified image is obtained from the interactions of the electrons and the specimen onto the imaging device such as a fluorescent screen or a charge-coupled device (CCD) camera. The resolution of the microscope changes with different functions of the instrument such as electron source, electromagnetic lenses in the microscope column, and so on.

The comparison between light microscope, SEM, and TEM are shown in Figure 3.25 and Table 3.2.

Imaging Modes of TEM

Bright-field TEM: Images are generated by transmitted electrons. Information on the size, shape, uniformity, and dispersity of nanomaterials can be obtained.

Dark-field TEM: Diffracted electrons from crystal planes are used to image and locate defects in crystals.

H. F. Yin *et al.* showed Au-Fe_3O_4 nanoparticles from their studies with bright-field and dark-field TEM images (Figure 3.26).

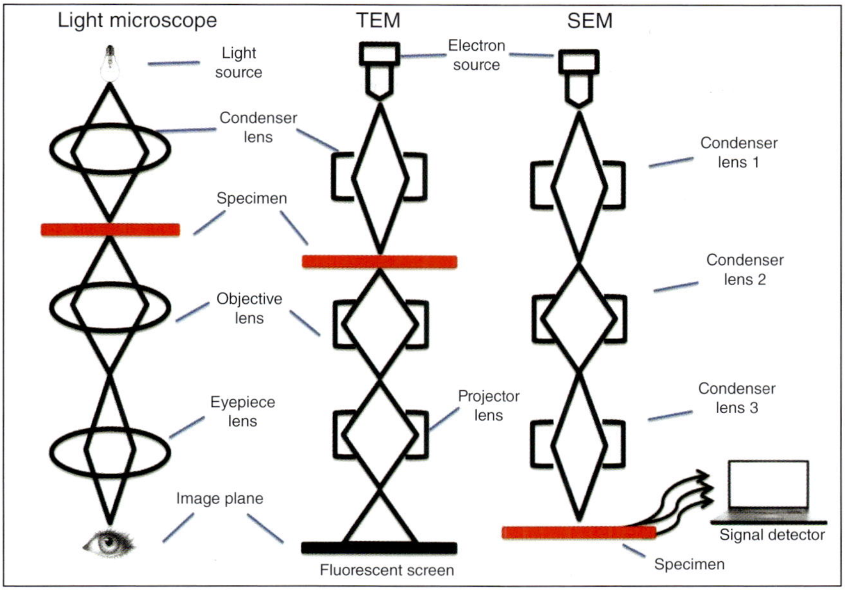

Figure 3.25 The schematic comparison between light microscope, SEM, and TEM [58].

High-resolution transmission electron microscope (HRTEM): This technique is used to study the lattice structure of crystals. When the kinetic processes of the nanomaterials are studied, the crystallinity and crystal defects can be investigated at the atomic level when changes usually begin.

Electron diffraction (ED): The symmetry and spacing of diffraction patterns can be shown by the crystal structure and phase of the sample.

In the study by X. J. Li *et al.*, the microstructure of B doped diamond-like carbon (B-DLC) film is investigated mainly by HRTEM. By the preparation of proper cross-sectional specimens, HRTEM can provide more precise analysis of the structure and give direct microstructural information of the doped element in the film (Figure 3.27).

Scanning transmission electron microscope (STEM): The image is generated by raster scanning of an electron beam over the sample.

High-angle annular dark-field scanning transmission electron microscope (HAADF-STEM): It is known as *Z-contrast imaging*. STEM coupled with HAADF detector can image by the atomic number of the elements in the sample.

The study by H. Akamine *et al.* shows that the anti-phase boundaries (APBs) in the stable CoPt alloy are characterized by two atomic planes using *HAADF-STEM*. Even in TEM images, it is not possible to see something about *Z*-contrast. However, the *Z*-contrast of HAADF-STEM is highly dependent on the number and types of atoms in each atomic column (Figure 3.28).

Energy-dispersive X-ray spectroscopy (EDX or EDS): This technique is used for qualitative and quantitative elemental analysis.

Table 3.2 Comparison between light microscope–SEM–TEM.

	Light microscope	SEM	TEM
Operating environment	Air	Vacuum	Vacuum
Resolution	0.2 μm	3 nm	0.2 nm
Magnification	≤1000×	≤1.000.000×	≥1.000.000×
Source	Light	Electron gun	Electron gun
Lens type	Glass lenses	Electromagnetic lenses	Electromagnetic lenses
Viewing the sample	Eyepiece (ocular)	Digital or CCD camera	Fluorescent screen, digital or CCD camera
Usage	Simple to use	Users require technical skills	Users require technical skills and need theoretical knowledge
Electron beam		Beam focused to fine point; sample is scanned line by line	Broad, static beams
Voltages needed		Accelerating voltage much lower; not necessary to penetrate the sample (0.2–30 kV)	Need high voltage for transmission of the sample (20–1000 kV)
Specimen type	Can view both live and dead samples	Wide range of samples allowed but can view only dead samples	Sample must be very thin, can view only dead specimens
Sample preparation	Easy	Simplifies	Complicated and time consuming
Create image	An image of the sample magnified by the ocular	Beam is scanned along the surface of the sample and build up the image	Transmitted electrons are jointly focused by the objective lens and magnified to create a real image
Cost	Cheap	Expensive	Extremely expensive

Electron energy loss spectroscopy (EELS): It is more powerful than EDX. Composition of the materials and map data showing the actual element distribution in a nanostructure can be generated.

Cryo-TEM: The biological samples be imaged under cryogenic temperature. The technique allows analysing the samples at their native hydrated state of the samples at high resolution, nano to the micro-scale. For instance, Hollander and Danino used Cryo-TEM to investigate the structures of the cochleates made of dioleoylphosphatidylserine (DOPS) that are negatively charged lipid microstructures and calcium, which induce the binding of lipid layers as well as complex formation [62] (Figure 3.29).

Three-dimensional transmission electron microscope (3D-TEM): 3D images can be obtained with proper detectors and software [55].

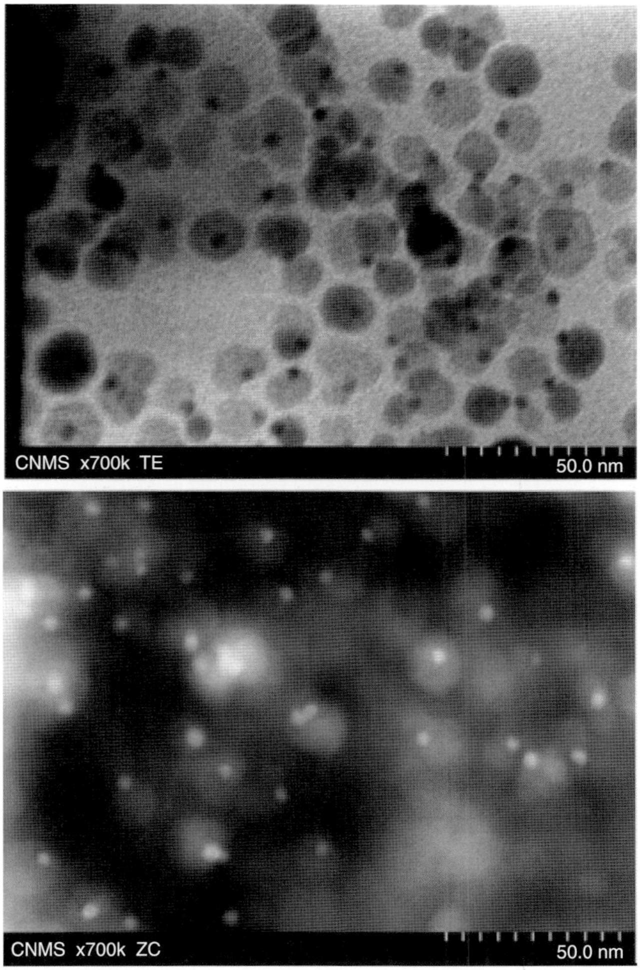

(a)
(b)

Figure 3.26 (a) Bright-field TEM image of Au-Fe_3O_4 dispersed in hexane. (b) Dark-field TEM image of Au-Fe_3O_4 dumbbells supported on SiO_2. (Yin 2008 [59]. Reproduced with permission of Royal Society of Chemistry.)

Sample Preparation of TEM The samples must be very thin (40–150 nm) for transmitted electrons. Some techniques of sample preparation before analyzing the specimens are given below. After these stages, the specimen places the TEM grids, support films, and SiN membranes (Figure 3.30).

- *Ultramicrotome*: To cut materials into extremely thin sections
- *Chemical and Cryo fixation* : To stabilize the materials
- *Dehydration*: To remove free water in the specimen by using organic solvent (alcohol or ethanol)
- *Embedding*: To immobilize the specimen in a hard resin-based matrix
- *Sectioning*: To obtain by slicing

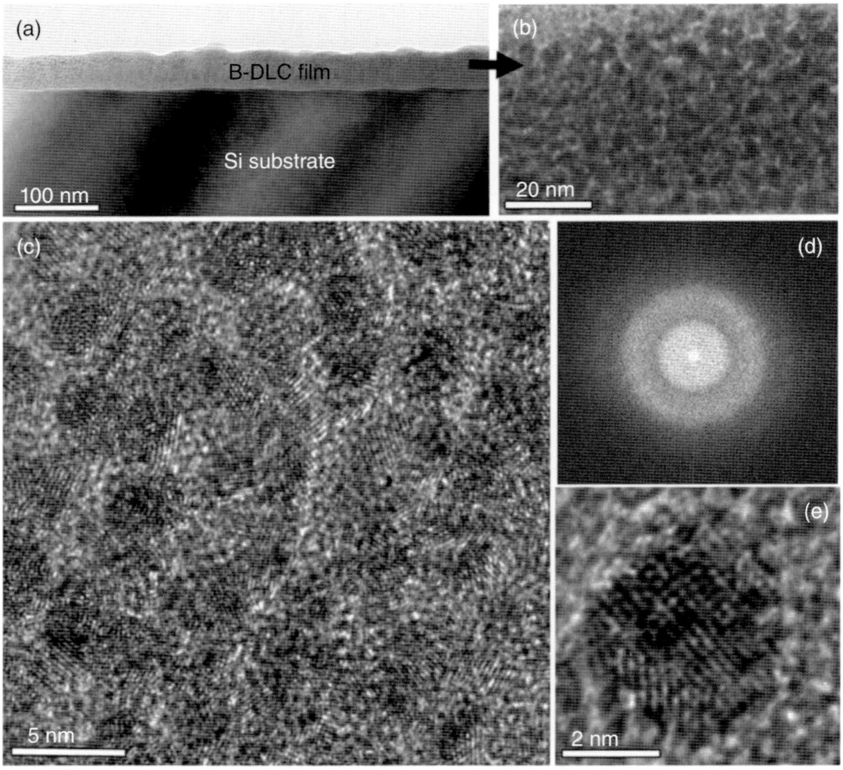

Figure 3.27 (a) Cross-sectional bright-field (BF) image of the B-DLC film; (b) the enlarged BF image of the film in (a); (c,d) HRTEM image of the film and its corresponding fast Fourier transform (FFT), showing many crystalline nanoparticles uniformly dispersed and embedded in the amorphous matrix film; and (e) HRTEM image of one of the nanoparticles in the film. (Li 2015 [60]. Reproduced with permission of American Chemical Society.)

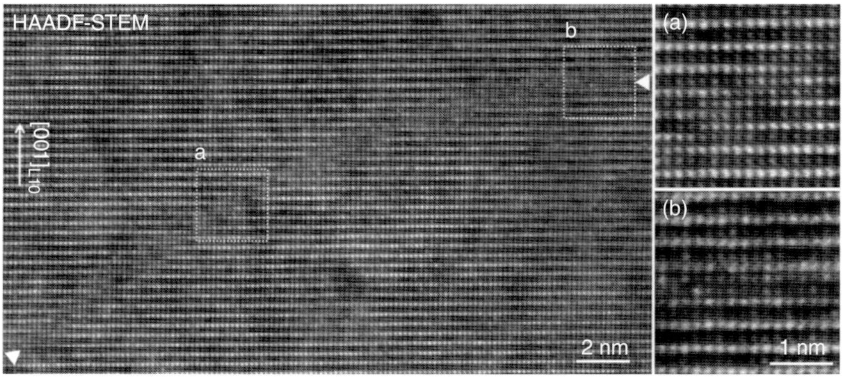

Figure 3.28 [100]L10 HAADF-STEM image of single-variant of CoPt including a single APB. Dark and bright columns correspond to Co and Pt columns, respectively. The APB starts from the lower-left corner (at white arrowhead) and curves toward the top-right corner where it follows the horizontal cubic plane. Panels (a) and (b) are enlargements of the corresponding dotted rectangles. (Akamine 2015 [61]. Reproduced with permission of Elsevier.)

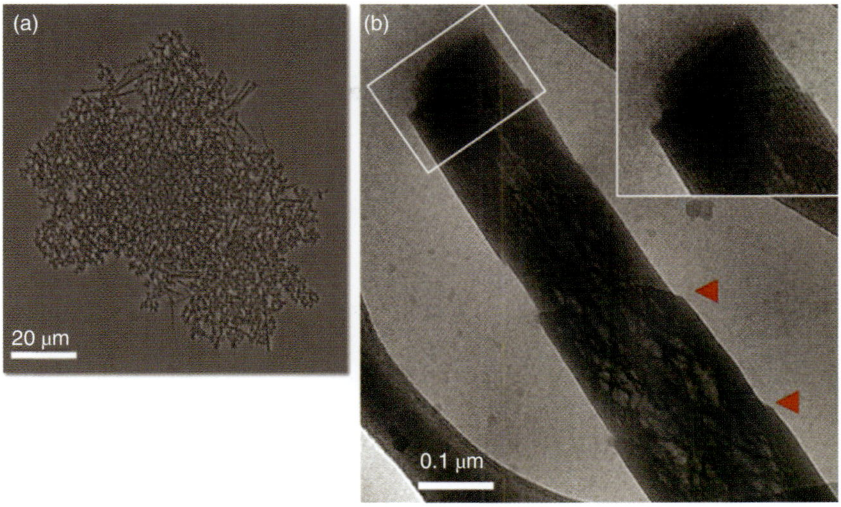

Figure 3.29 (a) Needle-like structures formed by DOPS with $CaCl_2$ detected by light microscopy. (b) Cryo-TEM image of the complex, unraveling the multilamellar structure (see also insert) and several wrapping folded layers (red arrowheads) (For interpretation of the references to color in this figure legend, the reader is referred to the web version of this article.). (Hollander 2015 [62]. Reproduced with permission of Elsevier.)

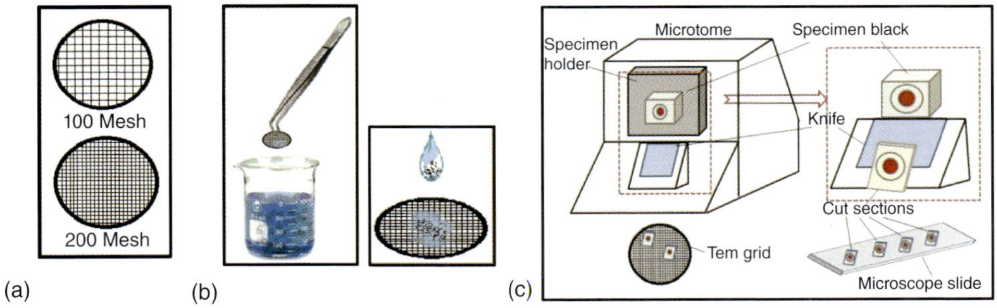

Figure 3.30 (a) Some TEM grids, (b) samples in the solution can drop directly on to the TEM grid or the TEM grid can dip into the solution, and (c) simple presentation of sample preparation with ultramicrotome [63].

- *Staining*: To enter preferentially high atomic weight metal ions into or arround the specimen that is composed of low atomic weight elements (especially biological samples)
- *Grinding and polishing*: To flatten the sample surface to be removed any damage during sectioning
- *Etching*: To generate contrast between the microstructural features of the sample surface [55, 64].

3.2.5 Scanning Probe Microscopy

Scanning probe microscope (SPM) belongs to a category of microscopes that have the ability to get atomic scale resolution images by scanning flat surfaces.

The first member of the SPM family, scanning tunneling microscope (STM) was invented by Gerd Binnig and Heinrich Rohrer at the IBM Zurich Research Laboratories in 1981 [65]. Binnig and Rohrer received the Nobel Prize in Physics in 1986 [66] for designing the STM. After the invention of atomic force microscopy (AFM) by Binnig, Quate, and Gerber in 1986 [67], SPM became one of the fundamental techniques for the surface characterization of soft materials (polymers, biological samples, etc.) and hard materials (metals, thin films, etc.)

3.2.5.1 Working Principle

Unlike optical microscopes that use light waves and electron microscopes that use electrons for imaging, SPM measures attractive or repulsive atomic forces between a sharp needle and sample surface using a spring. The spring is a microfabricated cantilever that is usually made of silicon or silicon nitride with an integrated tip. A laser beam is used to measure the deflection of the cantilever due to lowered and raised features on the sample; as the tip scans the surface in x- and y-directions over a region of interest, the focused laser beam on the cantilever is reflected due to the deflection of the cantilever and hits a position sensitive quadrant photo diode. The deflection causes slight position changes, and hence, light intensity changes on the photo diode. The difference in light intensities on the photodiode is sent to the computer control feedback loop. The feedback loop keeps the cantilever deflection constant by maintaining a constant distance between the cantilever and the sample by adjusting the voltage that is applied to the piezo scanner. The applied voltage, is then converted to a cantilever deflection in the Z-direction. Finally an accurate topographic map of the surface features is acquired. In Figure 3.31 SPM parts are seen.

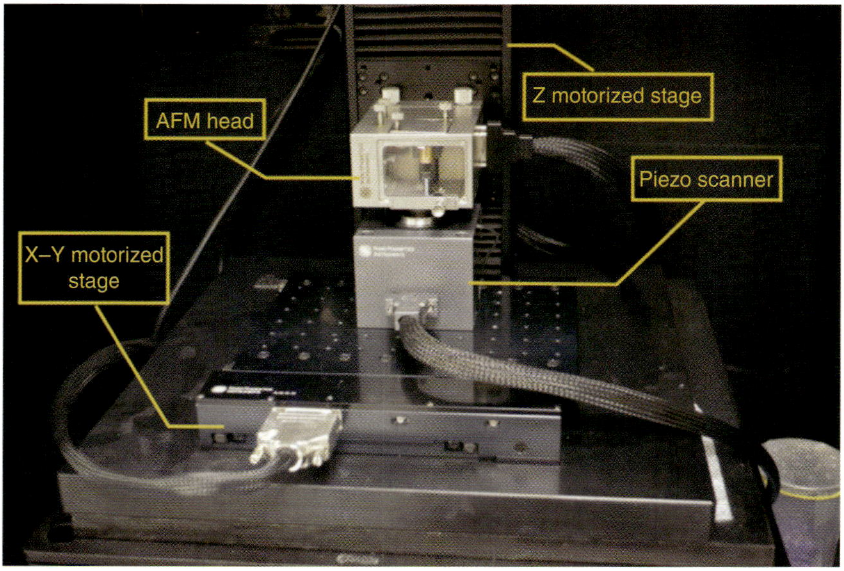

Figure 3.31 SPM scanner and stages.

3.2.5.2 Operating Modes of SPM

An SPM can be operated in the following modes:

AFM

- Contact mode AFM
- Non-contact mode AFM
- Intermittent contact mode AFM
- Closed cell liquid AFM.

STM

- Magnetic force microscope (MFM)
- Electrostatic force microscope (EFM)
- Lateral force microscope (LFM)
- Nanoindentation.

These modes of operation are being applied to a wide variety of application areas, from semiconductors to biology, from polymers to measurement of forces between particles and surfaces.

3.2.5.3 Contact Mode AFM

In the contact mode AFM, the cantilever is in physical contact with the surface. A soft cantilever with a low spring constant is used for sensitivity and low damage to the sample. Strong repulsive forces cause the cantilever to deflect while it passes over lowered and raised features. In Figure 3.32, an image acquired by contact mode AFM is seen. In contact mode AFM, hard samples and soft samples can be studied. Roy and Desai [68] determined the mechanical properties of spatially heterogeneous breast tissue specimens using contact mode AFM.

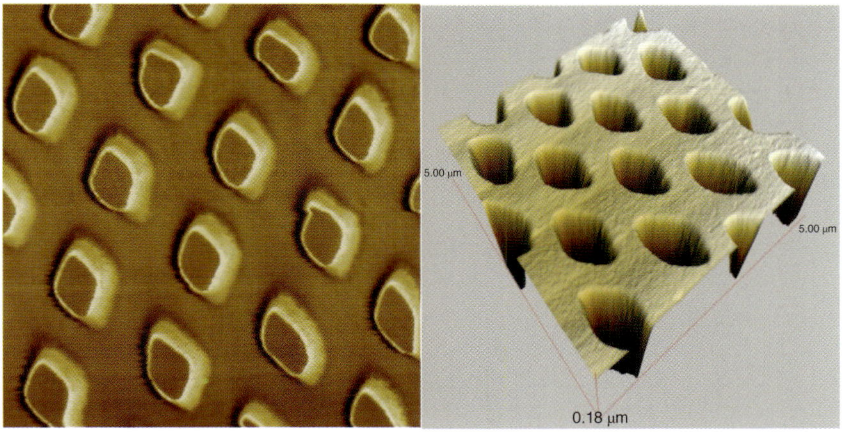

Figure 3.32 2D and 3D contact mode AFM image of platinum coated array. (Image Courtesy of Gökhan Erdoğan.)

3.2.5.4 Noncontact Mode AFM

In the noncontact mode AFM, the cantilever is not in physical contact with the surface. For sensitivity and low damage to the sample, stiff cantilevers with a high spring constant are used. Cantilever oscillates at its resonance frequency using a feedback loop. The feedback loop prevents the cantilever from crashing into the surface. The oscillation amplitude of the cantilever decreases when the tip approaches the sample surface. The feedback loop corrects these amplitude deviations in order to retrieve topographical information from the sample surface. Guillermet *et al.* [69] imaged single decastarphene molecules, adsorbed on Cu(1 1 1) and on a bilayer of NaCl/Cu(1 1 1) by using noncontact mode AFM.

3.2.5.5 Intermittent Contact Mode AFM

In the intermittent contact mode AFM, the cantilever is occasionally in physical contact with the surface. For sensitivity and low damage to the sample, stiff cantilevers with a high spring constant is used. The cantilever oscillates at its natural resonance frequency using a fixed frequency signal source. The oscillation amplitude is higher than that of a noncontact mode. The forces between the sample and the surface change the oscillation amplitude and the relative phase of the resultant oscillations. The bigger oscillations make the deflection signal large enough for the feedback circuit, which keeps the root mean square (RMS) amplitude constant by measuring the phase shifts and RMS amplitude of the vibrations. In Figure 3.33, intermittent contact mode AFM image of a CCD camera sensor is seen. Turkan and Guden [70] investigated the effects of alkali and nitric acid surface treatment and acid etching on the CaP deposition used for biomedical applications. They measured and compared the surface roughness of the samples using intermittent contact mode AFM. 2D and 3D AFM images of surface-treated samples are seen in Figure 3.34.

3.2.5.6 Closed Cell Liquid AFM

Apart from other modes of operation, closed cell liquid AFM is the most important mode due to its ability to operate under aqueous solutions. Attractive forces are eliminated due to surface tension, which enables the sample surface to be examined with a minimum cantilever tip force. It is an ideal case for imaging biological specimens and soft materials. There are some difficulties in this method such as performing good sealing, avoiding air bubbles that cause trouble in focusing the laser beam properly, maintaining the immobility of the sample, and minimizing laser refractions while passing through air-liquid interfaces. The procedures for imaging samples under liquid are the same as that for Contact Mode AFM and Intermittent Contact Mode AFM. Campen *et al.* [71] studied the *in situ* formation and properties of self-assembled films formed on mica surfaces by octadecylamine from alkane solution using liquid cell AFM.

3.2.5.7 STM

STM is based on electron tunneling that is a quantum mechanical phenomenon. In this method the sample must be conductive because a bias voltage is applied between the tip and the sample surface. The biased tip scans the surface while the tunnel current is present. There are two modes in this operation. The first one is

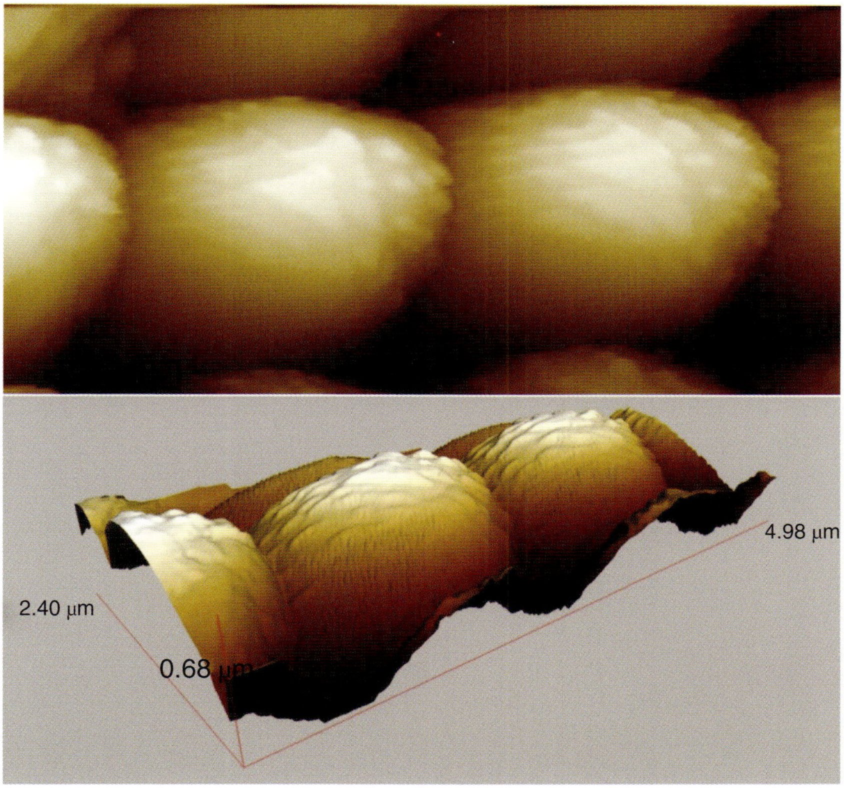

Figure 3.33 2D and 3D intermittent contact mode AFM image of a CCD camera sensor. (Image Courtesy of Gökhan Erdoğan.)

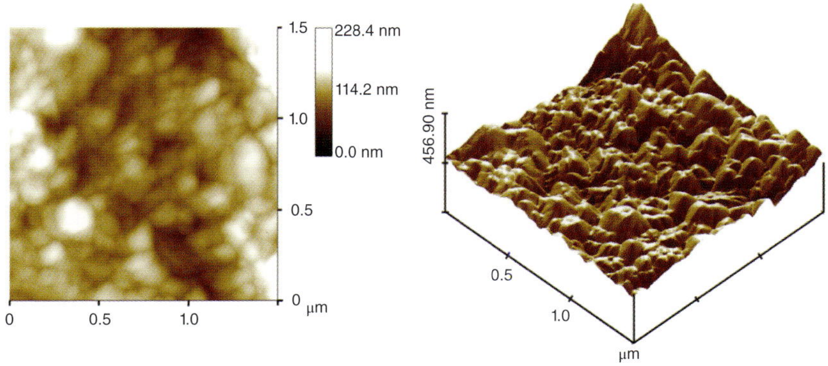

Figure 3.34 2D and 3D intermittent mode AFM images of surface topologies of foam specimen. (Image Courtesy of Uğur Türkan.)

the constant height mode in which the distance between the tip and the surface is kept constant using a feedback circuit. The second one is the constant current mode in which the tunneling current is kept constant using a feedback circuit. The feedback circuit supplies appropriate voltage to the scanner piezo, which moves the tip with picometer resolution up and down to keep the tunneling current constant. Shin *et al.* [72] studied the geometric and electronic structures of Fe islands on MgO film layers with STM.

3.2.5.8 MFM

MFM is a mode that maps the magnetic force gradient above the sample surface. In this mode, the cantilever must be coated with a thin magnetic film and the sample must have magnetic properties. There are two scans in this mode. The first scan is used to determine the topography while the second one is used to determine the magnetic force gradient. Varghese *et al.* [73] reported a new methodology to obtain reliable correlation between magnetic cluster details by using MFM.

3.2.5.9 EFM

EFM is a mode that maps conductive and nonconductive gradient above the sample surface. The cantilever must be coated with a conductive film. A voltage is applied between the sample and the tip in order to adjust the strength of the electric interaction between the tip and the sample. The first scan is used to determine the topography while the second one is used to determine the electrostatic state of the sample surface. Kazakova *et al.* [74] presented local electrical characterization of epitaxial graphene grown on both Si- and C-faces of 4H-SiC using electrostatic force microscopy. In another study, Pruneanu *et al.* [75] organized metallic nanoparticles into one-dimensional assemblies by using DNA as a "template" and allowing the preparation of various silver nanostructures. They demonstrated this study by employing EFM.

3.2.5.10 LFM

LFM measures frictional forces between the tip and the sample surface by measuring the tip deflection in lateral direction and, therefore, it can give information about the areas with varied friction. Taran *et al.* [76] used LFM techniques to determine the molecular-scale characteristics of friction forces between silica particles and silica wafers in aqueous solutions.

3.2.5.11 Nanoindentation

Nanoindentation measures the mechanical properties and wear characteristics of soft and hard materials such as thin films and polymers. It measures these properties by indenting the tip into the sample and measuring the force and distance curves. Enax *et al.* [77] characterized the mechanical properties such as stiffness and hardness of shortfin mako shark teeth using nanoindentation. Tekaya *et al.* [78] used nanoindentation and nanoscratch tests to characterize the mechanical and tribological properties of Ti/TiN multilayer and monolithic TiN nanocoatings.

3.2.6 Contact Angle

Wetting is an important parameter for a wide range of industrial and daily applications including self-cleaning, anti-icing, anti-wetting, and more. Contact angle measurement is used to determine surface wettability and the contact angle represents the value of wettability. Theoretically, the contact angle values must be between 0° and 180°. If the contact angle is less than 90°, these surfaces are called *hydrophilic*. If the wetting angle is higher than 90°, these kind of surfaces can be classified as water hydrophobic. When the contact angle approaches to 180°, the surface is considered superhydrophobic.

It is well known that the contact angle depends on surface chemistry (Cassie model [79]) and surface topography (Wenzel model [80, 81]). Therefore, apart from wetting properties of the surface, some other chemical, physicochemical, and morphological properties can be learned/perceived from the contact angle measurement. For instance, the surface free energy of solids can be calculated using contact angle measurements [82–84].

There are several ways to measure the contact angle. Among them, the contact angle goniometer is the most common technique (Figure 3.35). In this method, a drop is placed on the solid surface by a syringe that can be operated manually or automatically. Then, the image of the drop on the surface is magnified and captured by the camera. Finally, the angle of incidence between the liquid–surface interface and the liquid–vapor interface, namely, the contact angle, is measured by means of software. A contact angle goniometer device enables a direct, easy, and efficient contact angle measurement. However, if the sample surface area is smaller than that of the water drop or if the substrate has a porous structure, this method cannot be used. In such cases, other measurement methods such as Washburn's equation capillary rise and thin-layer wicking methods can be used in order to overcome these handicaps [85, 86]. Therefore, depending on the physical properties of the sample, the most suitable contact angle measurement technique should be chosen.

The dynamic contact angle is another important definition, which is related to wettability. It refers to the *advancing angle* and the *receding contact angle* (Figure 3.36). As liquid is added to the drop on the surface, the contact angle approaches a maximum value. This maximum angle is called *advancing contact angle*. When liquid is withdrawn from the drop, the contact angle decreases to a minimum value. This minimum contact angle is called *receding contact*

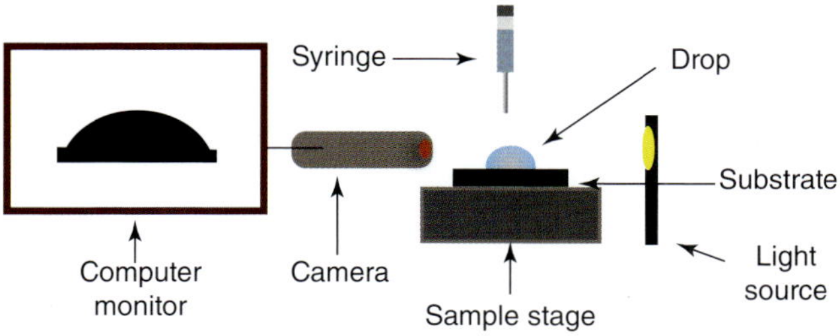

Figure 3.35 Schematic diagram of contact angle goniometer device.

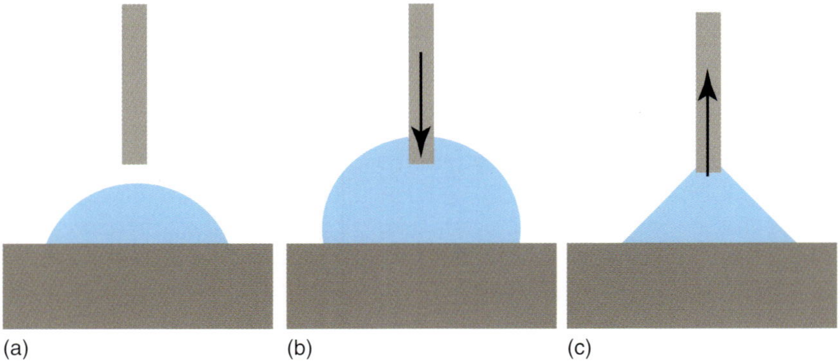

Figure 3.36 (a) Static contact angle, (b) advancing contact angle, and (c) receding contact angle.

angle. Contact angle hysteresis is the difference between the advancing contact angle and receding contact angle. Contact angle hysteresis gives a rich source of surface properties such as surface chemistry, morphology, contamination, adhesive properties, and so on.

3.2.7 BET (Brunauer–Emmett–Teller) Analysis

The Brunauer, Emmett, and Teller (BET) analysis, which is the most common method for measuring the specific surface area of a material, involves calculating the amount of adsorbate gas against the relative vapor pressure of inert gases at a variety of conditions. Nitrogen gas is generally used as the gas adsorptive molecule because of its inert nature property. In the analysis, nitrogen gas is exposed to solid materials under investigation at liquid nitrogen conditions (i.e., 77 K).

In the BET analysis, a sample is degassed prior to the surface analysis to help remove water and possible absorbed molecules. The degassed sample is weighed and then employed for analysis in an analysis compartment. Nitrogen gas is used for the analysis because of its inert nature property. The entire sample tube is immersed in a coolant bath of liquid nitrogen exposing it to lower temperature. A purge nitrogen gas is slowly adsorbed on the solid material and the relative vapor pressure of nitrogen is recorded against the volume of adsorbate by using the computer program of the BET analyzer. The connection between the relative vapor pressure and the amount of adsorbed at constant temperature is called *adsorption isotherm*. In BET analysis, the interpretation of adsorption isotherm is useful for the calculating the surface area of the material and understanding the pore structure of the material. The types of isotherm, which have been classified by UIPAC in 1985, are presented in Figure 3.37 [87]. Type I is characterized by a microporous solid (pore widths below 2 nm). Type II isotherm, which represents unrestricted monolayer–multilayer adsorption, is an indicative of nonporous or macroporous materials. Type III and V isotherms result from weak interactions between solid and gas molecules. Type IV isotherms occur on materials that possesses mesopore structure (pore widths from 2 to 50 nm). The most characteristic

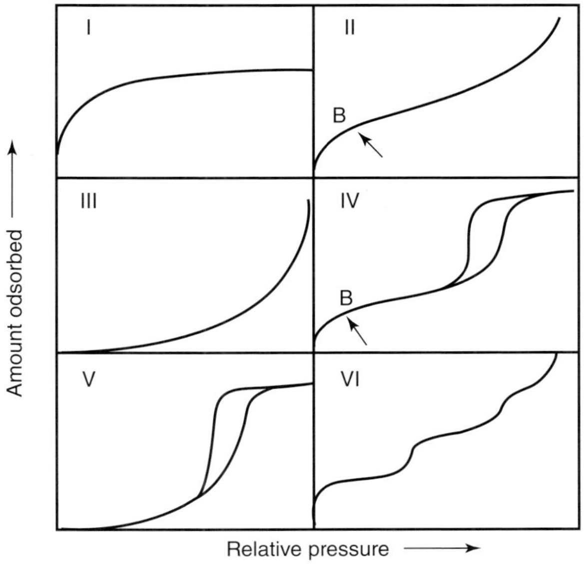

Figure 3.37 The six types of adsorption isotherm according to the 1985 IUPAC classification [87].

features of the Type IV and V isotherms are the hysteresis loops which result from the capillary condensation effect [88]. Type VI exhibits a stepwise multilayer adsorption on a nonporous surface [89].

The surface area of the material is determined from the evaluation of the adsorption isotherm data that is obtained from the BET analysis. Langmuir and BET model equations are applied to the isotherm data for this purpose.

The Langmuir model equation describes microporous material exhibiting Type I isotherms [90]. This model assumes that:

- Monolayer adsorption
- Energetically uniform surface
- No interactions between adsorbed species.

The form of the Langmuir equation is:

$$P/P_0V = 1/bV_m + P/P_0V_m \tag{3.11}$$

where V is the volume of adsorbed vapor, V_m is the volume of monolayer, P is the partial pressure of the adsorbate, P_0 is the saturation vapor pressure of the adsorbate, and b is the constant.

A plot of P/P_0V versus P/P_0 gives slope = $1/V_m$ and intercept = $1/bV_m$. The values of V_m and Langmuir constant b are determined from the slope of intercept of the Langmuir equation plot.

The surface area of the material is determined from

$$S = V_m \times N \times A_x/M \tag{3.12}$$

where A_x is the cross sectional adsorbate area, N is the Avagadro number, V_m is the volume of monolayer, and M is the molar volume of gas.

BET equation model, which is the extension of the Langmuir theory, is applicable to Type II and IV isotherms [91]. Type II isotherm describes unrestricted multilayer formation whereas Type IV isotherm is indicative of restricted multilayer formation. BET equation model assumes the following [91]:

1) Gas molecules physically adsorb on a solid in layers infinitely
2) There is no interaction between each adsorption layer
3) The Langmuir theory can be applied to each layer.

The form of the BET model equation is:

$$\frac{P}{V[P_0 - P]} = \frac{1}{V_{\mathrm{m}}C} + \frac{C-1}{V_{\mathrm{m}}C}\frac{P}{P_0} \tag{3.13}$$

where V is the volume of adsorbed vapor, V_{m} is the volume of monolayer, P is the partial pressure of the adsorbate, P_0 is the saturation vapor pressure of the adsorbate, and C is the constant.

A plot of $P/(P_0 - P)V$ versus P/P_0 gives slope $= C - 1/V_{\mathrm{m}}C$ and intercept $= 1/CV_{\mathrm{m}}$. The values of V_{m} and BET constant C are determined from the slope of intercept of the BET equation plot. For BET surface area measurements, relative pressures are in the range of $0.05 \leq P/P_0 \leq 0.30$. At higher relative pressures BET equation is invalid for occurrence of the capillary condensation effect. At lower relative pressure values ($P/P_0 < 0.05$), the amount of adsorbed gas is too small to be measured. The surface area of the material is determined from the Eq. (3.5).

In the literature, BET analysis was performed to measure the surface area of different materials, which is important information for industrial applications such as catalyst, adsorbent, ion exchanger, heat insulator, and so on. Sand, clay, organo-clays, and cotton fibers that were used in the clean-up operation of oil/fuel spills were characterized by N_2 adsorption–desorption and their isotherms are given in Figure 3.38 [92]. The adsorption isotherm of the materials exhibited a Type II and IV isotherm, which described unrestricted and restricted multilayer formation, respectively. The BET surface areas of the materials calculated from BET model equation and their values varied from 1.2 to 27.1 $\mathrm{m^2\,g^{-1}}$.

3.2.8 Terahertz Time Domain Spectroscopy

Terahertz time domain spectroscopy (THz-TDS) is a very promising method for a wide range of applications. This spectroscopic technique uses THz waves generated and detected by far IR pulses. The electromagnetic waves have the frequencies 0.1–10 THz known as *THz waves* or *T-rays*. This regime falls between the IR and microwave regions of the electromagnetic spectrum (Figure 3.39). THz beams can be manipulated by optical components such as mirror and lens and the phase of the electric field can be measured; therefore, THz region combines electronic and optic spectrum.

Terahertz radiation is nonionizing and can pass through plastic, paper, wood, and ceramics. But it cannot penetrate metal or liquid water. There are two common methods used in the generation of pulsed THz beams. They are photoconductivity (PC) and optical rectification (OR) methods. In the photoconductivity method, THz pulses are generated by illuminating the semiconductor material

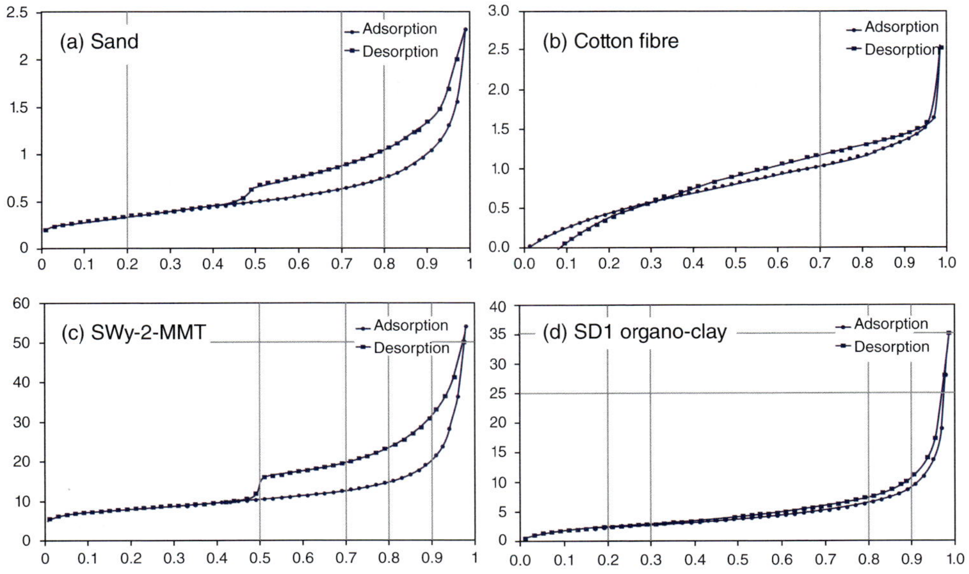

Figure 3.38 N_2 adsorption isotherms for (a) sand, (b) cotton fiber, (c) Swy-2-montmorillonite (MMT) (clay), and (d) SD1 organo-clay, vertical axis: volume adsorbed ($cm^3\ g^{-1}$ STP); horizontal axis (P/P_0). (Carmody 2007 [92]. Reproduced with permission of Elsevier.)

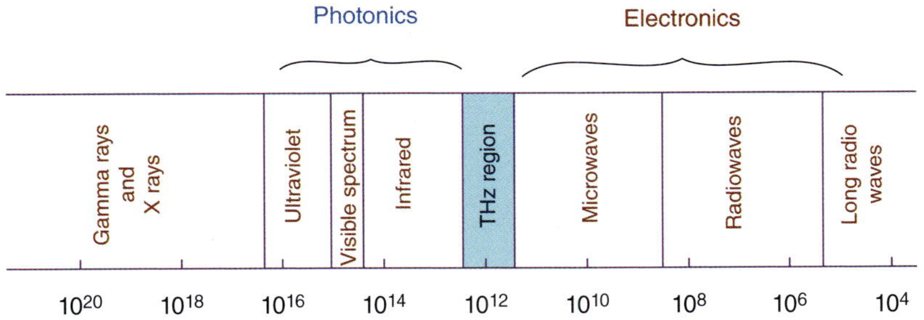

Figure 3.39 The electromagnetic spectrum. THz radiation is in the transition region between electronics and photonics.

with a femtosecond laser pulse [93]. This technique is also used in the detection of THz waves. Low-temperature grown GaAs (LT-GaAs) have short carrier life time, high mobility, and high resistivity and, therefore, are attractive for THz generation and detection [94, 95]. THz wave generation via optical reflection (OR) is method based on difference frequency mixing and materials that have large second order susceptibility $\chi^{(2)}$ is used for this method [96]. ZnTe is the most favorable electro-optic material for generation of the THz beam because of its large coherence length of 0.8 μm, which is important for the OR method [96]. But also DAST [97], GaP [98], $LiNbO_3$, ZnSe, CdS [99] are used to generate THz radiation. The peak electric field is higher in the PC method than in OR. However, terahertz beams generated by the OR technique has an extraordinary bandwidth.

The frequencies of the electromagnetic waves are too high to be measured digitally using electronic counters and there are two techniques that are frequently used to measure the wave form of terahertz pulses: the PC method [100] and electroopting sampling (EO). The PC method detects THz waves by a similar mechanism used in the generation by the PC method. EO sampling is based on the Pockells effect and was investigated by many researchers [96, 101]. A comparison of these two detection methods can be found in [102].

Pulsed terahertz waves generated and detected by these methods are used in the THz-TDS. Before the advent of THz-TDS, the most widely used method was the Fourier transform spectroscopy (FTS). FTS is only sensitive to the amplitude of the wave but THz-TDS is a phase-sensitive method and, therefore, both the amplitude and the phase of the electric field can be measured directly. The real and imaginary parts of the dielectric function can be obtained without using Kramers–Kronig relations. Another advantage of the THz-TDS technique is the high signal to noise ratio (S/N) in the range of 10 GHz to 4 THz. The field amplitude S/N is 10^4 in this range while FTS is about 300 [103]. There are two different configurations that are generally used for THz-TDS: transmission and reflection. In the case of strongly reflective and optically thick samples, reflection measurements are necessary. In terahertz time domain reflection spectroscopy [104], the complex refractive index of the material is obtained by taking the ratio of Fourier transformed electric field of the sample and the reference mirror. The surface of the mirror must be positioned at the same place as that of the sample. Any difference between the sample and reference mirror (>1 μm) results in a severe error in the phase measurement and, thereby, in the complex refractive index [105].

If the sample is optically thin and transparent or semi-transparent in the THz region, it is convenient to use terahertz time domain transmission spectroscopy [106]. In terahertz time domain transmission spectroscopy, the electric field of the THz waves is measured in the absence of the sample $E_{\text{ref}}(t)$ and in the presence of the sample $E_{\text{sam}}(t)$. $E_{\text{ref}}(t)$ and $E_{\text{sam}}(t)$ are Fourier transformed and the transmission function is calculated by taking the ratio of $\tilde{E}_{\text{sam}}(\omega)$ and $\tilde{E}_{\text{ref}}(\omega)$ [107].

$$\begin{aligned}\frac{\tilde{E}_{\text{sam}}(\omega)}{\tilde{E}_{\text{ref}}(\omega)} &= |\sqrt{T(\omega)}|\ \exp\left\{-i\left[\Delta\varphi(\omega) - \frac{\omega}{c}d\right]\right\} \\ &= \frac{4\tilde{n}(\omega)}{[\tilde{n}(\omega)+1]^2}\ \frac{\exp\left\{-i[\tilde{n}(\omega)-1]\frac{\omega}{c}d\right\}}{1-\frac{[\tilde{n}(\omega)-1]^2}{[\tilde{n}(\omega)+1]^2}\exp\left[-i2\tilde{n}(\omega)\frac{\omega}{c}d\right]}\end{aligned} \tag{3.14}$$

where $\tilde{n}(\omega) = n(\omega) - ik(\omega)$ is the complex refractive index of the sample, c is the speed of light, d is the thickness of the sample, $T(\omega)$ and $\Delta\varphi$ are the power transmittance and the phase shift due to the sample respectively. $n(w)$ and $k(w)$ can be determined by using experimentally obtained $\sqrt{T(\omega)}$ and $\Delta\varphi$. The complex dielectric constant $\tilde{\varepsilon}(\omega) = \varepsilon(\omega) - i\varepsilon(\omega)$ which is related to the complex refractive index by the relation $\tilde{\varepsilon} = \tilde{n}^2(\omega)$ can be calculated.

The experimental setups for terahertz time domain transmission spectroscopy is shown in Figure 3.40. In the first system (Figure 3.40a), a THz beam is generated and detected by photoconductive antenna, which consists of two metal striplines

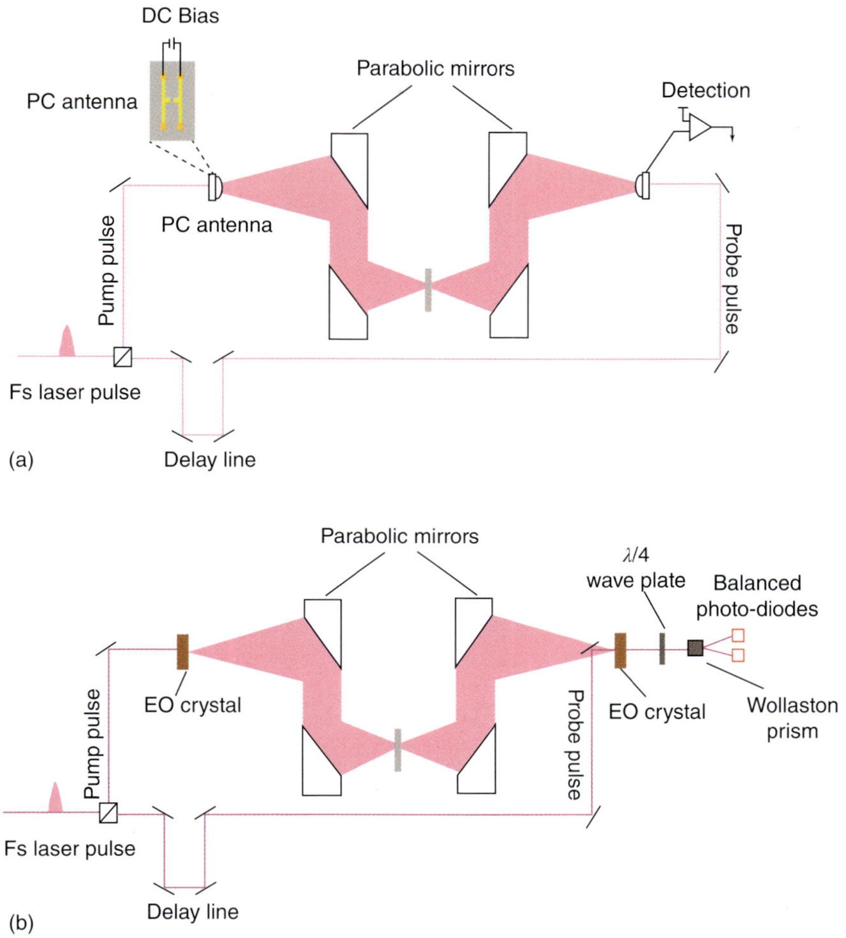

Figure 3.40 The experimental setups for terahertz time domain transmission spectroscopy. Terahertz pulses are generated and detected by (a) photoconductive antenna and (b) electro-optic material.

deposited on a semiconductor material. Laser pulse is splitted into the probe and pump pulse by the beam splitter. Pump pulse incident onto the PC antenna and generates electron-hole pairs. Acceleration of the charge carriers under the DC bias field gives rise to a transient current, and this time-dependent current radiates THz waves. The emitted radiation is focused onto the sample by parabolic mirrors. After passing through the sample, THz pulses are incident onto the second PC antenna and accelerate the photocarriers generated by the probe pulse. The transient current is proportional to the electric field of the terahertz pulse. By changing the time delay between the probe and the pump pulse THz wave form is obtained.

In the second system (Figure 3.40b) electro-optic crystals are used for generation and detection of THz beams. A femtosecond laser pulse that contains a range of frequencies interacts with the electrooptic material and THz waves are generated by frequency beating. After passing through the sample THz waves are focused onto another electrooptic crystal with the laser pulse at the same time. The frequency of the optical pulse is higher than the THz pulse and, therefore, the electric field of the terahertz beam serves as a static field and induces a small birefringence in the crystal. Linearly polarized probe beams gain a small elliptical polarization that is proportional to the electric field of the terahertz waves. The quarter wave plate induces a phase shift and the Wollaston prism splits the probe beam into the s and p polarization components. The difference between these components is measured by a balanced photodetector. Terahertz waveform is obtained by changing the time delay between the probe and the pump pulse.

THz-TDS is a powerful technique and it is widely used in material characterization [108], biomedical research [109], imaging [110], security [111], and communication applications [112].

References

1 History http://www.history.com/this-day-in-history/german-scientist-discovers-x-rays (accessed 20 April 2016).

2 Niemantsverdriet, J.W. (2007) *Spectroscopy in Catalysis*, John Wiley & Sons, Inc..

3 Robinson, J.W. (1974) *Handbook of Spectroscopy*, CRC Press, Cleveland, OH.

4 Kelsall, R.W., Hamley, I.W., and Geoghegan, M. (eds) (2005) *Nanoscale Science and Technology*, John Wiley & Sons, Ltd., Chichester.

5 Bubert, H., Rivière, J.C., Arlinghaus, H.F., Hutter, H., Jenett, H., Bauer, P., Palmetshofer, L., Fabry, L., Pahlke, S., Quentmeier, A., and Hinrichs, K. (eds) (2002) *Surface and Thin-Film Analysis*, Wiley-VCH Verlag GmbH & Co. KGaA.

6 Tang, W.-W., Dong, F., Wong, K.-H., and Wang, Y. (2015) Preparation, characterization and in vitro release of zein–pectin capsules for target delivery. *Curr. Drug Delivery*, **12**, 397–405.

7 Gupta, A., Kaur, C.D., Saraf, S., and Saraf, S. (2016) Formulation, characterization, and evaluation of ligand-conjugated biodegradable quercetin nanoparticles for active targeting. *Artif. Cells Nanomed. Biotechnol.*, **44**, 960–970.

8 Vidigal, G.M., Groisman, M., de Sena, L.A., and de Almedia Soares, G. (2009) Surface characterization of dental implants coated with hydroxyapatite by plasma spray and biomimetic process. *Implant Dent.*, **18**, 353–361.

9 Liu, K.-Z., Shi, M.-H., and Mantsch, H.H. (2005) Molecular and chemical characterization of blood cells by infrared spectroscopy: a new optical tool in hematology. *Blood Cells Mol. Dis.*, **35**, 404–412.

10 Güler, G., Gärtner, R.M., Ziegler, C., and Mäntele, W. (2016) Lipid-protein interactions in the regulated betaine symporter BetP probed by infrared spectroscopy. *J. Biol. Chem.*, **291**, 4295–4307.

11 Movasaghi, Z., Rehman, S., and Rehman, I.U. (2008) Fourier Transform Infrared (FTIR) spectroscopy of biological tissues. *Appl. Spectrosc. Rev.*, **43**, 134–179.

12 Belfer, S., Fainchtain, R., Purinson, Y., and Kedem, O. (2000) Surface characterization by FTIR-ATR spectroscopy of polyethersulfone membranes-unmodified, modified and protein fouled. *J. Membr. Sci.*, **172**, 113–124.

13 Mäntele, W. (2012) *Biophysik*, 1st edn, Eugen Ulmer Verlag, Stuttgart.

14 Barth, A. (2007) Infrared spectroscopy of proteins. *Biochim. Biophys. Acta*, **1767**, 1073–1101.

15 Güler, G. (2013) Investigation of structural and functional properties of the betaine transporter BetP from Corynebacterium glutamicum by using infrared spectroscopy. Doctora thesis, pp. 1–165.

16 Dubal, D.P., Lee, S.H., Kim, J.G., Kim, W.B., and Lokhande, C.D. (2012) Porous polypyrrole clusters prepared by electropolymerization for a high performance supercapacitor. *J. Mater. Chem.*, **22**, 3044–3052.

17 Chung, I.-M., Park, I., Seung-Hyun, K., Thiruvengadam, M., and Rajakumar, G. (2016) Plant-mediated synthesis of silver nanoparticles: their characteristic properties and therapeutic applications. *Nanoscale Res. Lett.*, **11**, 40.

18 Gürsoy, M., Uçar, T., Tosun, Z., and Karaman, M. (2016) Initiation of 2-hydroxyethyl methacrylate polymerization by tert-butyl peroxide in a planar PECVD system. *Plasma Processes Polym.*, **13**, 438–446.

19 Ding, B., Jasensky, J., Li, Y., and Chen, Z. (2016) Engineering and characterization of peptides and proteins at surfaces and interfaces: a case study in surface-sensitive vibrational spectroscopy. *Acc. Chem. Res.* doi: 10.1021/acs.accounts.6b00091

20 Hu, Q., Suzuki, H., Gao, H., Araki, H., Yang, W., and Noda, T. (2003) High-frequency FTIR absorption of SiO2/Si nanowire. *Chem. Phys. Lett.*, **378**, 299–304.

21 Noda, T., Suzuki, H., Araki, H., Yang, W., Shi, Y., and Tosa, M. (2005) Microstructures and IR spectra of long amorphous SiO2/Si nanowires. *Appl. Surf. Sci.*, **241**, 231–235.

22 Nagai, N. and Hashimoto, H. (2001) FT-IR-ATR study of depth profile of SiO2 ultra-thin films. *Appl. Surf. Sci.*, **172**, 307–311.

23 Laroche, G., Fitremann, J., and Gherardi, N. (2013) FTIR-ATR spectroscopy in thin film studies: the importance of sampling depth and deposition substrate. *Appl. Surf. Sci.*, **273**, 632–637.

24 Gunasekaran, S., Anbalagan, G., and Pandi, S. (2006) Raman and infrared spectra of carbonates of calcite structure. *J. Raman Spectrosc.*, **37**, 892–899.

25 Boronat, A., Silvestre, S., and Castañer, L. (2013) Optical and compositional characterization of GaAs(Ti) thin films deposited by R.F. magnetron sputtering. *J. Non-Cryst. Solids*, **359**, 21–26.

26 Pusep, Y., Milekhin, A., and Poropov, A. (1993) FTIR-spectroscopy of (GaAS)n/(AlAs)m superlattice. *Superlattices Microstruct.*, **13**, 115–123.

27 Pusep, Y., Milekhin, A., and Poropov, A.I. (1994) FTIR spectroscopy of longitudinal confined phonons and plasmon-phonon vibrational modes in GaAsn/AlAsm superlattices. *Solid State Electron.*, **37**, 613–616.

28 Juang, F.-S., Su, Y.-K., Yu, H.H., and Liu, K.-J. (2003) Characterization of the InAsSb/GaSb superlattices by Fourier transform infrared spectroscopy. *Mater. Chem. Phys.*, **78**, 620–624.

29 Catauro, M., Papale, F., and Bollino, F. (2015) Characterization and biological properties of TiO2/PCL hybrid layers prepared via sol–gel dip coating for surface modification of titanium implants. *J. Non-Cryst. Solids*, **415**, 9–15.

30 Tamosaityte, S., Galli, R., Uckermann, O., Sitoci-Ficici, K.H., Later, R., Beiermeister, R., Doberenz, F., Gelinsky, M., Leipnitz, E., Schackert, G., Koch, E., Sablinskas, V., Steiner, G., and Kirsch, M. (2015) Biochemical monitoring of spinal cord injury by FT-IR spectroscopy--effects of therapeutic alginate implant in rat models. *PLoS One*, **11**(2), e0150237.

31 Malakauskaite-Petruleviciene, M., Stankeviciute, Z., Niaura, G., Garskaite, E., Beganskiene, A., and Kareiva, A. (2016) Characterization of sol-gel processing of calcium phosphate thin films on silicon substrate by FTIR spectroscopy. *Vib. Spectrosc.*, **85**, 16–21.

32 Güler, G., Džafić, E., Vorob'ev, M.M., Vogel, V., and Mäntele, W. (2011) Real time observation of proteolysis with Fourier transform infrared (FT-IR) and UV-circular dichroism spectroscopy: watching a protease eat a protei. *Spectrochim. Acta, Part A*, **79**, 104–111.

33 Güler, G., Vorob'Ev, M.M., Vogel, V., and Mäntele, W. (2016) Proteolytically-induced changes of secondary structural protein conformation of bovine serum albumin monitored by Fourier transform infrared (FT-IR) and UV-circular dichroism spectroscopy. *Spectrochim. Acta, Part A*, **161**, 8–18.

34 Fabian, H. and Mäntele, W. (2002) Infrared spectroscopy of proteins, in *Handbook of Virational Spectrosopy* (eds J.M. Chalmers and P.R. Griffiths), John Wiley & Sons, Ltd., Chichester.

35 Korkmaz, F., Köster, S., Yildiz, Ö., and Mäntele, W. (2008) The role of lipids for the functional integrity of porin: an FTIR study using lipid and protein reporter groups. *Biochemistry*, **47**, 12126–12134.

36 Tyagi, G., Jangir, D.K., Singh, P., and Mehrotra, R. (2010) DNA interaction studies of an anticancer plant alkaloid, vincristine, using Fourier transform infrared spectroscopy. *DNA Cell Biol.*, **29**, 693–699.

37 Tyagi, G., Pradhan, S., Srivastava, T., and Mehrotra, R. (2014) Nucleic acid binding properties of allicin: spectroscopic analysis and estimation of anti-tumor potential. *Biochim. Biophys. Acta*, **1840**, 350–356.

38 Bourassa, P. and Tajmir-Riahi, H.A. (2015) Folic acid binds DNA and RNA at different locations. *Int. J. Biol. Macromol.*, **74**, 337–342.

39 Wehrli, F.W. and Wirthlin, T. (1976) *Interpretation of Carbon-13 NMR Spectra*, Heyden & Son Ltd., London.

40 Zerbe, O. and Jurt, S. (2014) *Applied NMR Spectroscopy for Chemists and Life Scientists*, Wiley-VCH Verlag GmbH & Co. KGaA, Weinheim.

41 Günther, H. (2013) *NMR Spectroscopy: Basic Principles, Concepts and Applications in Chemistry*, Wiley-VCH Verlag GmbH, Weinheim.

42 Che, M. and Vedrine, J.C. (eds) (2012) *Characterization of Solid Materials and Heterogeneous Catalysts: From Structure to Surface Reactivity*, Wiley-VCH Verlag GmbH & Co. KGaA, Weinheim.

43 Laws, D.D., Bitter, H.-M.L., and Jerschow, A. (2002) Solid-state NMR spectroscopic methods in chemistry. *Angew. Chem. Int. Ed.*, **41**, 3096–3129.

44 Jackowski, K. and Jaszuński, M. (eds) (2016) *Gas Phase NMR*, Royal Society of Chemistry, London.
45 Lambert, J.B. and Mazzola, E.P. (2004) *Nuclear Magnetic Resonance Spectroscopy: An Introduction to Principles, Applications, and Experimental Methods*, Pearson Education, Upper Saddle River, NJ.
46 Ablett, S. (1992) Overview of NMR applications in food science. *Trends Food Sci. Technol.*, **3**, 246–250.
47 Preston, C.M. (1996) Applications of NMR to soil organic matter analysis: history and prospects. *Soil Sci.*, **161**(3), 144–166.
48 Marcone, M.F., Wang, S., Albabish, W., Nie, S., Somnarain, D., and Hill, A. (2013) Diverse food-based applications of nuclear magnetic resonance (NMR) technology. *Food Res. Int.*, **51**(2), 729–747.
49 Demco, D.E. and Blümich, B. (2001) NMR imaging of materials. *Curr. Opin. Solid State Mater. Sci.*, **5**(2-3), 195–202.
50 Diercks, T., Coles, M., and Kessler, H. (2001) Applications of NMR in drug discovery. *Curr. Opin. Chem. Biol.*, **5**, 285–291.
51 Goldstein, J., Newbury, D., Joy, D., Lyman, C., Echlin, P., Lifshin, E., Sawyer, L., and Michael, J. (2012) *Scanning Electron Microscopy and X-Ray Microanalysis: a Text for Biologists, Materials Scientists, and Geologists*, Springer Science & Business Media.
52 Kimseng, K. and Meissel, M. (2001) *Short Overview about the ESEM: The Environmental Scanning Electron Microscope*, Produced by CALCE Electronic Products and Systems Centre, University of Maryland, Maryland.
53 Yang, D.S., Mohammed, O.F., and Zewail, A.H. (2013) Environmental scanning ultrafast electron microscopy: structural dynamics of solvation at interfaces. *Angew. Chem. Int. Ed.*, **52**(10), 2897–2901.
54 Microscope Master http://www.microscopemaster.com/transmission-electron-microscope.html (accessed 10 February 2016).
55 Kumar, C.S. (2013) *Transmission Electron Microscopy Characterization of Nanomaterials*, Springer Science & Business Media.
56 Williams, D.B. and Carter, C.B. (1996) *The Transmission Electron Microscope*, Springer, New York.
57 FEI https://www.fei.com/documents/introduction-to-microscopy-document (accessed 10 February 2016).
58 The University of Iowa http://cmrf.research.uiowa.edu/transmission-electron-microscopy (accessed 10 February 2016).
59 Yin, H., Wang, C., Zhu, H., Overbury, S.H., Sun, S., and Dai, S. (2008) Colloidal deposition synthesis of supported gold nanocatalysts based on Au–Fe 3 O 4 dumbbell nanoparticles. *Chem. Commun.*, **36**, 4357–4359.
60 Li, X.J., He, L.L., Li, Y.S., Yang, Q., and Hirose, A. (2015) Characterization of boron doped diamond-like carbon film by HRTEM. *Appl. Surf. Sci.*, **357**, 2141–2146.
61 Akamine, H., van den Bos, K.H.W., Gauquelin, N., Farjami, S., Van Aert, S., Schryvers, D., and Nishida, M. (2015) Determination of the atomic width of an APB in ordered CoPt using quantified HAADF-STEM. *J. Alloys Compd.*, **644**, 570–574.

62 Hollander, A. and Danino, D. (2015) Cochleate characterization by cryogenic electron microscopymethods: Cryo-TEM and Cryo-SEM. *Colloids Surf., A*, **483**, 187–192.
63 http://webmediinfo.blogspot.com.tr/2012/02/microtome.html (accessed 30 May 2016).
64 Leng, Y. (2013) *Materials Characterization: Introduction to Microscopic and Spectroscopic Methods*, 2nd edn, John Wiley & Sons, Ltd.
65 Binnig, G., Rohrer, H., Gerber, C., and Weibel, E. (1982) Surface studies by scanning tunneling microscopy. *Phys. Rev. Lett.*, **49**(1), 57.
66 Nobelprize.org http://www.nobelprize.org/nobel-prizes/physics/laureates/1986/press.html (accessed 5 January 2016).
67 Binnig, G., Quate, C.F., and Gerber, C. (1986) Atomic force microscope. *Phys. Rev. Lett.*, **56**(9), 930.
68 Roy, R. and Desai, J.P. (2014) Determination of mechanical properties of spatially heterogeneous breast tissue specimens using contact mode atomic force microscopy (AFM). *Ann. Biomed. Eng.*, **42**(9), 1806–1822.
69 Guillermet, O., Gauthier, S., Joachim, C., De Mendoza, P., Lauterbach, T., and Echavarren, A. (2011) STM and AFM high resolution intramolecular imaging of a single decastarphene molecule. *Chem. Phys. Lett.*, **511**(4), 482–485.
70 Turkan, U. and Guden, M. (2010) The effect of surface treatment on CaP deposition of Ti6Al4V open cell foams in SBF solution. *Ceram. Int.*, **36**(6), 1805–1816.
71 Campen, S., Green, J.H., Lamb, G.D., and Spikes, H.A. (2015) *In situ* study of model organic friction modifiers using liquid cell AFM: self-assembly of octadecylamine. *Tribol. Lett.*, **58**(3), 1–15.
72 Shin, H.J., Kim, S.H., Yang, H., and Kuk, Y. (2014) Modified gap states in Fe/MgO/SrTiO 3 interfaces studied with scanning tunneling microscopy. *Curr. Appl. Phys.*, **14**(12), 1692–1695.
73 Varghese, B., Piramanayagam, S.N., Lee, W.K., and Tan, H.K. (2014) Noise characterization of perpendicular recording media by cluster size measurements. *IEEE Trans. Magn.*, **50**(5), 1–6.
74 Kazakova, O., Panchal, V., and Burnett, T.L. (2013) Epitaxial graphene and graphene–based devices studied by electrical scanning probe microscopy. *Crystals*, **3**(1), 191–233.
75 Pruneanu, S., Olenic, L., Al-Said, S.A.F., Borodi, G., Houlton, A., and Horrocks, B.R. (2010) Template and template-free preparation of one-dimensional metallic nanostructures. *J. Mater. Sci.*, **45**(12), 3151–3159.
76 Taran, E., Kanda, Y., Vakarelski, I.U., and Higashitani, K. (2007) Nonlinear friction characteristics between silica surfaces in high pH solution. *J. Colloid Interface Sci.*, **307**(2), 425–432.
77 Enax, J., Janus, A.M., Raabe, D., Epple, M., and Fabritius, H.O. (2014) Ultrastructural organization and micromechanical properties of shark tooth enameloid. *Acta Biomater.*, **10**(9), 3959–3968.
78 Tekaya, A., Benameur, T., Labdi, S., and Aubert, P. (2013) Effect of Ti/TiN multilayer protective nanocoatings on Zr-based metallic glasses mechanical performance. *Thin Solid Films*, **539**, 215–221.

79 Cassie, A.B.D. and Baxter, S. (1944) Wettability of porous surfaces. *Trans. Faraday Soc.*, **40**, 546–551.
80 Wenzel, R.N. (1936) Resistance of solid surfaces to wetting by water. *Ind. Eng. Chem.*, **28**(8), 988–994.
81 Wenzel, R.N. (1949) Surface roughness and contact angle. *J. Phys. Chem.*, **53**(9), 1466–1467.
82 Fox, H.W. and Zisman, W.A. (1950) The spreading of liquids on low energy surfaces. *J. Colloid Sci.*, **5**, 514–531.
83 Owens, D.K. and Wendt, R.C. (1969) Estimation of the surface free energy of polymers. *J. Appl. Polym. Sci.*, **13**(8), 1741–1747.
84 Janczuk, B. and Bialopiotrowicz, T. (1989) Surface free-energy components of liquids and low energy solids and contact angles. *J. Colloid Interface Sci.*, **127**, 189–204.
85 Washburn, E.W. (1921) The dynamics of capillary flow. *Phys. Rev.*, **17**, 273–283.
86 Van Oss, C.J., Giese, R.F., Li, Z., Murphy, K., Norris, J., Chaudhury, M.K., and Good, R.J. (1992) Determination of contact angles and pore sizes of porous media by column and thin layer wicking. *J. Adhes. Sci. Technol.*, **6** (4), 413–428.
87 Sing, K.S.W., Everett, D.H., Haul, R.A.W., Moscou, L., Pierotti, R.A., Rouquerol, J., and Siemieniewska, T. (1985) Physical and biophysical chemistry division commission on colloid and surface chemistry including catalysis. *Pure Appl. Chem.*, **57**(4), 603–619.
88 Lowell, S., Shields, J.E., Thomas, M.A., and Thommes, M. (2012) *Characterization of Porous Solids and Powders: Surface Area, Pore Size and Density*, vol. **16**, Springer Science & Business Media.
89 Hill, T.L. (1949) Physical adsorption and the free volume model for liquids. *J. Chem. Phys.*, **17**(6), 590.
90 Langmuir, I. (1918) The adsorption of gases on plane surfaces of glass, mica and platinum. *J. Am. Chem. Soc.*, **40**(9), 1361–1403.
91 Brunauer, S., Emmett, P.H., and Teller, E. (1938) Adsorption of gases in multimolecular layers. *J. Am. Chem. Soc.*, **60**(2), 309–319.
92 Carmody, O., Frost, R., Xi, Y., and Kokot, S. (2007) Surface characterisation of selected sorbent materials for common hydrocarbon fuels. *Surf. Sci.*, **601**(9), 2066–2076.
93 Smith, P.R.D., Auston, H., and Nuss, M.C. (1988) Subpicosecond photoconducting dipole antennas. *IEEE J. Quantum Electron.*, **24**, 255–260.
94 Piao, Z., Tani, M., and Sakai, K. (2000) Carrier dynamics and terahertz radiation in photoconductive antennas. *Jpn. J. Appl. Phys.*, **39**, 96–100.
95 Dragoman, D. (2004) Terahertz fields and applications. *Prog. Quantum Electron.*, **28**, 1–66.
96 Nahata, A., Weling, A.S., and Heinz, T.F. (1996) A wideband coherent terahertz spectroscopy system using optical rectification and electro-optic sampling. *Appl. Phys. Lett.*, **69**, 2321–2323.
97 Schneider, A., Neis, M., Stillhart, M., Ruiz, B., Khan, R.U.A., and Günter, P. (2006) Generation of terahertz pulses through optical rectification in

organic DAST crystals: theory and experiment. *J. Opt. Soc. Am. B*, **23**, 1822–1835.

98 Ding, Y.J. and Shi, W. (2006) Efficient THz generation and frequency upconversion in GaP crystals. *Solid-State Electron.*, **50**, 1128–1136.

99 Yajima, T. and Takeuchi, N. (1970) Far-infrared difference-frequency generation by picosecond laser pulses. *Jpn. J. Appl. Phys.*, **9**, 1361–1371.

100 Tani, M., Herrmann, M., and Sakai, K. (2002) Generation and detection of terahertz pulsed radiation with photoconductive antennas and its application to imaging. *Meas. Sci. Technol.*, **13**, 1739–1745.

101 Wu, Q. and Zhang, X.-C. (1996) Ultrafast electro-optic field sensors. *Appl. Phys. Lett.*, **68**, 1604–1606.

102 Cai, Y., Brener, I., Lopata, J., Wynn, J., Pfeiffer, L., Stark, J.B., Wu, Q., Zhang, X.-C., and Federici, J.F. (1998) Coherent terahertz radiation detection: direct comparison between free-space electro-optic sampling and antenna detection. *Appl. Phys. Lett.*, **73**, 444–446.

103 Han, P.Y., Tani, M., Usami, M., Kono, S., Kersing, R., and Zhang, X.C. (2001) A direct comparison between terahertz time-domain spectroscopy and far-infrared Fourier transform spectroscopy. *J. Appl. Phys.*, **89**, 2357–2359.

104 Jeon, T.-I. and Grischkowsky, D. (1998) Characterization of optically dense, doped semiconductors by reflection THz time domain spectroscopy. *Appl. Phys. Lett.*, **72**, 3032–3034.

105 Nashima, S., Morikawa, O., Takata, K., and Hangyo, M. (2001) Measurement of optical properties of highly doped silicon by terahertz time domain reflection spectroscopy. *Appl. Phys. Lett.*, **79**, 3923–3925.

106 Kužel, P. and Petzelt, J. (2000) Time-resolved terahertz transmission spectroscopy of dielectrics. *Ferroelectrics*, **239**, 949–956.

107 Nishizawa, S., Sakai, K., Hangyo, M., Nagashima, T., Takeda, M.W., Tominaga, K., Oka, A., Tanaka, K., and Morikawa, O. (2005) in *Terahertz Optoelectronics* (ed. K. Sakai), Springer, Berlin-Heidelberg, Germany, pp. 203–271.

108 Naftaly, M. and Miles, R.E. (2007) Terahertz time-domain spectroscopy for material characterization. *Proc. IEEE*, **95**(8), 1658–1665.

109 Wallace, V.P., Taday, P.F., Fitzgerald, A.J., Woodward, R.M., Cluff, J., Pye, R.J., and Arnone, D.D. (2004) Terahertz pulsed imaging and spectroscopy for biomedical and pharmaceutical applications. *Faraday Discuss.*, **126**, 255–263.

110 Duling, I. and Zimdars, D. (2009) Terahertz imaging: revealing hidden defects. *Nat. Photonics*, **3**, 630–632.

111 Shen, Y., Lo, T., Taday, P.F., Cole, B.E., Tribe, W.R., and Kemp, M.C. (2005) Detection and identification of explosives using terahertz pulsed spectroscopic imaging. *Appl. Phys. Lett.*, **86**, 241116.

112 Yang, Y., Mandehgar, M., and Grischkowsky, D. (2015) THz-TDS characterization of the digital communication channels of the atmosphere and the enabled applications. *J. Infrared Millimeter Terahertz Waves*, **36**, 97–129.

4

Surface Modification of Polymeric Membranes for Various Separation Processes

Woei-Jye Lau, Chi-Siang Ong, Nik Abdul Hadi Md Nordin, Nur Aimie Abdullah Sani, Nadzirah Mohd Mokhtar, Rasoul Jamshidi Gohari, Daryoush Emadzadeh, and Ahmad Fauzi Ismail

4.1 Introduction

Membrane is defined as a selective barrier between two phases that separate based on the different permeation rates of subjected components. Among different classes of membrane materials, polymeric-based membrane is the main interest of research owing to its reliability, ease of fabrication, excellent mechanical strength, high rigidity, and tailorability toward specific separation demand. The breakthrough in polymeric membrane history was made by Loeb and Sourirajan in the 1960s with the introduction of asymmetric cellulose acetate (CA) membrane that was developed by phase inversion technique. This membrane consisted of a thin selective layer supported by a porous sublayer, providing minimum transport resistance for water molecules while maintaining promising separation efficiency against sodium chloride (NaCl). Nowadays, most of the commercial membranes are made from this technique and have been found applicable in a wide range of industrial processes such as in water and wastewater treatment processes, gas separation and purification, organic solvent recovery, and so on.

Although membranes can be industrially implemented without any major problems, continuous improvement in membrane science and technology is still needed in order to enhance its competitive advantages. For instance, membranes used in ultrafiltration (UF), nanofiltration (NF), and reverse osmosis (RO) still suffer from fouling that causes water permeability to decline as a function of operation time. Depending on the feed characteristics, fouling that is caused by organic and/or inorganic foulants cannot be completely avoided during aqueous medium separation processes. Membranes for gas separation, on the other hand, are bounded by the Robeson trade-off limit between permeability and selectivity. Furthermore, exposure of polymeric membrane to pressurized acid gas (e.g., carbon dioxide and hydrogen sulfide) tends to let these gases pass through the membrane easily with less discrimination, affecting gas pair selectivity.

The limitations encountered by polymeric membranes have encouraged researchers to find alternatives to further improve membrane properties. To date, various strategies have been explored to tackle the problems. Some of them

Surface Treatments for Biological, Chemical and Physical Applications, First Edition.
Edited by Mehmet Gürsoy and Mustafa Karaman.

are new polymeric materials synthesis, hybrid organic–inorganic membrane development, membrane pretreatment, and membrane surface modification. Development of new polymeric materials for specific applications has shown great potential. For example, in gas separation, polymer with intrinsic microporosity and thermally rearranged polymer have potential to surpass the Robeson trade-off limit. However, factors such as plasticization resistance, processability, mechanical strength, and chemical stability of the newly developed polymers are still under investigation and these new polymers are not likely to be commercially available for large-scale membrane production in the next few years.

Fluid separation during the membrane process is mainly governed by the characteristics of the membrane top layer. Thus, modification of membrane surface properties is the practical yet effective approach to achieve desired separation efficiency. In general, membrane surface modification can be performed by techniques such as interfacial polymerization (IP) (for thin film composite (TFC) membranes), organic–inorganic blending (for mixed matrix membranes (MMMs)), layer-by-layer (LbL) assembly, photo-initiated polymerization, and conventional spray coating. Herein, this chapter aims to overview the surface modification methods available for polymeric membranes development and discuss the importance of surface-modified membranes for various applications, covering aqueous, solvent, and gas phases.

4.2 Methods of Membrane Surface Modification

4.2.1 Blending

4.2.1.1 Polymer–Polymer Blending

Polymeric additive is frequently used in a membrane formation process as it can play a role in suppressing macrovoid formation in the membrane, promoting pore formation, improving pore interconnectivity, and/or modifying membrane surface hydrophilicity. For instance, adding hydrophilic additive into hydrophobic polyvinylidene fluoride (PVDF) solution can make the resultant membrane suitable for UF membrane, improving water permeability and reducing fouling propensity.

There are many works reported using polymeric additives of different molecular weights (MWs) at various concentrations during polymeric solution preparation. Of the additives (usually hydrophilic polymers) studied, poly(ethylene glycol) (PEG) and polyvinylpyrrolidone (PVP) are among the two most frequently used materials. These two additives are significantly different in terms of MW in which PEG displays average MW in the range of 200–20 000 g mol^{-1} while PVP shows much higher MW, ranging between 10 000 and 360 000 g mol^{-1}. Self-assembly amphiphilic copolymer is also considered in recent years as a very promising approach in the preparation of membranes with improved characteristics as it offers modification on membrane surface and internal pores while maintaining desirable bulk properties. Some examples of amphiphilic copolymers are polyacrylonitrile-*block*-polyethylene glycol (PAN-*b*-PEG), polyacrylonitrile-*graft*-poly(ethylene oxide)

(PAN-*g*-PEO), polystyrene-*block*-polyethylene glycol (PS-*b*-PEG), Pluronic, and polyacrylonitrile-*random*-*N*,*N*-dimethyl-*N*-methacryloxyethyl-*N*-(3-sulfo propyl) (PAN-*r*-DMMSA). Amphiphilic copolymers can be purposely designed to selectively self-migrate to a polymer–water interface during the phase inversion process to impart membrane surface hydrophilicity, offering facile operation for separation processes.

Besides the characteristics of polymer additives, other variables that are important during the polymer–polymer blending process include the properties of main membrane forming material (e.g., MW and concentration in polymeric solution), solvent type (e.g., solubility parameter and viscosity), fabrication conditions (e.g., shear rate/take up speed, coagulation medium, evaporation temperature, humidity, etc.), and post treatment (e.g., drying procedure). The variation of these conditions can influence the final morphology of the membrane and thus its separation performance. Because of this, in certain cases, opposite effects by adding the same polymeric additive might be observed in membrane properties.

4.2.1.2 Polymer–Inorganic Blending

Polymer/inorganic membranes are heterogeneous membranes consisting of inorganic filler embedded in a polymer matrix. Various combinations of polymer and inorganic materials have been attempted for making this type of membrane in the configurations of both flat sheet and hollow fiber. Mostly, they are used for the pervaporation of aqueous mixtures and gas separation. Studies on their potentials for aqueous solvent systems for UF applications have also been conducted.

Figure 4.1 presents a simple fabrication process of polymer/inorganic flat sheet membrane via blending method [1]. The homogenous polymeric/inorganic solution can be prepared by dispersing inorganic nanomaterials into suitable organic solvents such as *n*-methyl-2-pyrrolidone (NMP) and dimethylformamide (DMF). A small amount of polymeric additive-PVP is normally added, which serves as pore former for microporous membrane making. The suspension is then sonicated for several hours with the aim of creating uniform solution with well-dispersed nanomaterials. Finally, the main membrane-forming material (in this case, it is polyethersulfone PES) is added slowly into the mixture under vigorous stirring. The mixture is continuously stirred overnight to form a homogeneous mixed matrix slurry before it is used for casting process for flat sheet membrane making or for spinning process for hollow fiber membrane synthesis. This type of membrane that is made of polymer/inorganic solution is generally known as MMM.

Although most of the studies on the MMMs used large particles with the size in micron range, it is reported that the addition of such micron-size particles can cause the resultant membranes to be extremely brittle and they tend to crack at higher particles loading even though these materials are effective for enhancing the overall performance of the membranes. In view of this, the incorporation of smaller inorganic particles, which are nano-sized particles, is more sensible for the commercialization of membranes, because the smaller particles yield higher

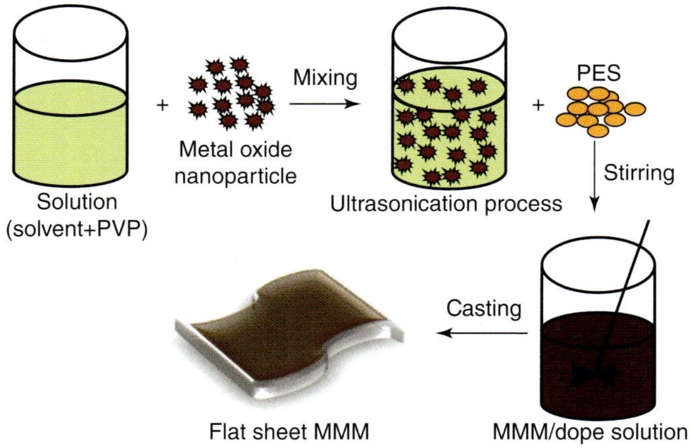

Figure 4.1 Fabrication process of polymer/inorganic membranes via blending method [1].

polymeric/particle interfacial area and have relatively lower impact on the physical changes on membrane surface roughness. It also provides the opportunity to introduce higher percentage loading of the inorganic nanomaterials into the polymer matrix.

4.2.2 Surface Coating

4.2.2.1 Interfacial Polymerization

The concept of IP was first introduced by Morgan in the 1960s [2]. By employing this technique, an ultrathin selective polyamide (PA) layer with thickness in the range of 100–500 nm is able to form over the surface of microporous support membrane, producing a TFC membrane with a good combination of permeability and selectivity. PA is a polymer where the monomer units are connected together by an amide group (—C(=O)—NH—). It can be easily synthesized by IP between polyfunctional amine monomer and polyacyl chloride monomer at the interface between two immiscible solvents with hydrochloride acid created as byproduct.

Figure 4.2 depicts the schematic diagrams of the TFC membrane preparation using typical IP technique. In order to establish a very thin PA selective layer on top of a support membrane, an aqueous solution containing amine monomer of 1–3 w/v% (prepared in the unit of w/v%, that is, g amine monomer/100 ml pure water) is first poured onto the support surface and allowed to come into contact with the polymeric surface for several minutes before draining the excess aqueous solution. The PA layer will start to develop as soon as an organic solution (usually *n*-hexane or cyclohexane) that contains a secondary monomer (acyl chloride monomer with concentration of 0.05–0.2 w/v%) is poured on top of the support membrane. Once the interaction between the two active monomers is completed, the organic solution is drained and the resultant composite membrane is rinsed using the same organic solution. For some cases, heat treatment at 70–90 °C is required before storing the membrane in a water container. This

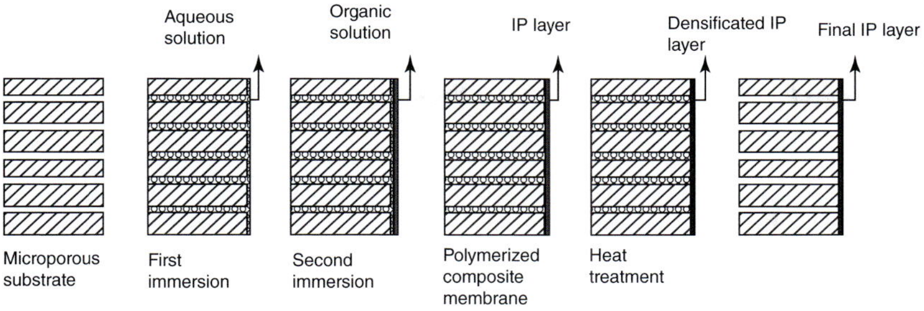

Figure 4.2 Preparation process of TFC membrane by interfacial polymerization technique.

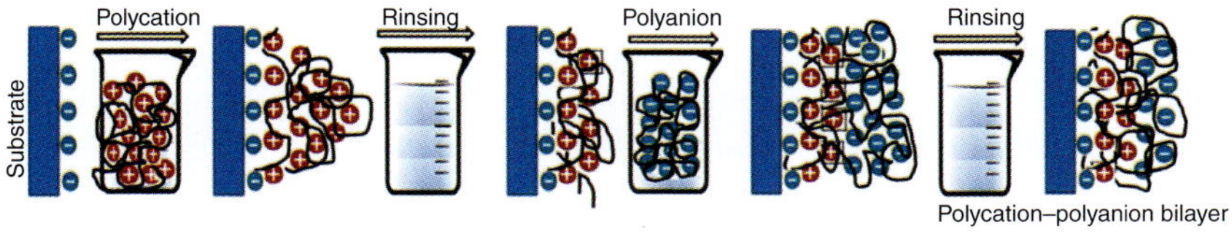

Figure 4.3 Preparation of polyelectrolyte multilayer membranes by the layer-by-layer assembly method (dipcoating). (Joseph 2014 [3]. Reproduced with permission of Royal Society of Chemistry.)

step is important to densify the polymerization properties of the PA layer on the surface of the support membrane.

4.2.2.2 Layer-by-Layer Coating

LbL assembly method is an attractive technique for the fabrication of ultrathin polyelectrolyte multilayer (PEM) film for diverse applications since its re-discovery almost two decades ago. The ability of this technique to provide conformal LbL coating of various functional materials on different substrates has opened the possibility of its use for the preparation of multilayer membranes. The LbL method involves alternating sequential adsorption of polycations and polyanions on a substrate, followed by a rinsing step after each adsorption to remove weakly associated polymer chains, as illustrated in Figure 4.3 [3]. Electrostatic interaction between the positive and negative charges leads to the formation of thin multilayer films.

The main advantage of the LbL method is the nanometer-scale thickness control during the deposition process, which is governed and tunable by the number of sequential adsorption steps. The number of bilayers in the multilayer film determines the thickness of the polyelectrolyte multilayer membrane (PEMM) and, thus, selectivity and flux can be finely tuned by the number of LbL cycles as well as by the chemical composition of the utilized polyelectrolytes. PEMMs have been considered as a good choice for the preparation of separation membranes possessing high selectivity and permeability. Selection of a suitable support material, the right pair of polyelectrolytes, and the appropriate LbL deposition conditions are the main steps, which control the properties of PEMMs.

It is well known that adsorption of a polyelectrolyte is strongly influenced by the properties of the substrate. A wide variety of substrates have been used for membrane preparation such as PE, hydrolyzed PA, polyethylene terephthalate (PET), PES, polystyrene, CA, polycarbonate, and so on. The characteristics of a good substrate for LbL PEM assembly are low surface roughness, high surface charge density as well as possessing minimal surface charge for the polyelectrolytes to adhere on it. This charge can be inherent or created by surface treatment. The influence of the substrate is restricted to the first few layers and the interaction between both polyelectrolytes governs the adsorbed amount in the higher layers. Moreover, the substrates should be chemically, mechanically, and thermally stable and possess sufficient permeability and hydrophilicity. To perform LbL coating on different substrates, methods such as dip coating, spray coating, and spin coating are used with dip coating most commonly employed.

Although PEMM has potential for separation and purification applications, it is still faced with great challenges. One major restriction of the LbL-assembled membrane is the poor stability of the bilayer constructed by the electrostatic interaction between oppositely charged PEs. Studies have shown that the LbL assembled structures can be easily destroyed under severe conditions, such as high ionic strength and chlorine treatment which are always involved in water purification processes. Although glutaraldehyde cross-linking process can be performed on an LbL-assembled membrane, the membrane stability is still insufficient under some conditions [4]. Besides, the membrane water flux tends to decrease significantly with increasing bilayer numbers, although higher rejection is achieved. Xu *et al.* [5] further stated that LbL-assembled membranes face great challenges to achieve the same level of currently used commercial TFC membrane (prepared via IP technique), that is, good balance between water permeability and solute selectivity. The utilization of this assembly method might be more suitable for surface modifications of TFC membranes.

4.2.2.3 Sol–Gel Coating

Sol–gel process is a wet-chemical technique widely employed recently in the fields of materials science and ceramic engineering. This technique is used primarily for the fabrication of materials (typically a metal oxide) starting from a chemical solution which acts as the precursor for an integrated network (or gel) of either discrete particles or network polymers.

In general, sol–gel process defines the transition of a liquid solution "sol" into a solid "gel" phase. The sol–gel process is a low temperature processing route for preparing complex oxides [6] and complex functional oxide nanostructure [7]. It is a multistep processing technique, involving both physical and chemical reactions such as hydrolysis, condensation, drying, and densification. A different size and shape of micro/nanoparticles and powders, fibers, membranes, porous materials, and coatings can be produced through a sol–gel method as illustrated in Figure 4.4 [8]. The following is the brief description of the three main steps involved in sol–gel thin film coating:

1) *Preparation of a sol:* Sol is prepared via hydrolysis and condensation reactions at room temperature in a designed chemistry with controlled rates.
2) *Coating:* Deposition of sol onto a substrate surface using either spray coating, dip coating, or spin coating.

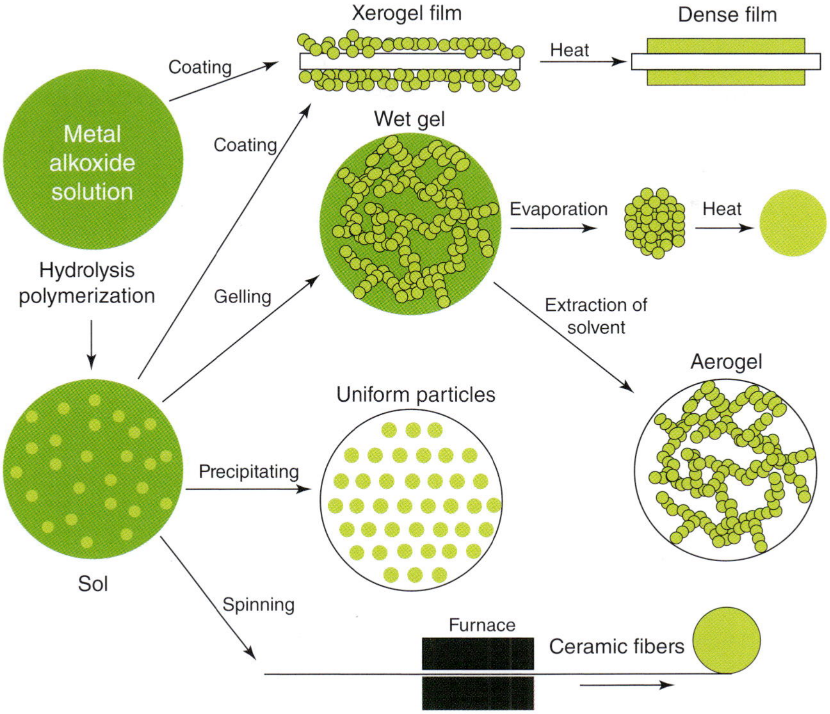

Figure 4.4 Schematic summary of sol–gel process [8]. (Reproduced with permission of Centexbel.)

3) *Heating:* Heat treatment at temperature of around 500 °C is applied to achieve dense and crystallized coating film on the substrate surface. This step is also called as *annealing*, *sintering*, or *calcination*.

The major advantages of the sol–gel process are its low processing temperatures in producing oxide products, ability to produce multicomponent materials, high purity, and stoichiometry control. Table 4.1 summarizes the procedure of three different coating techniques together with their pros and cons.

There are many research works reporting the potential of sol–gel method for membrane surface modification. For instance, Ma *et al.* [13] deposited Si-doped titanium dioxide (TiO_2) sol–gel onto a tubular alumina membrane to create an additional dense layer that could perform separation and photodegradation simultaneously. The Si-doped TiO_2 layer exhibited an anatase structure with a grain size of 8–10 nm and was able to degrade some of the reactive azo dyes dissolved in solution under UV irradiation. Choi *et al.* [14] reported that, besides exhibiting the potential for degrading organic pollutants, the membrane coated with sol–gel TiO_2 layer achieved better photodegradation and higher separation efficiency of organic matter pollutants. This can be attributed to the photocatalytic activity of TiO_2 coating layer that creates an additional selective layer to the membrane barrier and improves separation efficiency.

Table 4.1 Summary of different coating techniques together with their respective advantages and disadvantages.

Type of coating method	Procedure	Advantages	Disadvantages
Spray coating [9]	Sol is sprayed onto the substrate surface with controlled scanning speed and spray pressure	The coating process is contactless and the substrate topography is not influenced by the coating weight	Hard brittle materials like ceramics cannot be sprayed without using ductile binders. Gas consumption is quite high
Dip coating [10]	There are three stages: • *Immersion*: the substrate is dipped into coating sol with a controlled speed • *Holding time:* Substrate is holding in the coating sol to allow penetration of sol species • *Withdrawal*: substrate is withdrawn with a controlled speed to avoid vibrations, and so on	Any size and curved substrates can be used. Simple, low cost, low contamination risk, and operation atmosphere can be flexibly adjusted	Large volume of coating sol is required if the substrate size is too large. This is not economical and practical as precursor-based sols are expensive. Selective coating on one side of substrate is difficult
Spin coating [10–12]	There are four stages: • *Deposition*: An excess amount of sol is deposited on the substrate surface. The sol should completely wet the surface • *Spin-up*: The substrate is rotated at high speeds for desired time. During this rotation period spiral vortice is formed and deposited sol spread on substrate surface by centrifugal force • *Spin-off*: In this stage acceleration of rotation speed is kept constant and spinning reaches a defined rate. Viscous forces dominate the film formation behavior and edge effects are generally seen • *Evaporation*: In a regular sol formulation, solvent is usually volatile and causes simultaneous evaporation which is the dominant coating thinning mechanism at this stage. After evaporation of the solvent green, coating starts to transform into a gel	Film thickness can be easily manipulated by controlling spin speed. The film layer is usually much thinner and uniform compared to other techniques	The deeply etched features cause a physical obstruction to the solution flow, preventing complete coverage. Lower reproducibility and lack of material efficiency. Significant amount of sol (95–98%) is flung off to the coating bowl and wasted

4.2.2.4 Spin Coating

The technique of spin coating is widely used in many industrial processes, in particular in the microelectronics industry to apply a uniform photoresist and photolithographic film onto silicon wafers. This coating technique is also used in the preparation of composite membranes but only in the laboratory scale. The membranes prepared from spin coating could exhibit thickness in the range of 0.5–30 μm and are much thinner than those made from the casting technique using a knife-like tool (80–250 μm). Since preparation of membranes is a pressure-driven process, the thinner the membrane thickness the lower the transport resistance of the membrane, which, as a consequence, can lead to greater water permeability.

Figure 4.5 illustrates a schematic of a lab-scale spin coater (also known as *spinner*) [15]. Prior to spin coating, an excess of polymeric solution is applied to the center of a circular plate made of either glass or stainless steel material. It is followed by rapid acceleration of the plate to the desired rotation rate (angular velocity) at a given duration of spinning. In the spin coating process, polymeric solution flows radially due to the action of centrifugal force, causing the excess solution to be ejected off the edge of the plate. As the solvent used to dissolve polymer is volatile, an ultrathin layer of film is thus produced on the plate surface due to a dramatic rise in film viscosity from solvent evaporation. After the coating process, the nascent membrane together with the plate is immersed into a water coagulation bath in which phase inversion takes place and a membrane is produced. The resultant membrane is further immersed in another water bath in order to completely remove the residual solvent from the membrane prior to use.

There are many factors governing the thickness and morphology of the membrane produced during spin coating. The properties of the membrane prepared from spin coating can be affected by both polymeric solution and the spinning conditions. For instance, the thickness of film increases with decreasing spin speed or decreases with decrease in viscosity of solution [15, 16]. The solution

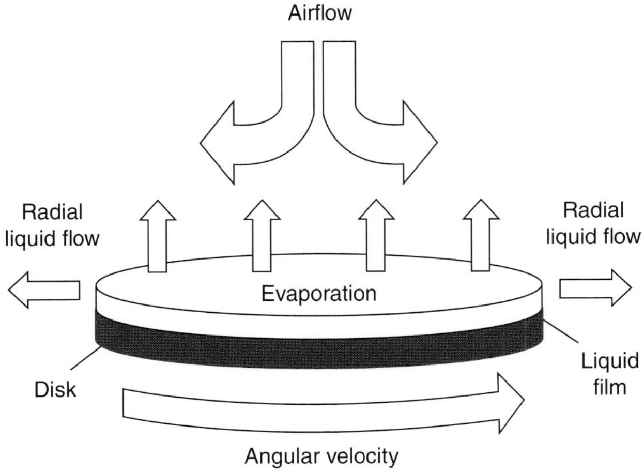

Figure 4.5 Schematic of a spin coating process. (Norrman 2005 [15]. Reproduced with permission of Royal Society of Chemistry.)

viscosity is in general dependent on the polymer concentration and type of polymer used in the preparation of solution in which the higher the polymer concentration and the MW of polymer used in the solution preparation, the greater the viscosity of the solution produced. Increase in spinning time on the other hand has potential to produce thinner membranes with more visible pores on surface. Wang *et al.* [17] reported that the PVDF membrane layer tended to become thinner while its surface pore became visible with increasing spinning time from 60 to 240 s. For prolonged spinning time (240 s), the cross-sectional structure of the membrane was found to be more porous and disordered and this could be the result of the evaporation-induced phase separation instead of nonsolvent (water)-induced phase separation.

4.2.3 Photo-Initiated Polymerization

4.2.3.1 UV-Initiated "Grafting-to" Membrane Surface

The UV-initiated polymerization modifies membrane top surface by grafting polymer chains onto the surface and/or in the membrane pores. UV irradiation can cross-link polymer chains and cleave polymer bonds, thereby forming functional groups such as hydroxyls, carbonyls, or carboxylic acids on the surface [18]. This method has several significant advantages such as (i) the ability to modify the polymer surface to have distinct properties through the choice of different monomers, (ii) the controllable introduction of graft chains with a high density and exact localization to the surface, without affecting the bulk properties, and (iii) greater chemical stability, which is assured by covalent attachment of graft chains [19].

The "grafting-to" method of a polymeric membrane is based on the UV-reaction between membrane materials and functionalization agents, which requires special UV-reactive groups. Depending on the location of UV-reactive moieties, two routes can be classified. They are via UV-reactive membrane polymer and via UV-reactive functionalization agents as illustrated in Figure 4.6 [20]. Since this approach is relatively independent on the chemical composition of materials, various functionalization agents can be attached onto a membrane surface with UV-reactive moieties, or a variety of membranes (nonporous and porous, different membrane polymers) can be modified without any pre-treatment using UV-reactive functional molecules.

An example is the functionalization of UF polyacrylonitrile (PAN) membrane with low MW aromatic azide derivatives comprising different hydrophilic and hydrophobic substituents. The separation characteristics and protein fouling tendency were altered and the extent of change was found to be dependent on the type of functional groups introduced. In brief, the improved membrane performance was explained by the increased surface hydrophilicity or presence of the desirable pores in the active membrane layer [20]. Amphiphilic graft copolymers, composed of a poly(cyclooctene) backbone with grafted PEG and photo-reactive phenyl azide pendant groups, were also synthesized and applied as coating to commercially available PVDF UF membranes [21]. Photo-induced cross-linking of the graft copolymer coatings prevented delamination from

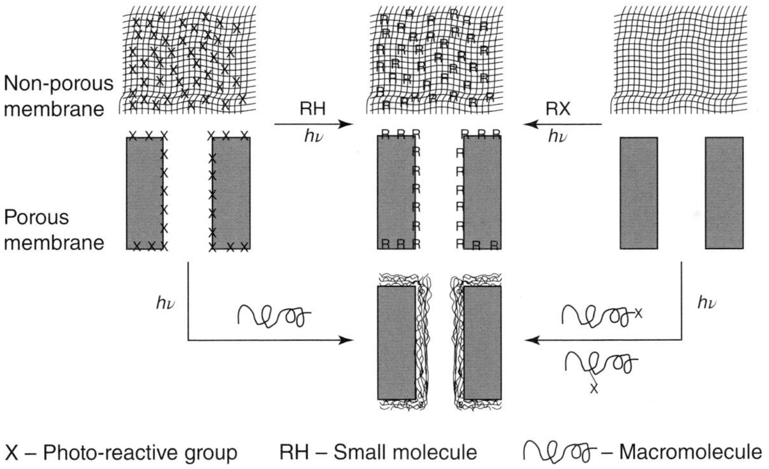

Figure 4.6 Depiction of principal mechanisms for photo-functionalization of polymeric membranes: via photo-reactive membrane polymer (from left) and via photo-reactive functionalization agents (from right). (He 2009 [20]. Reproduced with permission of Elsevier.)

the underlying membrane and led to composite membranes with improved antifouling properties in the treatment of oil-in-water emulsions.

Although this approach allows a precise control of grafter chain structure, its efficiencies are limited due to the difficulty in achieving complete grafting densities and homogeneity on the surface of membrane caused by the sterical hindrance effect. The effect is significant with increasing size of the functionalization agent. Moreover, photo-reactive polymer conjugates might react not only with the membrane surface but also with the other adjacent molecules. In certain cases, accessibility of membrane small pores for macromolecular functionalization agents may be limited [20].

4.2.3.2 UV-Initiated "Grafting-from" Membrane Surface

The "grafting-from" method has been used to overcome the limitation of the "grafting-to" method with respect to grafting density. Most composite membranes prepared via "UV-grafting-from" are pore surface-functionalized. Depending largely on the initial membrane pore size (micro/mesopores vs macropores) and its pore size distribution, membranes with different grafted layer properties can be achieved. Membrane modification via "grafting-from" approach in general can be performed with or without using a photo-initiator.

The "grafting-from" method in the absence of photo-initiator involves the direct generation of free radicals from the base membrane polymers under UV irradiation. Such a method requires either a photosensitive base polymer (photo-reactive side group or part of polymer backbone) or the introduction of photo-sensitive groups onto the membrane surfaces prior to graft copolymerization. Poly(arylsulfone) (PAS) membranes such as PES and polysulfone (PSF) are examples of polymeric membranes that are light sensitive in the UV (200–320 nm) and hence do not need a photo-initiator for radical production. These membranes are self-initiated and can produce sufficient radicals for vinyl grafting of various functional groups. Although both PES and PSF are photosensitive, PES is far more sensitive to UV-initiated graft polymerization and requires far less energy to attain a desired degree of grafting as compared to PSF. However, the UV irradiation should be carefully performed because in certain cases, severe degradation of membrane pore structure could occur under UV irradiation, affecting membrane surface integrity.

For surface modification with an added photo-initiator, the initiating radical sites should be generated on the membrane surface by the reaction of a photo-initiator with the base membrane polymer under UV irradiation. Benzophenone (BP) and its derivatives are the most often used photo-initiators for initiation of the UV-initiated graft polymerization of vinyl monomers on the surface of polymer membranes. This type of photo-initiator undergoes photo-reduction by hydrogen atom abstraction from surrounding chemical species, which leads to the generation of initiating radicals. By means of UV-induced graft polymerization with photo-initiators, a number of vinyl monomers were successfully used to modify and functionalize a wide range of polymeric membranes. These include, PSF, PAN, PVDF, polypropylene (PP), polyurethane (PU), PET, polyethylene, nylon, and styrene-butadiene-styrene membranes.

The photo-initiator may be loaded on the membrane surface by dissolving in the monomer solution. PET nucleopore membranes, for example, were modified using BP dissolved in the monomer solution followed by UV-irradiation of the solution and the immersed membrane [22]. It was shown that the photo-grafting occurred only on the top membrane surface rather than in the membrane pores. Although this approach is relatively simple,; its main drawback is a low local concentration of BP on the membrane surface because BP moves to the membrane surface only by diffusion. This results in low grafting efficiency, whereas high bulk BP concentration may cause a side reaction such as homopolymerization. In order to improve the grafting efficiency, the photo-initiator adsorption method has been proposed. In this way, the local BP concentration on the membrane surface was increased, while the BP concentration in the bulk of monomer solution was kept very low to reduce the homopolymerization process [19].

In general, it should be noted that very detailed studies have been carried out on surface membrane modification via UV-initiated graft polymerization, including the grafting mechanisms, dependence of grafting efficiency on grafting time, monomer type, UV-wavelength, and intensity. The attractive features of UV grafting are easy with controllable introduction of graft chains with a high density and their exact localization to the membrane surface. Furthermore, covalent attachment of graft chains onto a polymer surface avoids their delamination, and assures the long-term chemical stability of introduced chains in contrast to physically coated polymer layers. These advantages have resulted in this modification technique gaining interest from membrane manufacturers, for example, in continuous modification of PS hollow fiber membranes with an anionic grafted polymer layer to obtain NF membrane [23]. The disadvantages of the modified membranes are usually their reduced fluxes compared with those of the unmodified membranes because the grafting layer adds extra hydraulic membrane resistance. Grafting may also increase manufacturing costs due to additional use of organic solvents, monomers, and UV equipment.

4.2.4 Other Surface Modification Methods

Apart from the above-mentioned modification methods, heat treatment can also be employed to modify the surface (and/or structure) of a polymeric membrane. Heat treatment is used to improve the mechanical and tribological properties of the polymers. Upon heat treatment, the elastic part of the polymer tends to increase, causing an increase in compressive strength and heat conductivity. For certain cases, heat treatment process is claimed to improve the chemical stability of the modified membrane by altering the polymer chain mobility.

Generally, heat treatment process can be carried out in two different ways, either by subjecting the membrane to direct heating in an oven or by post-treating the membrane in a water bath under different temperatures. In terms of applications, heat treatment process is more preferred in the membrane gas separation process. It is proven that a heat-treated membrane has the potential to improve the gas permeability, permselectivity, and antiplasticization properties. Studies by Koros and Woods [24] confirmed that the modified membrane exhibited significantly improved hydrogen permeance (300 GPU) when treated at 175° C. The surface pore size of the heat-treated membrane was reported to reduce upon heat treatment, resulting in the formation of a denser structure. From gas separation and pervaporation perspectives, it is desirable to have denser membrane for both applications as their main separation mechanism is governed by a solution-diffusion mechanism.

Other than the heat treatment, membrane post-treatment using chemical solution is also known to have an impact on altering membrane surface properties. Previous work has shown that post-treatment with ethanol was useful to improve both the surface hydrophobicity and the effective porosity of the resulting PVDF-based hollow fiber membrane, leading to greater performance for ammonia removal [25]. Recently, surface modification by chemical reaction has gained great interest in producing a membrane with improved surface property, especially in terms of surface hydrophilicity and membrane permeability. Surface grafting and cross-linking are examples of surface modification with a chemical process. In the chemical process, the surface modification involves surface reaction in order to form a covalent bonding interaction between the modified surface and a bulk of the polymer membrane. Generally, the polymer chains are first activated either by a chemical reaction (strong alkali or strong acid) or high-energy radiation (plasma treatment) to generate active groups, followed by the grafting of hydrophilic/hydrophobic modifiers. The membrane surface properties can be improved while the membrane bulk is not significantly affected. Moreover, the covalent attachment of modifiers on the membrane surface avoids their delamination and offers a long-term chemical stability, in contrast to physical surface coating.

There is also a special surface modification technique that does not belong to any of the above methods such as molecular imprinting technology (MIT). Generally, MIT allows preparing polymeric materials with selectivity toward specific

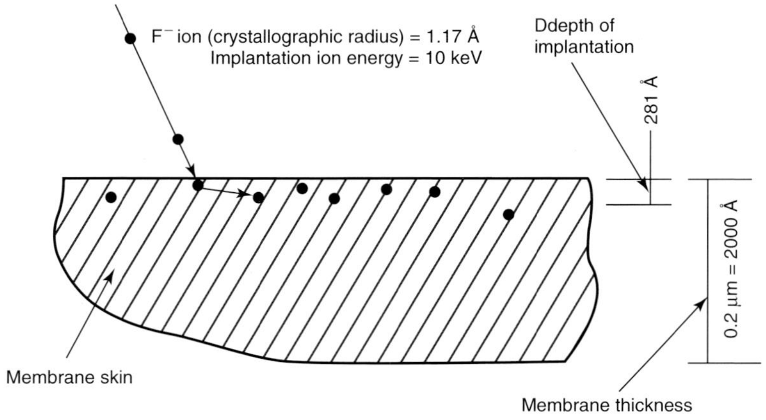

Figure 4.7 Schematic diagram for ion implantation. (Mukherjee 2005 [26]. Reproduced with permission of Elsevier.)

molecules through polymerization or phase inversion in the presence of a template. Mukherjee *et al.* [26] studied ion implantation on the surface of commercial PA NF membrane (NF90 from Filmetec) and a CA NF membrane (SP28 from Osmonics). The surfaces of both NF90 and SP28 were modified by implanting with F-ions at two different intensities as illustrated in Figure 4.7 [26]. In this work, they have observed that compared to the unmodified membranes, the zeta potential of the modified membranes tended to show higher negativity with an increased intensity of ion implantation, resulting in an increase in salt rejection property. This promising result can be attributed to the greater Donnan exclusion effect.

4.3 Advancements of Surface-Modified Membranes for Various Separation Processes

4.3.1 Wastewater Treatment

4.3.1.1 Ultrafiltration and Forward Osmosis for Oily Wastewater

UF membrane has been applied in oily wastewater treatment over the past decades due to its low energy consumption, minimal chemical agents required, easy automation, less production of oil sludge, and compact process [27, 28]. It has been previously reported that for oil and gas produced treatment process, membrane technologies possess significantly lower energy consumption and operating cost (i.e., US$ 0.08–0.34 bbl^{-1}) than those of conventional technologies (US$ 0.1–5 bbl^{-1}) [27, 28]. The low energy consumption in membrane technologies, in particular, UF is mainly attributed to low operating pressure (several bars) applied throughout the operation period. Furthermore, UF membranes are characterized by their surface pores between 0.02 and 0.1 μm, which can be used to efficiently remove emulsified oil and suspended solids from oily wastewater.

Numerous studies have been focused on enhancing the performance of UF membrane in treating oily wastewater over the past two decades. Polymeric additive is frequently used in the membrane formation process as it can play a role in suppressing macrovoid formation in the membrane, promoting pore formation, improving pore interconnectivity, and modifying membrane surface hydrophilicity [29]. A brief overview of the performance of UF membranes modified by additives is provided in Table 4.2. For instance, Wu *et al.* [30] cross-linked cellulose-based microporous membrane with poly(vinyl) alcohol (PVA) via dynamic coating process. The results show that a remarkable water flux ($725\,l\,m^{-2}\cdot h$) and oil rejection (96.8%) were achieved when tested using

Table 4.2 Comparison of different UF membranes for oily wastewater treatment process.

Membrane configuration	Main membrane forming materials/ additives	Process conditions	Performances	References
Flat sheet	Cellulose ester/polyvinyl alcohol (PVA)	Oil concentration of 242.25 ppm oil at 4 bar	Permeate flux: $725\,l\,m^{-2}\,h^{-1}$ Oil rejection: 96.75%	Wu *et al.* [30]
Flat sheet	Polyacrylonitrile (PAN)/ polyacrylonitrile-graft poly(ethylene oxide) (PAN-*g*-PEO)	Industrial wastewaters with different COD values (14 300, 2 300, and 970 ppm) were evaluated at 0.07 MPa	Permeate flux: $1160\,l\,m^{-2}\,h^{-1}\,MPa^{-1}$ Oil rejection: 96%	Asatekin and Mayes [31]
Flat sheet	PAN	Oily solution with concentration of 78 ppm and pH 10 was tested at temperature of 50 °C and pressure of 3 bar	Permeate flux: $250.8\,l\,m^{-2}\,h^{-1}$ Oil rejection: 99%	Salahi *et al.* [32]
Flat sheet	Polysulfone (PSF)/polyethylene glycol (PEG) and polyvinylpyrrolidone (PVP)	Oil concentration of 100 ppm at 103.4 kPa and pH 6.0	*Cross-flow configuration* Permeate flux: $25-126\,l\,m^{-2}\,h^{-1}$ Oil rejection: 94.4% *Dead-end configuration* Permeate flux: $60-113\,l\,m^{-2}\,h^{-1}$ Oil rejection: 97.8%	Chakrabarty *et al.* [33]
Flat sheet	$PSF/SO_4^{2-}/ZrO_2-Y_2O_3$	Oil concentration of 80 ppm at 0.2 MPa	Permeate flux: $100\,l\,m^{-2}\,h^{-1}$ Oil rejection: 99.16%	Zhang *et al.* [34]

242.25 ppm feed oil solution at 4 bar. Qiu *et al.* [35], prepared the UF membrane by coating PVA onto the felt-metal filter. They obtained a permeate flux of $45\,l\,m^{-2}\cdot h$ and oil rejection of 90% using synthetic oily wastewater containing 500 ppm kerosene oil at 4 bar.

The incorporation of different MW of PVP (24 and 360 kDa) into PSF-based UF membrane was studied by Chakrabarty *et al.* [33] for the treatment of synthetic oily solution. The PSF-PVP24k was reported to exhibit the highest flux ($128\,l\,m^{-2}\cdot h$), with a greater degree of flux declination (43.1%) while for PSF-PVP360k, the flux is $70\,l\,m^{-2}\cdot h$ with lower water flux (31.6%). However, the rejection of oil is found to be quite similar, which is 77.5% and 78.8%, for both of PSF-PVP24k and PSF-PVP360k, respectively. It is explained that increasing the MW of PVP from 24 to 360 kDa could lead to lower flux due to the reduction of surface pore size caused by the swelling of PVP on the membrane top layer.

In recent years, incorporating inorganic fillers into the membrane matrix to prepare a type of composite membranes with enhanced surface properties has attracted a lot of attention owing to their small particle sizes, high surface area, and hydrophilic nature. For instance, Yuliwati *et al.* [36] modified the surface of hydrophobic PVDF hollow fiber membrane using hydrophilic TiO_2 nanoparticles in the presence of lithium chloride ($LiCl_2$) salt through blending method. The modified membrane was tested under submerged conditions in which air bubbles were created to minimize membrane fouling. Upon TiO_2 incorporation, the modified PVDF membrane achieved significantly higher water permeability ($82.5\,l\,m^{-2}\,h^{-1}$) and oil rejection (98.8%) than that of the unmodified PVDF membrane ($30\,l\,m^{-2}\,h^{-1}$ and 62.6%) when both membranes were evaluated under same conditions. The improvements are attributed to the improved membrane surface hydrophilicity (by 141.6%) upon addition of 1.95 wt% of TiO_2. Further increase in TiO_2 concentration to 3.8 wt% however did not make the membrane to become more hydrophilic. It is because TiO_2 tended to agglomerate when high loading was used.

Li *et al.* [37] reported that the oil separation performance of PVDF-based membrane could be further enhanced upon aluminum oxide (Al_2O_3) incorporation. In comparison to the unmodified PVDF membrane, the water flux of surface-modified membrane was reported to increase by 26–50% while oil removal efficacy remained at 98.7%. Similar outcomes were reported by Yi *et al.* [38] where a higher relative flux was obtained using Al_2O_3-modified PVDF membrane. Figure 4.8 shows that the modified PVDF membrane showed higher relative permeate compared to the control PVDF membrane in treating oily solution with concentration ranging from 20 to 400 ppm. This indicated that Al_2O_3 has played a role in improving membrane surface antifouling properties by reducing the deposition/adsorption of oil molecules.

Other inorganic fillers that have been recently used for improving UF membrane surface properties for better oil separation performances are silicon dioxide (SiO_2), hydrous manganese dioxide (HMO), phosphorylated Zr-doped hybrid silica (SZP), and phosphorylated TiO_2-SiO_2 (PTS). Zhang *et al.* [34] studied the performance of the PSF membrane incorporated with SZP particles for oily wastewater treatment and reported that water flux as high as $100\,l\,m^{-2}\,h^{-1}$ (at 1 bar) could be achieved by embedding 15 wt% SZP into the membrane made of 12 wt% PSF. The oil rejection of the membrane was not compromised by its

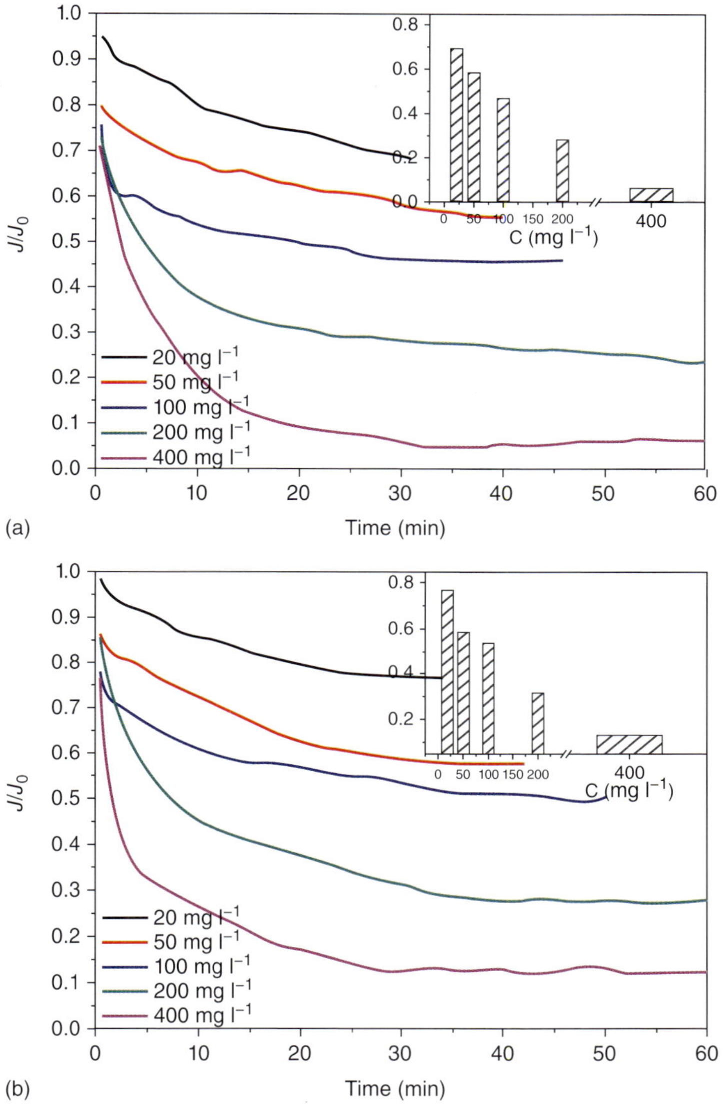

Figure 4.8 Relative flux of various initial oil concentration using (a) unmodified PVDF membrane and (b) Al_2O_3-modified PVDF membrane. (Li 2006 [37]. Reproduced with permission of Elsevier.)

high water permeability as almost complete elimination of oil molecules (99.2% rejection) was achieved at 80 ppm oily wastewater.

Further attempt was made by Zhang *et al.* [34] to investigate the performance of PSF-based membrane with the use of three different types of inorganic fillers as modifier, that is, SiO_2, SZP, and PTS. Of the composite membranes prepared, the membrane modified by PTS showed the highest permeate flux (116 l m^{-2} h^{-1}) followed by PSF/SiO_2 (100 l m^{-2} h^{-1}) and PSF/SZP (75 l m^{-2} h^{-1}). In terms of oil rejection, PSF/PTS achieved 92% rejection compared to 85.2%

and 90.8% recorded in PSF/SiO_2 and PSF/SZP, respectively. The greater filtration performance of the PSF/PTS membrane is mainly attributed to its better surface hydrophilicity coupled with greater antifouling and anticompaction properties.

A novel highly hydrophilic HMO was synthesized and incorporated with the PES membrane to treat synthetic oily solution containing 1000 ppm crude oil at 1 bar [39]. As the membrane was incorporated with very high amount of nanoparticles (at HMO:PES weight ratio of 2 : 1), it was possible to achieve extremely high water flux (573.2 $l\,m^{-2}\,h^{-1}$). It is also observed that almost complete oil rejection (99.99%) was achieved by this HMO-modified PES membrane. There are several factors contributing to the excellent performance of this novel membrane. The first one is due to the smallest surface pore size of the membrane that plays a main role in preventing oil particles from passing through the membrane. The second factor to be considered is the enhanced hydrophilicity of the PES/HMO membrane, which minimizes oil deposition and/or adsorption. The increase of membrane effective surface area that resulted from rougher surface also partly contributes to higher water production rate.

Apart from these hydrophilic inorganic fillers, some studies reported on the potentials of UF membrane made of copolymer in treating oily wastewater. Asatekin and Mayes [31] modified PAN-based membranes by incorporating amphiphilic PAN-*g*-PEO in treating oil-contaminated wastewater. It was reported that oil removal efficiency as high as 96.5% could be obtained with grafting modified membrane compared to 96.1% shown by the unmodified commercial membrane. Interestingly, a full recovery of initial permeate flux could be obtained when the studied membranes were subjected to simple backflushing process. This indicated that the fouling resistance of the modified membrane was greatly enhanced with the addition of copolymer. Using the same membrane forming material, Salahi *et al.* [32] revealed that permeate flux and oil rejection were recorded at 250.8 $l\,m^{-2}\,h^{-1}$ and 99%, respectively after 2.5-h operation at 3 bar.

Another method is using surface grafting in which hydrophilic polymer chains are immobilized onto the membrane surface either by chemical reaction or physical absorption to provide a stable fouling resistance layer. The introduced hydrophilic materials form compact hydrated layers to prevent the fouling of oil droplets on membrane surfaces and allow the fouled oils to be easily removed during cleaning process. Zhao and co-workers [40] grafted a low surface free energy molecule – pentadecafluorooctanoic acid onto the surface of an aminated PAN UF membrane via an acrylated reaction. The modified membrane exhibited excellent antifouling property (i.e., 100% flux recovery). Zhu *et al.* [41] reported a novel PVDF membrane grafted with zwitterionic polyelectrolyte brush poly(3-(*N*-2-methacryloxyethyl-*N*,*N*-dimethyl) ammonatopropanesultone) (PMAPS) via surface-initiated atom transfer radical polymerization process (see Figure 4.9). The modified membrane showed extremely high water flux and low contact angle upon 12-h polymerization time. They attributed the superior hydrophilicity and flux performance to the higher surface energy and hydrated behavior of zwitterionic polyelectrolyte polymers in water.

Although UF membrane could achieve an excellent oil separation rate, the removal of some recalcitrant, toxic, and non-biodegradable compounds are

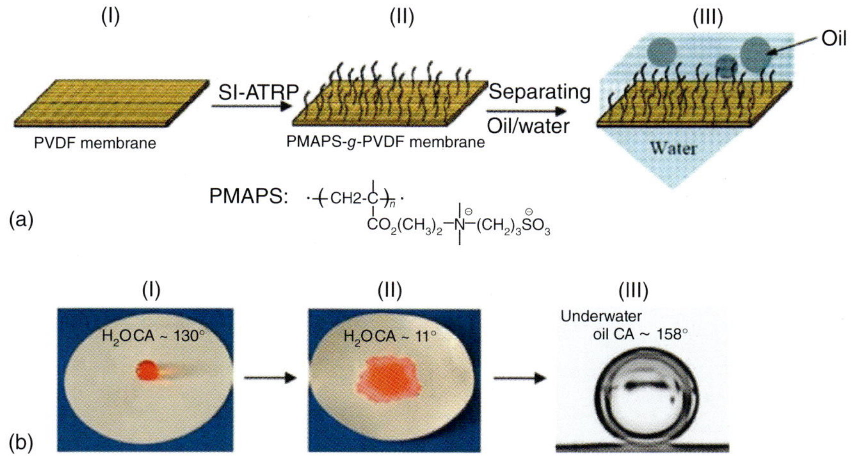

Figure 4.9 (a) Schematic illustration of the preparation of PMAPS-*g*-PVDF membranes for separating dispersed oil from water and (b) water and oil wettability. Photographs of an oil water droplet (3 ml) dropped onto the original PVDF membrane (left) and PMAPS-*g*-PVDF membrane (middle), and their contact angles. An underwater oil (1,2-dichloroethane) droplet on PMAPS-*g*-PVDF membrane showing a contact angle of 158° (right). (Zhu 2013 [41]. Reproduced with permission of Royal Society of Chemistry.)

found to be ineffective and require additional treatment processes to eliminate them. Furthermore, a typical UF membrane is susceptible to fouling at high feed oil concentration [37, 42, 43]. In view of this, various kinds of advanced treatment methods have been proposed to overcome the drawbacks of the existing UF treatment processes. Among the methods proposed, photocatalytic membrane, that contains photocatalyts within the membrane matrix and/or on top surface has drawn immense attention among scientists specializing in membranes as this hybrid process could deal with a broad range of organic pollutants via photocatalytic degradation and separation processes. Ong *et al.* [44] studied the separation and degradation of synthetic oily wastewater using submerged membrane photocatalytic reactor (SMPR). The hollow fiber membrane used consisted of PVDF and TiO_2 nanoparticles. It was found that average membrane flux, total organic carbon (TOC), and oil removal efficiencies of 73.0 $l\,m^{-2}\,h^{-1}$, 80% and >90%, were obtained, respectively, using a PVDF membrane embedded with 2 wt% TiO_2 when tested with 250 ppm oily solution. Another similar work conducted by Moslehyani *et al.* [45] using PVDF-multiwalled carbon nanotube (MWCNT) nanocomposite membrane with 200 ppm TiO_2 suspension revealed that optimum water flux (710 $l\,m^{-2}\,h^{-1}$), photodegradation (90%), and separation efficiency (99%) could be achieved when the membrane was evaluated using 100 ppm oily solution.

Other than UF membrane, Duong and Chung [46] also reported that oil-emulsion separation can be done in forward osmosis (FO) process using a dual-skin TFC membrane with a self-assembled sulfonated pentablock copolymer coating layer as illustrated in Figure 4.10. It was reported that pure water could be recycled from 200 000 ppm oily wastewater at a reasonable flow rate

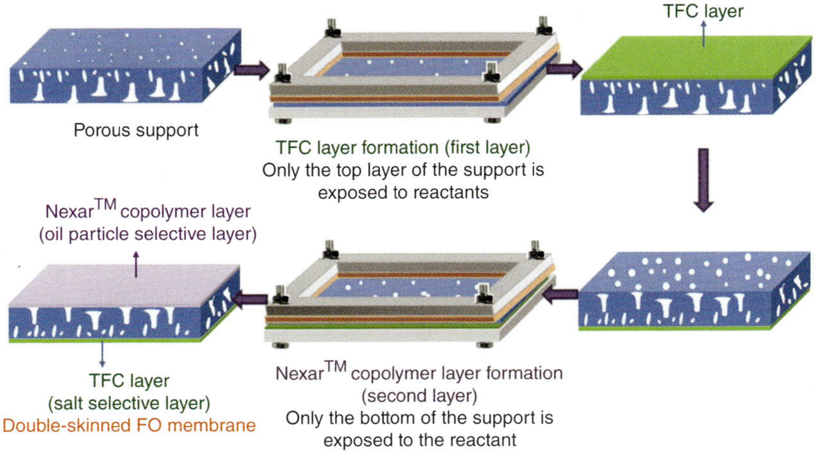

Figure 4.10 Laboratory-scale fabrication steps for the double-skinned TFC membrane [46].

of 11.9 $l\,m^{-2}\,h^{-1}$ and high quality pure water (with oil rejection of 99.88%) from various oil–water emulsions using 1 M NaCl as the draw solution. Furthermore, Han *et al.* [47] studied the FO performance using a novel hydrophilic cellulose acetate butyrate (CAB) as the membrane substrate (hollow fiber format) and coating its outer surface by polydopamine (PDA). Under the pressure retarded osmosis (PRO) mode, the newly developed TFC membrane exhibited a water flux of 37.1 $l\,m^{-2}\,h^{-1}$ with an oil rejection of 99.9% using a 2000 ppm soybean oil/water emulsion as the feed and 1 M NaCl as the draw solution.

To date, many studies have been performed to design advanced functional membranes for oil/water separation. For oily wastewater separation, the most important issue to be addressed is to decrease the adsorption of oil droplets on the membrane surface, which has been extensively investigated. In addition to oil droplets, membrane fouling caused by the adsorption of other organic molecules such as surfactants remains a problem but is often overlooked. More attention should be paid to this issue.

4.3.1.2 Nanofiltration and Membrane Distillation for Textile Wastewater

Today's world stands as a witness to the revolutionizing socio-economic influence of various industries including the textile industry. According to the World Trade Report 2008, the economic growth of textile industry has increased tremendously worldwide with 1.7% (or about US$ 238.1 billions) of the total world exports in 2007 [48]. Rapid industrialization and fast urbanization require adequate care at the source itself to attain a balanced sustainability between waste generation and economic growth. Despite the obvious benefits of this industrial progress, it contributes to environmental degradation from the both input and output sides of its activities. On the input side, the textile processing industry is regarded as a water-intensive sector as it uses a huge quantity of fresh water in the production process. Statistics revealed that textile industries consume an average of 0.06–0.40 m^3 water for each 1 kg of textile fabricated as water is used as the principal medium for applying dyes and finishing agents as well as for removing impurities. Meanwhile, on the output side, textile manufacturing process generates large amount of wastewater with a wide range of pollutant concentrations. Dasgupta *et al.* [49] reported that the textile industry discharges around 200–350 m^3 of wastewater per ton of finished product.

To date, membrane separation process has been extensively studied as one of the feasible alternative methods for textile wastewater treatment. The use of NF and membrane distillation (MD) for dye removal in particular is significantly better than those of the conventional treatment methods such as coagulation–flocculation and biological process. NF membranes with a molecular weight cut-off (MWCO) of 200–1000 g mol^{-1} exhibit one of the most competitive separation and purification processes in removing nano-sized dye components from textile wastewater. This is because dye-house effluent usually contains a mixture of dyes with MWs in the range of 300–800 g mol^{-1}. MD meanwhile, is very potential for textile wastewater treatment as it can exploit the heat energy from the hot effluent discharged (50–80 °C). Unlike NF membrane that strongly relies on external osmotic pressure to operate, the existence of vapor pressure difference between the hot solution (wastewater) and cold water across the membrane could act as a driving force, which potentially minimizes fouling tendency in the long run. Although both NF and MD demonstrate excellent separation efficiencies in dye removal, the deterioration of flux as a function of operation time still remains a main concern. Membrane flux decline is mainly caused by the dye adsorption on the membrane surface and/or inside the pores at the early stage of fouling. As the fouling continues to develop, cake formation and pore blocking will take place, leading to severe flux decline.

Development of a less-fouling-sensitive NF membrane surface is still the most sustainable solution to minimize the fouling tendency of NF membrane during a dyeing treatment process. Concerning the membrane characteristics, the hydrophobicity of the top layer is believed to cause the most flux decline. For charged organic compounds, electrostatic attraction or repulsion forces between the component and the membrane influence the degree of fouling. In view of that issue, hydrophilic groups and/or charged functional groups are introduced onto NF membrane surface via coating, blending, or grafting, aiming to produce a membrane capable of withstanding varying chemical composition of textile effluents and exhibiting less fouling sensitivity through decreased dye deposition.

There are many works reporting the improved surface characteristics of PA layer of TFC membranes for dye removal. Shao *et al.* [50] for instance, synthesized a PA selective layer using a novel combination of *m*-phenylenediamine (MPD) and piperazine (PIP) with TMC on a polyetherimide (PEI) hollow fiber support. The introduction of PIP into the MPD reaction system was reported to be capable of finetuning the sub-nano structure of the NF membrane formed via IP and obtaining a mean effective pore size (diameter) of 0.46 nm. Correspondingly, the MWCO values of the NF membrane with and without PIP incorporation were 273.4 and 257.1 g mol^{-1}, respectively. The newly developed NF membranes also showed good performance for dye removal and were suitable for the removal of both Safranin O and Aniline blue. Under specific operating conditions, the Safranin O and Aniline blue rejections were higher than 90%, and the flux of dye solutions increased with the operating pressure and declined with the feed concentration.

It is well known that the performance of NF membranes fabricated via IP depends largely on the choice of reactive monomers and the reaction process conditions. Liu *et al.* [51], for example, synthesized two types of novel sulfonated aromatic diamine monomers – 2,5-bis (4-amino-2-trifluoromethyl-phenoxy) benzenesulfonic acid (6FAPBS) and 4,4′-bis (4-amino-2-trifluoromethyl-phenoxy) biphenyl-4,4′-disulfonic acid (6FBABDS), aiming to modify the inherent property of a typical PA layer and increase water permeation of TFC membranes. The active skin layer of the TFC NF membranes was prepared through IP technique using amine solutions containing the novel 6FAPBS/6FBABDS and PIP as well as TMC solution on the microporous polyphenylsulfone (PPSU) support membrane. With increase of 6FAPBS or 6FBABDS content in the amine mixture, it was found that the water contact angle of TFC membrane surface declined significantly from >70° to <30° (Figure 4.11). The reduced water contact angle indicated the improved membrane surface hydrophilicity following the introduction of more hydrophilic sulfonic acid groups from the 6FAPBS or 6FBABDS in the PA layer. The high degree of hydrophilicity of the active skin layer increased the water diffusion (flux) through the membranes from 46.2 $l\,m^{-2}\,h^{-1}$ in the control membrane to 63.9 and 71.3 $l\,m^{-2}\,h^{-1}$ (evaluated at 0.5 MPa) for the membrane modified by 6FAPBS and 6FBABDS, respectively. Surprisingly, the water flux enhancement did not compensate the solute rejection. It was explained that the electrostatic repulsive interaction between dyes and negatively charged membrane surface is the main factor promoting the separation mechanism through Donnan exclusion effect.

Mondal and De [52] modified the properties of the PA layer of hollow fiber membrane through variation of TMC concentration on the PSF support layer.

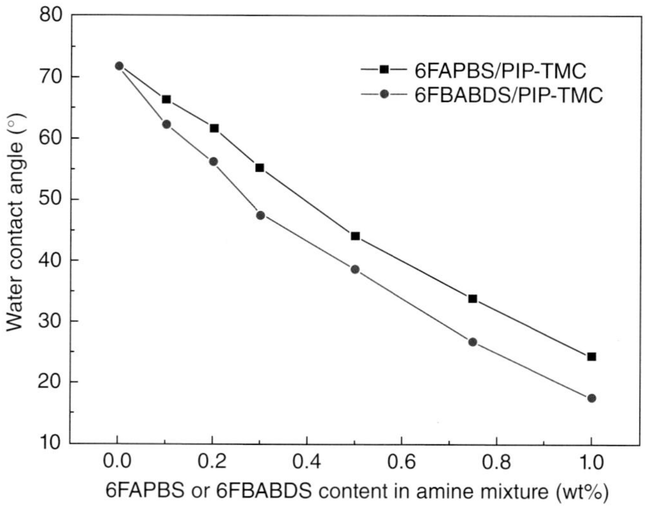

Figure 4.11 Water contact angle of the TFC NF membrane as a function of 6FAPBS or 6FBABDS amine content [51].

In this study, NF membranes were prepared by surface modification at a fixed MPD concentration (2 wt%) and three TMC concentrations (0.1, 0.5, and 1.0 wt%). The surface analysis using electron microscope showed that increasing TMC concentration tended to form PA layer with denser structure. The denser structure has led the membrane made of highest TMC concentration to exhibit highest conductivity (68%) and salinity (54%) rejection among the membranes studied. Besides its smallest surface pores, the larger effect of charge repulsion of the membrane made of highest TMC concentration was the other main factor enhancing separation efficiency.

Instead of using IP process, Yu *et al.* [53] developed negatively charged NF hollow fiber membranes by dip coating sodium carboxymethyl cellulose (CMCNa) on the outside of PP microporous hollow fiber membranes followed by cross-linking with $FeCl_3$. The fabricated CMCNa/PP composite hollow fiber membranes exhibited excellent dye rejections (more than 99.7% removal of Methyl Blue and Congo Red) throughout the studied period due to the combination effect of steric hindrance and Donnan exclusion. Meanwhile, the NaCl rejection at a concentration of 10 000 mg l^{-1} was recorded lower than 2%. The low NaCl rejection is the typical characteristic of an NF membrane that generally shows low separation performance for monovalent salt. The modified NF membrane however could provide an attracted avenue for salt recovery owing to its effectiveness in dye/salt fractionation.

Ismail and Lau [54] modified the surface of PES-based NF membrane by incorporating a small percentage of charged sulfonated poly(ether ether ketone) (SPEEK) via blending method. Figure 4.12 compares the outer surface of the control PES membrane with the PES/SPEEK blended membrane after a chemical cleaning process. Prior to the cleaning process, all these membranes were used for treating a synthetic dyeing solution that consisted of Reactive Black 5 (RB5), PVA, NaCl, and Na_2SO_4. With the introduction of negatively charged sulfonic acid groups by SPEEK, the membrane with surface charge similar to that of the dye foulant was developed, making the membrane less susceptible to fouling.

In addition to modifying the surface charge properties, the surface morphology (e.g., roughness) and hydrophilicity of the membrane can also be modified to become less susceptible to dye adsorption. The affinity of organic components to NF membranes can cause fouling of hydrophobic materials. Thus, blending of

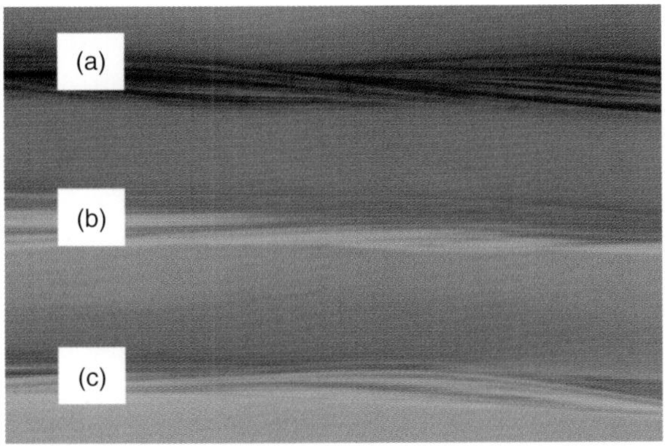

Figure 4.12 Comparison between the outer surface of control PES membrane and SPEEK-modified PES membrane after a chemical cleaning process, (a) control PES, (b) PES blended with 2 wt% SPEEK, and (c) PES blended with 4 wt% SPEEK. (Ismail 2009 [54]. Reproduced with permission of Taylor and Francis.)

original main membrane forming material with secondary hydrophilic polymer is likely to improve the hydrophilic characteristics of the developed membrane, minimizing the fouling tendency. To study the effects of surface roughness on dye adsorption, Yu *et al.* [55] developed two types of NF membranes through the phase inversion process and IP technique. It was found that the TFC PA membrane displayed significantly higher surface roughness in comparison to the asymmetric CA membrane and was more likely to promote dye adsorption on the membrane surfaces. As a consequence, this led to a low permeate flux, especially at high concentrations of dye. The distinct high roughness of the TFC membrane is mainly due to the presence of interfacially polymerized aromatic PA at the membrane top surface.

At present, most of the modified NF membranes are negatively charged, which are prepared primarily by IP and their applications are primarily oriented to the retention of small molecules and/or multivalent anions. Sun *et al.* [56] recommended an approach to the fabrication of positively charged NF membrane by introducing hyperbranched polyethyleneimine (HPEI) and isophthaloyl chloride (IPC) on a Torlon® polyamide-imide (PAI) dual-layer hollow fiber membrane surface. As a polyelectrolyte, HPEI has the ability to functionalize PAI membranes through cross-linking the carbonyl groups of imide rings in the polymer chains with amines to formamide groups. As a result, the resultant NF membrane with positively charged selective layer showed almost complete removal (99.8%) of Safranin O (positively charged dye). The promising result is mainly attributed to the charge repulsion effect between Na^+ and the positively charged membrane. Although using HPEI of higher MW tended to further increase membrane dye rejection, the membrane water flux was compromised. As evidenced from the field emission scanning electron microscopy (FESEM) images, higher MW HPEI could lead to the formation of thicker selective thin layer that increased mass transfer resistance of water molecules and thus reduced water permeability.

Compared to negatively charged membranes, positively charged NF membranes are more efficient in the cases where multivalent cations such as Ca^{2+} and Mg^{2+} and cationic dyes are required to be removed. Several approaches have been developed to prepare the positively charged NF membrane and among them, surface grafting, in particular, UV-initiated grafting has drawn great interest due to its easy operation and low cost. Moreover, the chemical bond between the substrate membrane and the active layer might increase the stability of the NF membranes during high pressure operation. Zhong *et al.* [57] fabricated novel positively charged NF membranes using sulfonated polyphenylenesulfone (sPPSU) support via UV-induced grafting. Two types of monomers [2-(methacryloyloxy)ethyl]trimethyl ammonium chloride (monomer A) and diallyldimethylammonium chloride (monomer B) were used during NF membrane fabrication. After UV-grafting positively charged vinyl monomers on the sPPSU substrate, the pore sizes of the resultant membranes were reduced to between 0.5 and 2 nm in their selective layers. Besides, the UV-modified NF membranes exhibited higher hydrophilicity in comparison to the initial substrate. With respect to filtration performance, the newly developed membranes could achieve pure water permeability of $9-14\,l\,m^{-2}\,h^{-1}$.bar and almost complete rejection (99.98%) against Safranin O dye due to their positive surface charges.

Zhang *et al.* [58] produced a novel positively charged composite NF membrane prepared by bio-inspired adhesion of PDA and surface grafting of poly(ethylene amine). The objective of selecting PDA is due to its potent adhesion property, making it less prone to fouling problems. PDA deposition, however is time consuming and not really practical. To solve this issue, grafting of poly(ethylene amine) to PDA surface can remarkably reduce the PDA deposition time down to as short as 4 h despite changing the structural properties into a dense and defect-free active layer. The work showed that the surface properties of the modified membrane could be facilely controlled by altering PDA deposition time, poly(ethylene amine) concentration, and reaction temperature and time. Based on experimental results, the novel NF membranes exhibited an acceptable $MgCl_2$ rejection (73.7%) with the pure water flux recorded at $14.4\,l\,m^{-2}\,h^{-1}$ at operating pressure of 0.20 MPa.

Liu *et al.* [59] recently developed positively charged loose NF membrane via UV-induced grafting polymerization of diallyl dimethyl ammonium chloride (DADMAC) onto a PSF support membrane. The optimized NF membrane was achieved by grafting 50 wt% DADMAC solution for 7 min at an irradiation distance of 22 cm. Surface chemistry occurrence of N and Cl, increasing thickness of dense barrier layer and roughness variation all verified the immobilization of poly(DADMAC) layer on the membrane surface. The optimized NF membrane showed good hydrophilicity and loose pore size of 8.6 nm. The rejection to different valence state followed the order of $AlCl_3 > CaCl_2 > MgCl_2 > NaCl > LiCl > MgSO_4 > Na_2SO_4$, and the rejections to divalent and multivalent cations were all higher than 90%. With respect to dye removal efficiency, the optimized membrane could show 99% and 100% rejection against Methylene Blue and Congo Red, respectively.

In view of different charge properties of textile dyes, Akbari *et al.* [60] developed NF membranes with surface charge identical with that of dye molecules using UV irradiation technique. Negatively charged and positively charged NF membranes were prepared using monomers of sodium *p*-styrene sulfonate (NaSS) and [2-(acryloyloxy)-ethyl]trimethylammonium chloride (AC), respectively. Upon surface modification, it was found that the membranes could be successfully used for the treatment of dye effluents, containing either anionic dyes or cationic dyes. Based on their study, it can be seen that the flux of dye solutions was quite stable irrespective of the concentration, indicating that fouling and/or osmotic pressure were limited. For a grafted negatively charged membrane, retentions of >98% were reported for all classes of dyes that were evaluated. The main separation mechanism of this membrane is due to the combination of charge repulsion and size exclusion effect.

In order to enhance water permeability, a great deal of study has been carried out through selection and modification of the microporous support membrane. A vast variety of polymers have been successfully used as porous supports for TFC membrane fabrication over the years. Generally, thermally stable polymeric substrate is highly desired for many industrial processes, especially in the textile industry, which originally discharge hot effluent. In view of this, Han *et al.* [61]

fabricated new NF membranes using poly(phthalazinone ether sulfone ketone) (PPESK) as the support substrate layer. They produced two types of NF (negatively charged and positively charged) membranes through IP and phase inversion methods, namely, PIP/PPESK and positively charged quaternized PPESK (QA PPESK). During the filtration process, QA PPESK membrane showed outstanding performance compared to PIP/PPESK in terms of flux and dye rejection. Over 14.5 $l\,m^{-2}\,h^{-1}$ flux, 92.3% dye rejection and 10% salt rejection were observed in the long-term operation at 60 °C, which indicated the outstanding thermal stability of the PPESK membrane toward hot effluent. The excellent performance of the QA PPESK NF membrane may be induced by the hydrophilic properties of the positively charged NF membranes that make it less susceptible to fouling.

Additionally, Wei *et al.* [62] focused on the application of a TFC NF hollow fiber membrane synthesized by IP on a PSF/PES blended support membrane for dye desalination and concentration. The fabricated negatively charged NF membranes yielded an MWCO of approximately 520 $g\,mol^{-1}$ and a pure water flux of 47.5 $l\,m^{-2}\,h^{-1}$ at 0.4 MPa. Rejections of different salts followed in the order of $MgSO_4 > Na_2SO_4 > MgCl_2 > NaCl$ at pH 6.8, with up to 96.2% rejection for $MgSO_4$. With respect to solute separation, the membrane demonstrated 99.9% dye removal owing to its negative charge properties.

Other than modifying the selective and substrate layer of the TFC NF membranes using the methods discussed earlier, Daraei *et al.* [63] chose to use a combination of inorganic nanoclay and chitosan to modify the active layer of the PVDF-based membrane in the absence of cross-linker. It was observed that synergistic effect from both organoclay and chitosan had led to higher removal rate of dyes, owing to their efficient adsorption performances. Addition of very low clay content (<2 wt%) into the chitosan solution could significantly improve the adsorptive property of the coating layer by enhancing the accessible active sites on the membrane surface. Indeed, increasing the amount of nanoclay promoted to denser thin film structure and decreased the permeate flux. Therefore, homogeneous and uniform clay particles dispersion on the membrane surface as well as contribution from clay intercalation interaction between the organoclay and PVDF membrane supported layer play more effective roles toward dye adsorption efficiency. Compared to Cloisite 15A (C15A), Cloisite 30B (C30B) presented superior adsorption capacity due to its modifier surfactant chemistry (ternary ammonium with two hydroxyl groups), which offers more positive charge. The steady performance of organoclay/chitosan nanocomposite membrane during 3 h of test indicated the stability of chitosan coating layer in spite of lacking cross-linker in the preparation procedure. The prepared membranes removed the dyes dominantly due to their adsorption potential along with other drastic mechanisms such as sieving, inertia effects, and concentration polarization.

NF membrane normally requires a hydrophilic surface for better performance, but not for the membrane used in an MD process. It is because hydrophobicity is very important in MD to ensure that a lower value of surface energy can be obtained to prevent liquid solution penetration at the membrane pores. As simple blending method is the common technique employed by many researchers to improve MD performances, the effects of different types of additives have been explored by Mokhtar *et al.* [64]. In this work, an attempt was made to prepare PVDF membranes incorporated with two different types of additives, that is, ethylene glycol (EG) and PVP for the MD process of dyeing wastewater application. Upon surface modification with the additives, it was found that the membrane blended with additive (both EG and PVP) had a greater overall porosity and mean pore size than that of the pristine PVDF membrane. However, in comparison to the PVDF membrane incorporated with EG, the PVP effect on the properties of the PVDF membrane was more pronounced as it could lead not only to decrease in membrane hydrophobicity but also reduction in membrane thermal stability and mechanical strength.

During MD experiments using $0.5\,g\,l^{-1}$ RB5 solution, PVDF–EG membrane demonstrated consistent separation efficiency with average dye rejection of 99.88%. Meanwhile, the rejection of PVDF–PVP membrane decreased gradually from initial 99.91% to 99.16% at the end of the experiment. The slight reduction of dye rejection in PVDF–PVP membrane could be related to the pore wetting and fouling that occurred in this membrane. Figure 4.13 compares the inner and outer layer of both PVDF–PVP and PVDF–EG membranes after being used for a dyeing treatment process. Obviously, the inner and outer layer of the PVDF–PVP membrane was severely stained with dye components in comparison to the marginal change on the whitish color of the PVDF–EG membrane. The direct observation on the physical color of the membranes has confirmed the penetration of dye molecules from the lumen side of the PVDF–PVP membrane

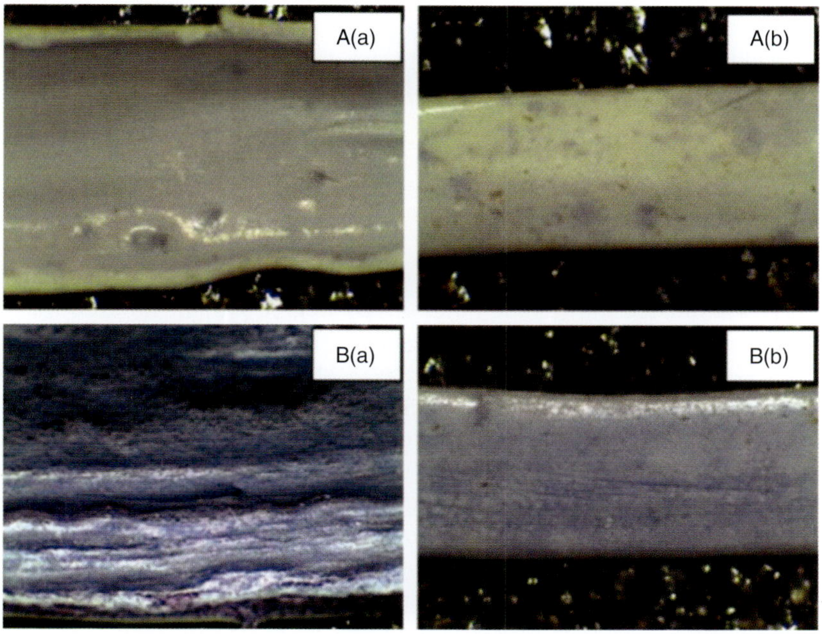

Figure 4.13 Stereoscopic images of (A) PVDF–EG and (B) PVDF–PVP hollow fiber membrane after MD tests, (a) inner surface and (b) outer surface. (Mokhtar 2015 [64]. Reproduced with permission of Taylor and Francis.)

toward the outer layer. The promising results of PVDF–EG membrane are mainly due to its greater wetting pressure and higher hydrophobicity degree.

Besides using polymeric additive to modify membranes, Mokhtar *et al.* [65] made an attempt to improve the surface hydrophobicity of PVDF-based membranes using inorganic clay – C15A. The PVDF–C15A membranes were prepared via a dry-jet wet spinning method and the clay content varied in the range of 0–10 wt%. It was reported that the incorporation of clay particles significantly improved membrane characteristics in terms of wetting pressure, hydrophobicity, and mean pore size. This is probably due to the disordered arrangement of polymer chain packing by the presence of clay particles in the membrane matrix. The higher wetting pressure obtained by the PVDF–C15A membranes showed that the membranes are able to prevent liquid penetration through the membrane pores due to the cross-membrane pressure difference that is sometimes caused by imbalance of feed and permeate side pressures. Prior to a dyeing separation process, the PVDF–3%C15A membranes recorded almost 70% flux enhancement compared to the pristine PVDF membrane when tested at 70 °C feed temperature and 20 °C permeate temperature. Among the resultant membranes, the PVDF–3%C15A showed the highest wetting pressure and contact angle value (greater hydrophobicity). These caused the membrane to have greater capability to prevent liquid intrusion into the membrane pores. Moreover, the mean pore size of this membrane was the smallest compared to other composite membranes which minimized wetting tendency. Meanwhile, the promising flux obtained by this membrane is mainly due to its lower membrane thickness and higher overall porosity, which decreased the mass transfer resistance across the membrane. The findings of this work revealed that C15A is a potential filler for PVDF membrane of MD process, improving not only the structural properties of the membrane but also enhancing the permeate flux without sacrificing dye rejection.

4.3.2 Drinking Water Production

4.3.2.1 Reverse Osmosis and Forward Osmosis for Brackish Water/Seawater Desalination

Desalination is a water treatment process during which, salts and non-ionic minerals are eliminated from saline water up to a degree suitable for customary drinking. Currently, there are various techniques that can be used to treat inland brackish water and seawater. But, the most commonly used method is RO technology. Engineered osmosis such FO and PRO on the other hand are the most interesting processes in recent years and have potential to overcome limitations encountered by an RO process.

Among the membrane structures available, TFC membrane that consists of a very thin selective layer coated on top of microporous substrate is the primary membrane used in modern desalination process. Although TFC membrane offers several distinct advantages in the water desalination process, it still requires surface modification in order to overcome the trade-off effect between water flux and solute selectivity as well as fouling and chlorine propensity.

Characteristics of the selective skin layer of the TFC membrane to a certain extent depends on the physicochemical properties of the support layer. An ideal support layer should possess the following features: thin and porous, low tortuosity factor, and appropriate surface pore size. The support layer is more important for the FO and PRO process compared with the RO process as the support layer would contact directly with either draw solution (FO mode) or feed solution (PRO mode) throughout the filtration process.

Improving the hydrophilic feature of support surfaces has been considered as a promising option to boost water flux as a hydrophilic surface tends to exhibit higher affinity against water molecules, improving its transport rate. In this regard, PEG and PVP have been added to the support layer of TFC membrane, aiming to ameliorate membrane hydrophilicity and porosity [66]. However, it was reported that the IP step was likely to create a dense PA layer inside the pores of modified support layer which later affected the water flux of TFC membrane. Misdan *et al.* [67] pointed out that by manipulating the surface pore size and porosity of the support layer, one could in fact produce desirable physicochemical properties of the PA layer. One of them is to restrict the penetration of PA into the pores of the support layer.

A review over the available literature indicates that a reliable method to prepare ideal support layers for TFC membrane is still lacking. It is generally agreed that the most ideal morphology in the support layer should possess a finger-like structure because this structure tends to provide high diffusion for both salts and water. Li *et al.* [68] and Widjojo *et al.* [69] however emphasized that a sponge-like structure is more important in order to obtain sufficient mechanical strength. They have further elucidated that if the support layer itself shows high hydrophilic behavior, it is not necessary to have a finger-like structure to overcome ICP for flux enhancement during an FO process.

Electrospun nanofibers have also been considered as the support layers for TFC membranes. Bui *et al.* [70] introduced a novel flat-sheet PA TFC membrane supported by an electrospun nanofibers. The nanofibers were electrospun onto polyester nonwoven fabric followed by PA layer coating onto the nanofiber support. Two polymeric materials (PSF and PES) were selected to prepare nanofiber substrate and the results showed that the PSF nanofibers demonstrated strong adhesion with top PA layer while PA layer delamination occurred on PES nanofibers. It was explained that the presence of bisphenol A moiety in PSF structure is the main factor contributing to better adhesion between PSF and PA derived from MPD and TMC through a specific chemical interaction as illustrated in Figure 4.14a. Figure 4.14b,c meanwhile shows the cross-section and top surface of TFC membrane made of PSF nanofiber substrate. With respect to performance, the best nanofiber supported-PA composite membranes exhibited two to five times higher flux than a standard commercial membrane. While using nanofiber membranes has improved the membrane performance in most cases, still there is uncertainty over the mechanical stability of the selective layers. To overcome this issue, Hoover *et al.* [71] deposited electrospun nanofiber on a backing layer in order to enhance the mechanical stability of the TFC membrane. As a result, the mechanical stability of the TFC membrane increased at the cost of slight reduction in water diffusion. This modification enabled this class of membranes to deal with real operational conditions where fluid shear forces need to be maintained. The promising findings of such studies

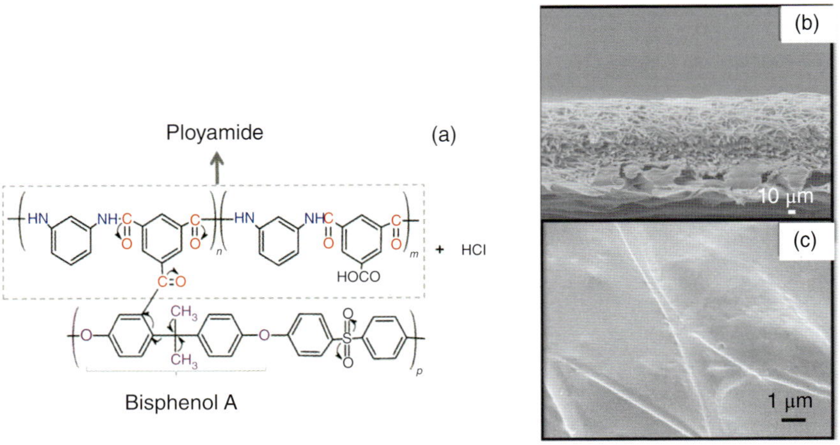

Figure 4.14 (a) Diagram of a possible cross-linking interaction between PA and the bisphenol A group of PSF (arrows show the proposed reaction mechanism) and (b,c) SEM images of cross-section and top surface of PSF-based TFC membrane. (Bui 2011 [70]. Reproduced with permission of Elsevier.)

has urged researchers to design new membrane structures suitable for industrial osmosis applications.

In addition to support layers, selecting suitable monomers, additives, inorganic nanofillers, and reaction conditions are also vitally important in governing surface properties of the PA selective layer formed. Wei *et al.* [72] studied the effect of monomer concentration on the characteristics of a PA selective layer deposited on a PSF support layer during an IP process. It was found that concentration of both MPD and TMC monomers had significant influence in PA film formation. A denser PA film structure was obtained by increasing MPD concentration, and increasing TMC concentration increased water flux at the cost of drop in the degree of cross-linking due to the formation of unreacted acyl chloride groups. From the results, it can be seen that enhancing MPD and TMC concentrations had dissimilar effects on the surface properties of TFC membranes and, therefore, should be properly controlled.

Newly synthesized monomer used to be the topic of study among the community of membrane scientists for TFC RO membrane making. Li *et al.* [73] synthesized a series of isomeric biphenyl tetraacyl chloride (BTEC) for the preparation of TFC RO membranes with MPD as amine monomer in aqueous solution. According to the authors, the organic phase reactant is likely to have greater impact on membrane performance compared to aqueous phase reactant because the IP process is generally diffusion-controlled in the organic layer. The experimental results showed that the membrane prepared from *op*-BTEC demonstrated the highest permeability ($54.2\,l\,m^{-2}\,h^{-1}$) followed by membranes prepared from *om*-BTEC ($50.0\,l\,m^{-2}\,h^{-1}$) and *mm*-BTEC ($31.7\,l\,m^{-2}\,h^{-1}$) when tested using 2000 ppm NaCl solution at 2 MPa. The reason for the flux enhancement might be due to the rougher and larger surface area of *op*-BTEC membrane produced, which led to greater contact with water molecules. Very interestingly, the membrane of *op*-BTEC did not suffer from a drawback of the "trade-off" between permeability and selectivity as NaCl rejection remained almost the same (~97%). Similar results were also reported in the work of Chen *et al.* [74] in which the incorporation of water-soluble amine reactants – sulfonated cardo poly(arylene ether sulfone) (SPES-NH_2) into aqueous solution containing MPD could overcome the trade-off effect between membrane permeability and selectivity. Under the optimum preparation conditions, the TFC membranes prepared from SPES-NH_2 showed remarkable increase in water permeability ($51.2\,l\,m^{-2}\,h^{-1}$) with slightly decrease in salt rejection (97.5% at 2000 ppm NaCl, 2 MPa) compared to the membranes prepared without SPES-NH_2 ($37.4\,l\,m^{-2}\,h^{-1}$ and 99%). The improved results are attributed to the incorporation of hydrophilic SPES-NH_2 to PA and/or higher degree of cross-linking formed in the thin selective layer. In view of the importance of hydrophilicity on TFC membrane performance, a novel amine monomer – 3,5-diamino-*N*-(4-aminophenyl)benzamide (DABA) with three amino groups was synthesized and used together with MPD in TFC RO membrane preparation [75] With increasing DABA content in aqueous phase from 0% to 0.25% (w/v), the membranes showed increase in water flux from 37.5 to $55.4\,l\,m^{-2}\,h^{-1}$ and maintained a high salt rejection (~98%) in filtering salt solution containing 2000 ppm NaCl at 2 MPa. It was revealed that the top membrane surface became more hydrophilic, smoother, and thinner as DABA concentration was increased in amine solution.

Benefiting from their optical, catalytic, and antimicrobial features, inorganic nanoparticles find extensive applications in many areas including membrane applications. The incorporation of inorganic particles during PA layer synthesis could produce hybrid PA-inorganic filler layer with improved surface characteristics. Most inorganic nanomaterials have positive contribution to the formation of selective layers [76]. Using inorganic nanomaterial during IP process can improve film formation by offering the following benefits: increasing the rate of monomers diffusion to the interface, expanding the wet zone on the top of the support layer, controlling the reaction pH (buffer agent), and so on. Moreover, inorganic additives (e.g., TiO_2, silica, and zeolite nanoparticles) are generally known to have positive impact on antifouling resistance, in addition to their nature of hydrophilicity.

In general, incorporation of inorganic nanoparticles results in improved membrane water permeability through either formation of a direct pathway for water diffusion or modification of a membrane matrix. By embedding a small amount of zeolite NaA nanoparticles within the PA layer (made of MPD and TMC monomers) of the RO membrane, a remarkable membrane flux enhancement was achieved with salt rejection remaining comparable to the typically prepared TFC membrane (without NaA particles) [77]. It was suggested that water molecules tend to flow preferentially through the superhydrophilic and molecular sieve nanoparticle pores that embedded on the top PA layer. However, the degree of flux enhancement varied depending on the zeolite type and loading as well as the conditions of the IP process. For instance, it was reported by Lind *et al.* [78] that smaller zeolites (~100 nm) tended to produce greater water flux due to larger characteristic pores while larger zeolite particles (~300 nm) produced more favorable surface properties by increasing the negative charge and lowering the contact angle, owing to its exposure at the PA surface. Exposure of nanofillers at the PA surface is highly possible to occur especially for the nanoparticles with large particle size. This is because of the irregular morphology of PA thin layer that is very often reported to be 100–500 nm in thickness.

Besides zeolite nanoparticles, other types of nanomaterials such as aluminosilicate single-walled nanotube (SWNT), SiO_2, titania nanotube (TNT), halloysite nanotube (HNT), silver (Ag), and metal alkoxides have been reported as potential fillers for PA layer synthesis. All these nanomaterials have shown positive features on TFC RO membrane performance, in particular water permeability, provided an appropriate amount of nanomaterial is added during the preparation process. For instance, Baroña *et al.* [79] reported that the water flux of the RO membrane incorporated with 0.59 wt% aluminosilicate SWNTs in PA layer increased to as high as 24.6 $l\,m^{-2}\,h^{-1}$ at 16 bar, that is, a 1.5-fold increase in comparison to the control RO membrane with little variation in NaCl rejection (95.6–96.2%). This significant surge in water flux can be attributed to the increased preferential flow of water molecules through the hydrophilic inner walls (with diameter ~1 nm) of the aluminosilicate nanotubes embedded within the PA matrix. Remarkable water enhancement in the RO membrane was also reported by Yin *et al.* [80] upon addition of 0.1 wt% porous MCM-41 nanoparticles into the PA selective layer without compromising NaCl separation efficiency (at least 97.7% rejection). The increase in water flux from 28.5 $l\,m^{-2}\,h^{-1}$ of control RO membrane to >46 $l\,m^{-2}\,h^{-1}$ of modified RO membrane at an operating pressure of ~20.7 bar is attributed to the internal pores of MCM-41 (about 3.85 nm), which create shorter flow paths for water molecules to pass through. In addition to the preferential flow mechanism, Baroña *et al.* [79] did not rule out that the reduction in membrane surface contact angle upon addition of nanotubes might partly contribute to better water permeability. Compared to solid nanomaterials such as TiO_2 and metal alkoxides, the existence of pore channels in mesoporous nanomaterials, coupled with unique surface characteristics could potentially achieve a synergistic effect for rapid diffusion of water molecules, resulting in much greater water permeability. However, the preferential flow channels created by mesoporous nanomaterials embedded in PA layer of RO membrane for water molecules deserves more research to validate this proposition.

The promising results obtained by Lind *et al.* [78] by incorporating zeolite nanoparticles in the PA layer of RO membrane motivated Ma *et al.* [81] to further explore the potential of zeolite NaY nanoparticles for FO membrane process. In the range of 0.02–0.4 wt/v% zeolite loading, the FO water flux of the modified membrane tended to increase with increasing loading from 0.02% to 0.1% but decreased with further increasing zeolite loading to 0.4%. Under specific testing conditions (feed solution: 10 mM NaCl; draw solution: 1 M NaCl), the most permeable zeolite-incorporated FO membrane (with 0.1% zeolite) demonstrated water flux of ~32 $l\,m^{-2}\,h^{-1}$ (for active-layer-facing-draw-solution (AL-DS) orientation) and ~15 $l\,m^{-2}\,h^{-1}$ (for active-layer-facing-feed-solution (AL-FS) orientation) in comparison to ~22 and ~10 $l\,m^{-2}\,h^{-1}$ recorded for the control TFC, respectively. The increase in water flux upon low loading of zeolite as well as the decrease in water flux at high zeolite loading are attributed to the porous nature of zeolite and formation of thicker PA layer, respectively, as elucidated by the authors.

Instead of using IP technique to prepare composite membrane for RO and FO desalination process, Saren *et al.* [82] fabricated composite membrane using LbL assembly method. In order to improve the negative charge density and hydrophilicity of the surface of the substrate made of PAN substrate, the substrate was subject to a post treatment using sodium hydroxide. It was followed by LbL assembly using poly(allylamine hydrochloride) (PAH) and poly(sodium 4-styrene-sulfonate) (PSS) as polycation and polyanion, respectively. Figure 4.15 shows the deposition of PSS and PAH layer via LbL assembly technique on the PAN substrate surface. The resultant LbL membrane in FO mode showed 15, 20, and $28\,l\,m^{-2}\,h^{-1}$ water flux at 0.1, 0.5, and 1.0 M $MgCl_2$, as draw solution respectively, and distilled water as feed solution.

Jin *et al.* [83] prepared a new type of RO membrane via LbL assembly method by alternating electrostatic adsorption of polyvinylamine and polyvinyl sulfate (PVS) on porous support. They reported that the rejections of NaCl and Na_2SO_4 were 84% and 96% at 5 bar, and 93.5% and 98.5% at 40 bar, respectively while the water permeance of the LbL composite membrane was recorded at $0.114\,ml\,m^{-2}\,h^{-1}\,bar^{-1}$. The results showed that at low and moderate pressure the membrane is suitable for water softening applications, while at a pressure of 40 bar or higher it can be used for water desalination. However, it must be pointed out that the water flux of LbL composite membranes is still significantly lower than those of composite membranes prepared by IP technique.

Figure 4.16 illustrates the surface modification of composite membranes in the configuration of flat sheet and hollow fiber via LbL assembly method for RO and FO application [84, 86]. The results of RO membrane after modifications revealed that the antifouling properties of the RO membrane was remarkably improved without significantly changing water flux [86]. For the hollow fiber modification, it has been reported that both inner [84] and outer [85] layers of membrane

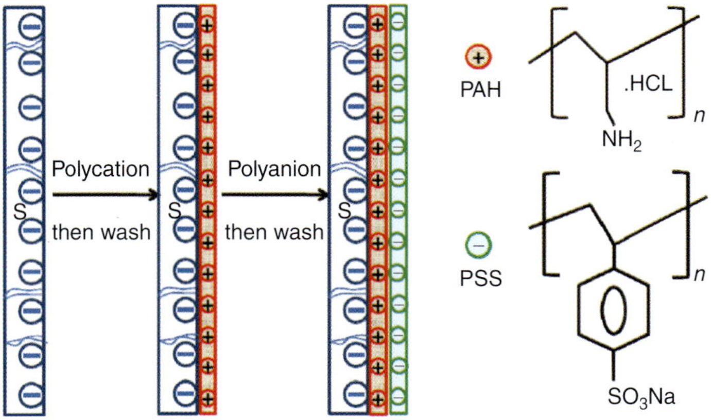

Figure 4.15 Schematic of LbL assembly of poly(allylamine hydrochloride) (PAH) and poly(sodium 4-styrene-sulfonate) (PSS) on PAN substrate. (Saren 2011 [82]. Reproduced with permission of American Chemical Society.)

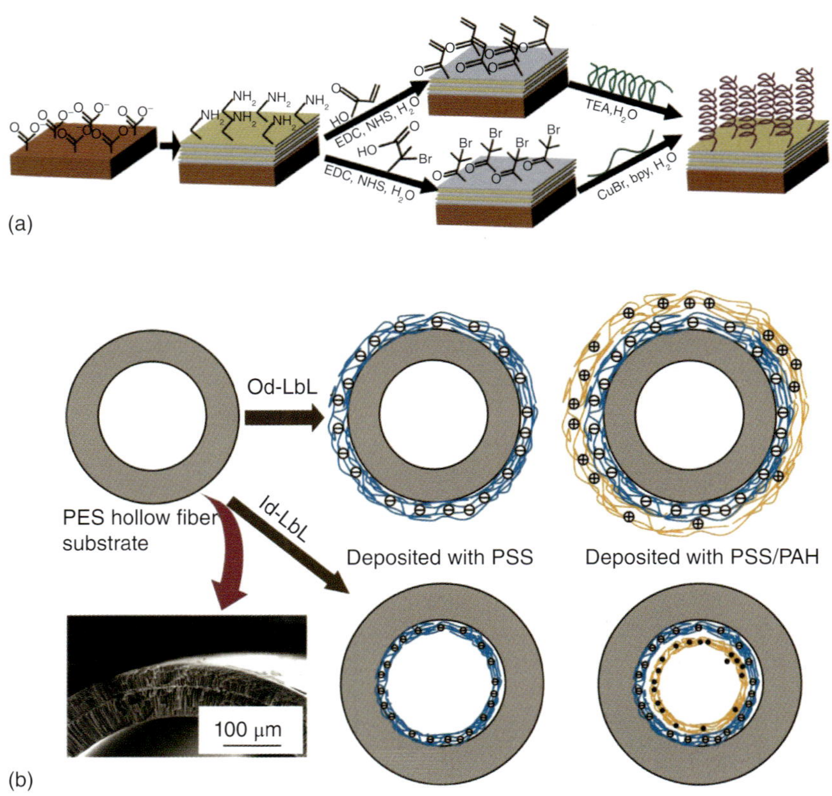

Figure 4.16 Schematic illustration of surface modification of membrane via LbL technique, (a) modification on top surface of flat sheet RO membrane [86] and (b) modification on lumen and outer side of hollow fiber FO membrane [84]. (Panel (a): Chen 2013 [86]. Reproduced with permission of Royal Society of Chemistry.)

could be modified via LbL assembly method. In PRO mode of operation, which used deionized (DI) water as feed solution, the water flux of outer-layer modified membrane with six-bilayers was recorded at 14.6, 25.9, and 40.5 $l\,m^{-2}\,h^{-1}$ throughout the time, salt to water flux (J_s/J_v) ratio of 0.034, 0.066, and 0.201 using 0.05, 0.1, and 0.5 M $MgCl_2$ as draw solutions, respectively. Although the inner-layer modified membrane with two layers deposition exhibited water flux (70 $l\,m^{-2}\,h^{-1}$) higher than that of the outer-layer modified membrane under specific testing conditions (feed solution: DI water; draw solution: 0.5 M), its salt leakage value was significantly higher (0.5 $g\,m^{-2}\,h^{-1}$). The better performance of the outer-layer modified membrane is most likely due to better deposition and formation layer of PVA and PVS on the top surface of the support layer. In fact, the top surface of the substrate has smaller pore size and smoother morphology which could provide better conditions to new layer deposition.

4.3.2.2 Adsorptive Ultrafiltration for Underground Water

Membrane technology is also very promising for underground water purification, in particular for removing heavy metals to the safety standard of drinking water.

RO process in general is able to provide safe drinking water with arsenic concentration (or other types of heavy meals) below World Health Organization regulation. Nevertheless, RO when applied for heavy metals decontamination removes all the inorganic materials that impart taste to drinking water and requires high operating pressure to produce minimum flux. Low pressure driven membranes like MF and UF on the other hand are not effective in removing heavy metals, mainly due to their porous structure, which offers minimal/none resistance against heavy metals. To promote the applicability of microporous membranes for heavy metal removal, embedding inorganic metal oxide nanoparticles (that have high adsorption capacity against heavy metal ions) into the membranes could create synergistic effect for the production of potable water from underground water source under low pressure.

As documented in the literature, Zheng *et al.* [1] and Jamshidi Gohari *et al.* [87] were among the pioneers in using inorganic adsorbents in polymeric host media for eliminating As(V) and As(III) from aqueous solution. Arsenic can exist in the underground water in four valence states: −3, 0, +3, and +5, but only arsenate (As(V)) and arsenite (As(III)) are mostly found. Jamshidi Gohari *et al.* [87] successfully produced adsorptive UF membrane by incorporating PES matrix with Fe–Mn binary oxide (FMBO) adsorbent through a phase inversion process for adsorptive removal of As(III) from aqueous solution. Figure 4.17 shows the changes in surface morphology of PES-based membrane upon blending with

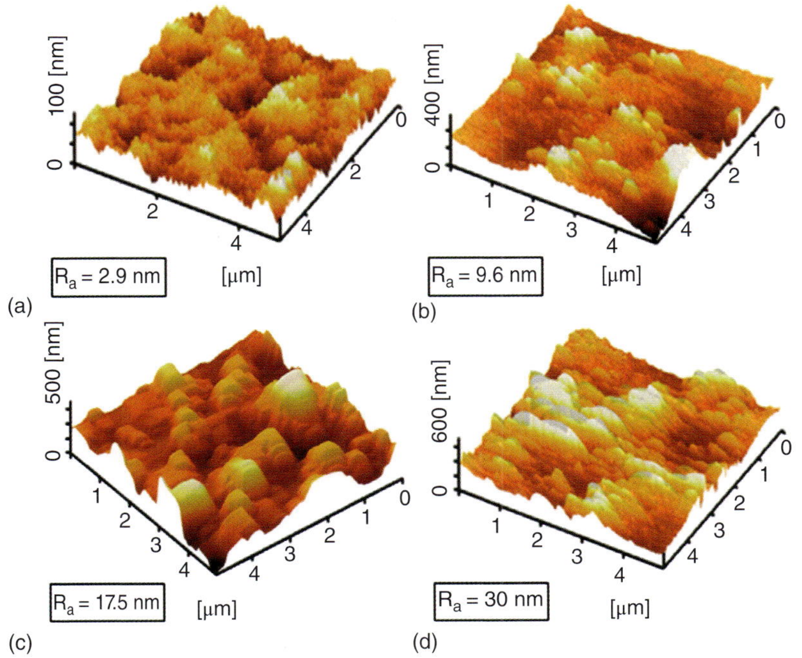

Figure 4.17 3D AFM images on the top surface of membrane made of different FMBO/PES weight ratio, (a) control (contained no FMBO), (b) 0.5, (c) 1.0, and (d) 1.5. (Jamshidi Gohari 2013 [87]. Reproduced with permission of Elservier.)

FMBO particles of different loadings. As can be clearly seen, the membrane surface roughness was significantly increased from mean roughness, R_a of 2.9 nm in the control PES membrane to >30 nm in the membrane blended with the highest amount of FMBO. The increased surface roughness is mainly attributed to the presence of FMBO particles in the membrane. Besides altering membrane surface roughness, the results have also shown that membrane surface hydrophilicity was also significantly improved, owing to the hydrophilic nature of FMBO. Most importantly, the FMBO-modified membranes showed promising results in eliminating As(III) with optimum adsorption rate recorded at 73.5 mg g^{-1}. The control PES membrane is not capable of removing As(III) because of the absence of FMBO particles that act as adsorbent for As(III) uptake. Although the increasing amount of inorganic adsorbents embedded in the membrane could lead to greater membrane adsorption capability, the amount that is added needs to be properly controlled as excessive use of adsorbents could lead to severe particle agglomeration in the resultant membrane, reducing not only its adsorption capability but also its mechanical properties.

A flat sheet composite membrane made of PAN polymer and iron oxide (laterite) adsorbent was also reported by Chatterjee and De [88] for the removal of arsenic from arsenic-contaminated ground water. The permeability and MWCO of the selected membrane were 3.4×10^{-11} m s^{-1} Pa^{-1} and 48 kDa, respectively. Morphological analysis showed macrovoids constricted by laterite particles. Surface characteristics assessed by atomic force microscopy revealed the increase in roughness with laterite concentration. With respect to the adsorption capacity, the laterite-modified membrane only showed maximum 1.4 mg g^{-1} of adsorption capacity, which was remarkably lower compared to other available adsorbents.

In addition to As(III), the flat sheet polymeric membranes incorporated with zirconia particles [1] and TNTs [89] were also reported to be effective for As(V) elimination. Upon incorporation with adsorptive inorganic materials, it was reported that the membrane surface became rougher and exhibited lower water contact angle (more hydrophilic), in addition to the improved adsorption capability against As(V). Figure 4.18 presents the contact angle of small water droplets on the top surface of PES membrane modified by different TNTs loadings. The contact angle is an important parameter for measuring surface hydrophilicity. In general, the smaller the contact angle, the higher is the hydrophilicity of the membrane surface. Higher membrane hydrophilicity is often associated with greater water permeability. Figure 4.19 illustrates the presence of abundant amount of —OH groups in the PES membrane matrix upon addition of TNTs [89]. With the presence of the highest amount of TNTs, the resultant PES/TNT blended membrane exhibited not only extremely high water permeability (1250 l m^{-2} h^{-1} bar^{-1}) but also excellent adsorption capability against As(V), that is, 124 mg g^{-1}.

The membrane in hollow fiber configuration could also be modified using inorganic nanoparticles for the purpose of As(V) removal. He *et al.* [90] demonstrated that upon incorporation with zirconia nanoparticles, the surface-modified PSF-based hollow fiber membrane could achieve promising

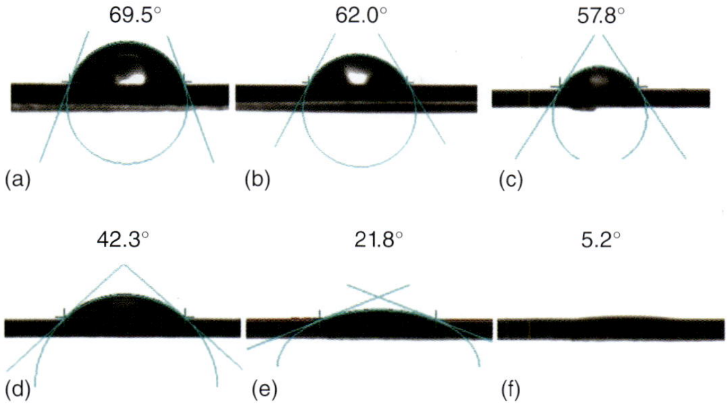

Figure 4.18 Shape of small water droplet (0.5 μm) on the surface of membrane made of different TNT/PES weight ratio, (a) control (contained no FMBO), (b) 0.125, (c) 0.25, (d) 0.5, (e) 1.0, and (f) 1.5. (Gohari 2015 [89]. Reproduced with permission of Royal Society of Chemistry.)

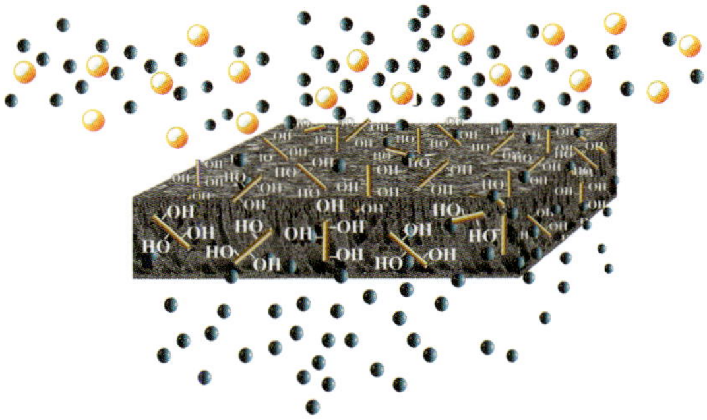

Figure 4.19 Schematic of the membrane surface modified with TNTs coupled with abundant amount of —OH groups. (Gohari 2015 [89]. Reproduced with permission of Royal Society of Chemistry.)

adsorption rate against both As(III) and As(V). Under optimized pH environment, the zirconia-modified membrane showed maximum adsorption capacity of 138 and 200 $mg\,g^{-1}$ for As(III) and As(V), respectively. The contact angle however decreased slightly when an excessive amount of nanoparticles was added. This is probably because the surface porosity of the hollow fiber membranes decreased with more additional nanoparticles in the membranes. The addition of hydrophilic NPs tends to decrease membrane surface contact angle, but the poor surface porosity increases the contact angle. The normalized pure water flux was reported to increase slightly when the content of zirconia nanoparticles was increased. Based on experimental results, the membrane water flux was affected by nanoparticles loading for the following several reasons: addition of nanoparticles enhances the surface hydrophilicity and overall porosity of the hollow fiber membranes, but decreases the number of pores on both the inner and outer surfaces of the membranes. Obviously, the first two effects should increase the pure water flux while the last effect should decrease it. The findings of this work clearly showed that the first two effects (improved hydrophilicity and increased porosity) were dominant in improving water flux.

A novel nano-scale polyaniline/Fe^0 composite (termed as *PANI/Fe^0*) was fabricated by Bhaumik *et al.* [91] via reductive deposition of nano-Fe^0 onto the PANI nanofibers matrix at room temperature. The resultant nanofibers were then evaluated for its performance for arsenic removal in batch experiments. The results showed that the maximum nanofibers adsorption capacities at pH 7.0 for As(III) and As(V) were 232.5 and 227.3 $mg\,g^{-1}$, respectively. These capacities were significantly higher than most of the reported adsorbents. A combination of surface complexation and electrostatics interaction between As species and surface sites of the PANI/Fe^0 could be the possible mechanisms involved in the removal process. Morphological analysis showed that the pure PANI fibers had relatively smooth surface. After formation of nano-Fe^0, the diameters of the PANI nanofibers were increased, leading to an increase in surface roughness compared to the pure PANI form.

Modification on the membrane properties using nanoparticles has also been attempted for lead (Pb) removal. Pb is a metallic element in group IV-A of the periodic table with an atomic number of 82 and a relative atomic weight of 207.19 $g\,mol^{-1}$. Lead can be found in the environment in oxidation states: 0, +2, and +4. In aqueous systems, Pb commonly forms complexes with hydroxides, chlorides, carbonates, sulfates, and sulfides; however, lead is predominately present in water as plumbous (Pb^{2+}) at near neutral to moderately acidic systems. Very similar to the work that was conducted for As(III) removal [1, 92], a separate work reported embedded selective HMO into PES-based membrane via blending method, targeting to remove Pb(II) from contaminated water. With respect to surface morphology, it was found that the R_a value of the membrane increased considerably from 2.96 nm in the control membrane (contained no HMO) to >35 nm in the membrane consisting highest HMO:PES weight ratio, that is, 2.0. The increase in the membrane surface roughness is very likely due to the alignment of HMO nanoparticles embedded in membranes. A large quantity of —OH functional groups were detected by FTIR spectroscopy on the surface of PES/HMO blended membrane, owing to the presence of homogeneously dispersed HMO nanoparticles, and these —OH groups are very likely responsible for the increased surface hydrophilicity. Besides being able to achieve high adsorption capacity against Pb(II) (204.1 $mg\,g^{-1}$), impregnating the HMO nanoparticles into the membrane matrix could overcome the challenge of separating nano-sized particles from water after the adsorption process, offering a synergistic effect to purify water in a simple yet efficient way. Furthermore, the PES/HMO blended membrane showed high recovery rate for Pb(II) removal after subjecting to a simple desorption process using HCl solution. The high recovery rate indicates that the membrane can be applied for multiple treatment cycles before losing its function as an adsorbent in a membrane matrix.

Recently, hydrous ferric oxide (HFO) has been considered by Abdullah *et al.* [93] as a potential adsorbent for Pb(II) removal. The particle was introduced into the membrane matrix using the same blending method as described by Jamshidi Gohari *et al.* [1, 92]. Very similar to other previous relevant works, the membrane surface roughness was increased while water contact angle decreased upon incorporation of nanoparticles. Although the PES/HFO membrane could achieve reasonably high pure water flux (942 $l\,m^{-2}\,h^{-1}\,bar^{-1}$), its adsorption capacity against Pb(II) was quite poor (13.2 mg g^{-1}). It is shown that proper selection of nanoparticle properties is required if high adsorption capacity were to be achieved. In another attempt, PVDF-based hollow fiber membrane incorporated with TiO_2 was prepared by Putri *et al.* [94] for Pb removal. The adsorption isotherm results indicated that the membrane demonstrated an adsorption capacity of 1.512 mg cm^{-2} at initial Pb(II) concentration of 100 mg l^{-1}. Although the number of surface pores appeared more in the membrane with increasing TiO_2 concentration, it did not lead to increased water flux. It is because TiO_2 nanoparticles tended to agglomerate at high concentration, leading to possible pores blockage.

4.3.3 Dense Membrane for Gas Separation Process

Membranes for gas separation processes have been successfully implemented worldwide since the 1970s by companies such as Monsanto, Cynara, Sparex, and Grace Membrane Systems. Some of its major industrial applications are air separation, CO_2 removal, O_2 enrichment, and hydrocarbon separation. According to a market research report [95], the global membrane market is projected to reach US$ 32.14 billion by 2020 from US$ 18.68 billion in 2014 with the fastest growth expected from gas processing membrane. Rubbery and glassy polymers are the two main polymeric materials that are used for membrane gas separations. Rubbery polymer, having flexible polymer backbones, allows high gas permeation with poor gas pair selectivity. In contrast, glassy polymer exhibits better gas pair selectivity but lower permeability, owing to its fixed chain structure that limits its segmental motion. Table 4.3 compares the gas permeability of several widely used polymers for different types of gases. By taking into consideration the balance between gas permeability and gas pair selectivity, glassy polymers are more preferable to be used in industrial applications.

Figure 4.20a shows that the separation factors of membranes are bounded by the trade-off limit between permeability and selectivity, which was first

Table 4.3 Pure gas permeability (barrer) of widely used polymers [96].

Gas	Rubbery polymer		Glassy polymer		
	Silicone rubber	Natural rubber	Cellulose acetate	Polysulfone	Polyimide
H_2	550	41	24	14	50
He	300	31	33	13	40
O_2	500	23	1.6	1.4	3
N_2	250	9.4	0.33	0.25	0.6
CO_2	2700	153	10	5.6	13
CH_4	800	30	0.36	0.25	0.4
C_2H_6	2100	—	0.2	—	0.08
C_3H_8	3400	168	0.13	—	0.015
C_4H_{10}	7500	—	0.1	—	—

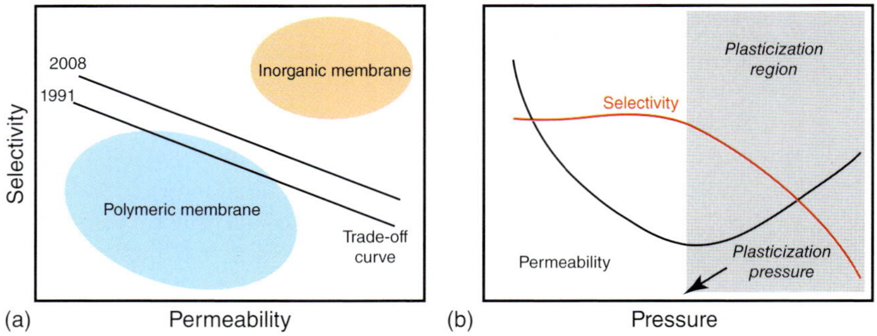

Figure 4.20 (a) The trade-off limit between permeability and selectivity of polymeric membrane and (b) the pressure-dependent plasticization phenomena.

introduced by Robeson [97] in 1991, and later updated in 2008. The trade-off depicted that highly permeable membranes are commonly accompanied by low gas pair selectivity and vice versa. Figure 4.20b shows the effect of operating pressure on the gas transportation behavior in membranes during separation process. At high operating pressure, highly condensable gas (e.g., CO_2 and H_2S) tends to dissolve in the membrane, causing excessive disruption in the polymer chain packing. This phenomenon is generally known as *pressure-induced plasticization*. Once the pressure exceeds the plasticization pressure of a polymer, sudden increase in permeability with poor selectivity is observed. The plasticization phenomenon still remains to be a great concern to industry, in particular in processes where high operating pressure is required. In case of O_2 enrichment process, the small difference between the kinetic diameter of N_2 (3.64 Å) and O_2 molecules (3.46 Å) makes the enrichment process very challenging and it is almost impossible for these two gases to be separated by a simple size-based diffusion rate difference. The present O_2/N_2 separation factor for the best commercially available polymer membranes is between 6 and 8, but selectivity of 15–20 is necessary in order for the membrane process to be commercially attractive. Many attempts have been made in the past to improve the separation characteristics of polymer membranes via surface modifications by means of cross-linking, polymer blending, coating, and so on.

As mentioned in the previous sections, polymer blending is among the methods that can be considered cost- and time-effective to improve membrane properties for water-based processes. This blending method is also widely employed to improve the separation performance of membranes for gas applications. Polymer blending is effective to produce membranes of enhanced performance by combining the unique characteristics of each polymer. For example, a polymer with high intrinsic separation factor, which is commonly expensive and is hardly available, is blended with a polymer that is relatively cheap, aiming to achieve a high separation factor with lower processing cost.

The study on gas separation of membranes made by polymer blending method has been reported by Visser *et al.* [98] in which different types of polymers were considered. In the work, polyimide Matrimid (BTDA-AAPTMI) was blended with either co-polyimide P-84 (BTDA-TDI/MDI) or PES. Compared to the control Matrimid membrane, the O_2 permeability and O_2/N_2 selectivity of the Matrimid membrane blended with co-polyimide P-84 were improved by 78% and 16%, respectively. Matrimid/PES blended membrane, on the other hand, showed lower permeability and selectivity in comparison to the control membrane. This is mainly due to the difference in the functional group introduced to the membrane. It is hypothesized that the presence of an imide group within the Matrimid/P-84 membrane provides good interaction with O_2 compared to the sulfone group present in the Matrimid/PES membrane. Since gas separation membrane is always sensitive against material chemistry, identifying suitable blending material is crucial to ensure high performance membrane development.

It should be also noted that the performance of the blended membrane is limited by the performance of the individual selected polymer. For example, Yong *et al.* [99, 100] have evaluated the performance of polyimide (PI) membrane blended with polymer of intrinsic microporosity (PIM-1) for gas separation process. PIM that is known to have high fractional free volume (FFV) is able to achieve high gas permeabilities coupled with moderate gas pair selectivities. PI meanwhile is highly selective and exhibits mild gas permeabilities. The improvement of membrane performance, relative to PI membrane, was observed after PIM-1 was incorporated. The CO_2 permeability increased linearly with increasing PIM-1 content due to high FFV of the resulted membrane. However, the CO_2/CH_4 selectivity shifted toward intrinsic selectivity of the PIM-1 owing to the trade-off effect between their parent polymers (Figure 4.21a). Interestingly, the O_2/N_2 separation trend of the blended membrane was similar to the CO_2/CH_4 separation (Figure 4.21b). At 5% PIM-1, it was found that both the O_2 permeability and O_2/N_2 selectivity were increased, contrary to the trade-off between their parent polymer performances. This is probably because of the formation of charge transfer complex (CTC) that offers different interaction sites of the functional groups to the O_2 and N_2. It is also speculated that the polymer chains are partially connected as a form of mild cross-linking and pack tightly with CTCs, thus enhancing the gas pair selectivity. Further discussion on the CTC will be provided in the following section.

Blending using low MW polymers such as PEG and poly(ethylene oxide) (PEO) was also attempted to improve membrane gas performance. This strategy has shown promising results when blended with rubbery polymers. Car *et al.* [101] reported that poly(amide-*b*-ether) block copolymer (PEBAX) was miscible with PEG even though PEG content was as high as 50 wt%. The presence of PEG in the PEBAX induces crystal structures with smaller and imperfect crystallite, leading to the formation of higher FFV. As a consequence, the CO_2 permeability increased linearly with PEG content. The permeability of the control PEBAX membrane increased from 24.8 to >50 barrer upon 50 wt% PEG incorporation. The increase in gas permeability of PEBAX/PEG membrane did not compensate gas pair selectivity.

In certain cases, a parent polymer is blended with a secondary polymer of higher plasticization resistance in order to maintain stability of the membrane against plasticization under high feed pressure operation. PI in general has good correlation between permeability and selectivity, but is associated with low CO_2

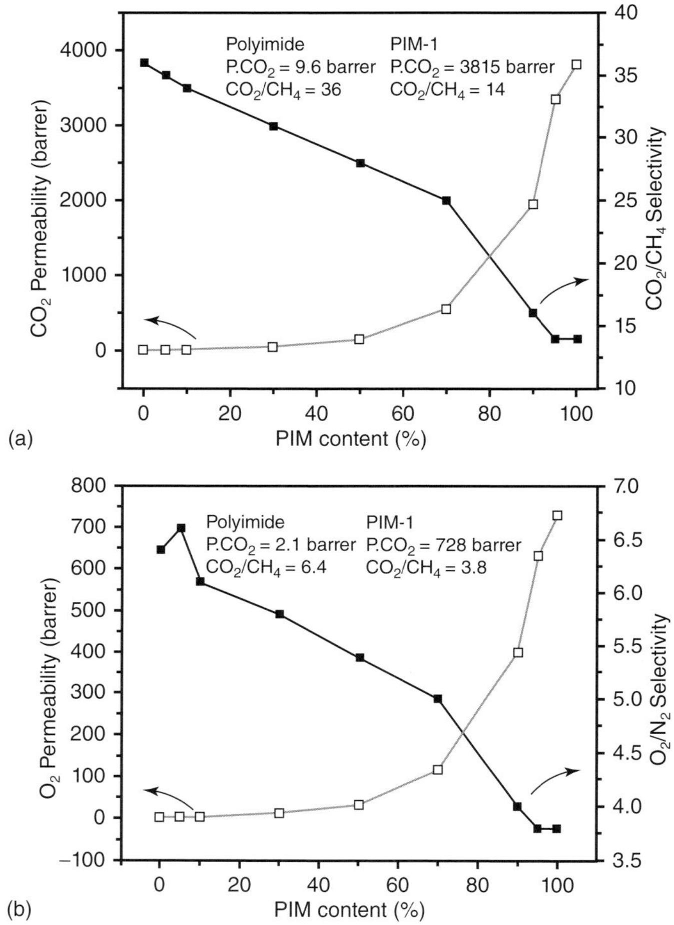

Figure 4.21 Gas separation performance of PI/PIM-1 blends at different PIM-1 content, (a) CO_2/CH_4 separation and (b) O_2/N_2 separation [99].

pressure of plasticization. To overcome this limitation, Visser *et al.* [98] incorporated higher plasticization-resistance polymers, that is, PES and co-polyimide P-84 into PI-based membrane. The resultant membrane showed positive features that improve plasticization resistance of PI-based membrane. In comparison, PI/PES membrane showed the highest plasticization resistance (16 bar) followed by PI/P84 membrane (~10 bar) and control PI membrane (4 bar) when all membranes were evaluated using CH_4/CO_2 (2/8, v/v) mixed gas as feed.

To improve membrane resistance against plasticization, chemical cross-linking can also be performed. Cross-linking plays a role in improving membrane plasticization resistance by tightening the interstitial space along polymer chains, protecting nodule integrity against CO_2 and restricting polymer chain vibration. Cross-linking agents are used to react with the specific functional groups on the membrane surface (e.g., carboxyl, biazides, and sulfone), forming new functional groups that are beneficial for targeted application.

Shao *et al.* [102, 103] cross-linked the surface of 6FDA-durene membrane with either 1,3-cyclohexanebis(methylamine) (CHBA) or ethylenediamine (EDA). Methanol was used to swell the membrane surface and rendering the polymer chains more accessible for the reaction with the diamine. During the cross-linking process, the cross-linkers interact with the imide group of the membrane and form an amide group. Prolonging cross-linking time would eventually result in the disappearance of the imide group (Figure 4.22). The results indicated that the 6FDA-durene/CHBA and 6FDA-durene/EDA membranes

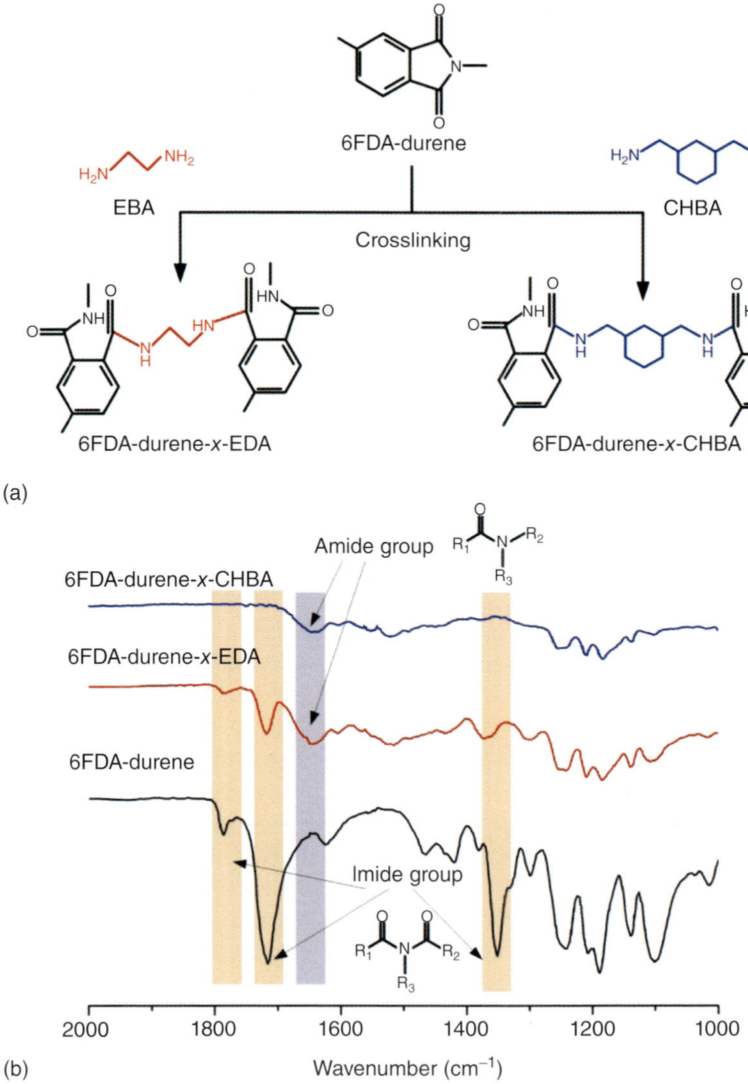

Figure 4.22 Chemical cross-linking of 6FDA-durene using 1,3-cyclohexanebis(methylamine) (CHBA) and ethylenediamine (EDA), (a) transformation of amine- and imide-group into amide-group and (b) FTIR spectra of cross-linked membranes [102, 103].

demonstrated higher CO_2/CH_4 selectivity than that of control 6FDA-durene membrane. Their selectivity was improved by 42% and 108%, respectively. The high selectivity of the membrane is caused by the presence of amide-group that eases CO_2 to be dissolved onto the membrane surface before diffusing to permeate inside. Increasing cross-linking time and temperature was reported to further induce amide group formation, leading to higher CO_2/CH_4 selectivity.

Khan *et al.* [104] carried out cross-linking of PIM-1 using alkoxy-biazide to improve its plasticization resistance. In this work, alkyl biazides (N_3—R—N_3) was used to produce highly reactive divalent nitrene (—N:) radical at elevated temperatures. The highly reactive nitrene radical generated was then reacted with the alkyl groups (such as —CH_2 and —CH_3) of the PIM-1 spirobisindane segment to form a covalent network. The cross-linking reaction is highly dependent on the cross-linking temperature as well as the ratio of cross-linker to the membrane. Upon completion of cross-linking, the FFV of the membrane decreased as flexible PEG chains tended to occupy the rigid and contorted PIM-1 polymer chains. This caused the membrane to become more size-selective and was able to offer higher gas pair selectivity. The cross-linked membrane on the other hand was less susceptible toward high pressure following its increased resistance against plasticization. It should be noted that by increasing the cross-linkers content, the permeability of the membranes against all gases tested were diminished. Larger gas molecules were severely affected compared to smaller gases. The reduced gas permeability is likely due to the reduction of FFV in the membrane.

In addition to chemical cross-linking, the membrane plasticization can also be improved using thermal cross-linking method. At elevated temperature, electron donors (e.g., carboxyl group) present in the membrane would begin to interact with electron acceptors (e.g., phenyl ring) within their chain (intra-molecular) as well as within the neighboring chain (inter-molecular). This phenomenon is known as *CTC*. The resulting CTC would suppress the mobility of polymer chains, providing high size and shape discrimination between gas molecules. Consequently, this leads to higher gas pair selectivity and superior plasticization resistance. In the case of CTC formed on the surface of PI/PIM-1 blended membrane, it is the nitrogen atom in Matrimid (PI) that acts as an electron donor to the acceptor (carbonyl group), shortening the donor–acceptor distance within Matrimid [99].

Although cross-linking could improve membrane plasticization resistance, at the same time it reduces membrane permeability. To compensate this drawback, using thermally labile material as cross-linking agents is recommended. The thermally labile materials tend to decompose and create micropores within the membrane (or surface) after undergoing a specific heat treatment process. Askari *et al.* [105] cross-linked the surface of 6FDA-durene/DABA co-polyimide membrane using cyclodextrin (CD) as thermally labile material. Thermally labile material was hypothesized to increase the FFV of the membrane and improve separation properties upon cross-linking. Compared to the smaller CD (α-CD with size of 4.7–5.7 Å and β-CD with 6.0–7.8 Å), the membrane that cross-linked with larger CD (γ-CD of 7.5–9.5 Å) showed the most promising results of CO_2 permeability (>4200 barrer). Its CO_2 permeability was 220% better than the control 6FDA-durene/DABA membrane. With respect to plasticization resistance, the membrane cross-linked with γ-CD also showed much higher resistance, that is, 30 atm compared to 5 atm recorded in the control membrane.

Another technique that can be used to modify a membrane surface is by establishing a thin selective layer on the top surface of a membrane. The thin layer can be formed via several major routes such as IP, co-precipitation, and coating. Zhao *et al.* [106] formed a thin PA layer on the top surface of PES substrate by IP technique that used triethylenetetramine (TETA) and TMC as active monomers. By subjecting the PA layer to heat treatment, the resultant membrane was seen to exhibit better stability against pressure and achieve greater CO_2/N_2 selectivity compared to the membrane without heat treatment. The heat-treated membrane recorded CO_2 permeance of 1.33×10^{-5} cm^3 (STP) cm^{-2} s^{-1} $cmHg^{-1}$ and CO_2/CH_4 selectivity of 94.1 at feed pressure of only 1.1 atm. The improvement in the gas separation performance is due to densification of the PA layer upon heat treatment that increases transport resistance and further hinders larger molecule (CH_4) to permeate across the membrane (Figure 4.23).

However, IP for membrane gas separation is still limited, owing to the possible defects formed on the PA layer during drying process. To overcome the defective surface formation, Du *et al.* [107] proposed solid–liquid interfacial reaction that could minimize defect formation during drying. Instead of using liquid phase amine, amine-based polymer (poly(*N*,*N*-dimethylaminoethyl methacrylate) (PDMAEMA) was used to minimize the penetration of amine solution into the pores of the membrane. The PDMAEMA was first coated onto the PSF substrate before the membrane was interfacially formed via cross-linking with *p*-xylylene dichloride (XDC) in heptane solution.

Developing composite membranes via coating is also commonly employed due to the simplicity in membrane surface modification. Polydimethylsiloxane (PDMS) is widely used as coating material in gas separation membrane to cover the defects on the membrane surface. Since PDMS is highly permeable with low gas pair selectivity, its coating only aims to eliminate convective flow through the defective surface. PDMS coating, therefore, does not significantly change the permeance or selectivity on the defect-free portion of the membrane. Since the inception of PDMS coating by Henis and Tripodi [108], this method is considered as a standard approach in preparing membrane for gas separation and has been widely practiced.

Zulhairun *et al.* [109] have reported that prior to coating, the control PSF membrane showed relatively low CO_2/CH_4 selectivity, that is, 2.96. Upon PDMS coating, the surface-modified membrane demonstrated significantly higher CO_2/CH_4 selectivity (i.e., 28.06) owing to the reduced surface defects. The additional coating layer however reduced the CH_4 permeance of the control membrane from 25.01 to 2.73 GPU as observed in the PDMS-modified membrane. CO_2 permeance of the membrane was reported to change marginally upon PDMS coating. Further improvement on membrane performance was done specifically on the surface layer by incorporating copper-based metal–organic framework (MOF) (Figure 4.24). The presence of MOF on the surface coating contributed to higher affinity toward CO_2 and increased gas permeation rates across the membrane. Promising results were obtained using the membrane

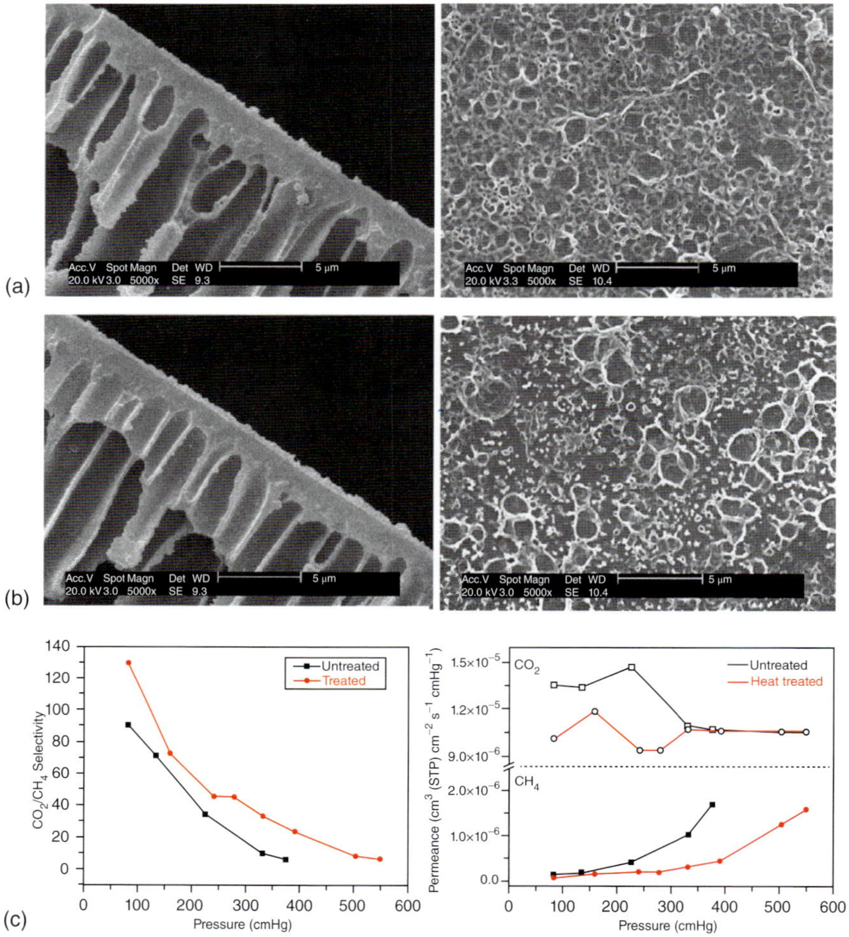

Figure 4.23 Membrane cross-section and surface morphology of (a) the untreated composite membrane, (b) the heat-treated composite membrane, and (c) gas separation performance of untreated and heat-treated membrane [106].

coated with composite material in which CO_2 permeance and CO_2/CH_4 selectivity recorded at 84.95 GPU and 33.27, respectively.

Poly(ether-block-amide) recently attracted wide attention as a coating medium. This polymer has been commercialized as PEBAX® and Polyactive®. Compared to Polyactive®, PEBAX is more commonly used for gas processing application. PEBAX consists of highly flexible ether chain can result in enhanced gas permeability without compromising good selectivity. Unlike PDMS coating that aims to eliminate surface defects on dense or asymmetric membrane, PEBAX itself acts as separation medium. The use of PEBAX for membrane making was first reported by Blume and Pinnau [110] for acid gas separation. Since then, PEBAX has been used as coating material on many types of support membranes. These include PES, PSF, polyaniline, and PEI.

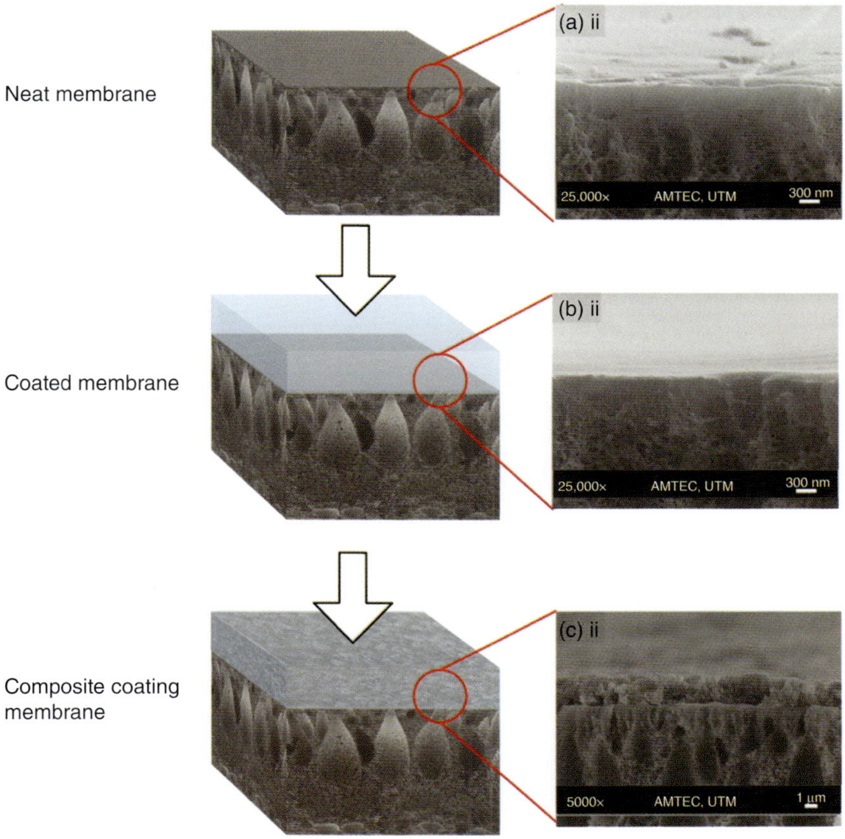

Figure 4.24 Asymmetric membrane with composite PDMS coating to simultaneously eliminate defects and improve gas separation [109].

Since the PEBAX layer is used solely for separation, incorporating inorganic particles within the matrix shows more prominent influence on gas separation compared to when using PDMS. It should also be noted that incorporating fillers within the selective layer is much effective to assist gas molecules to transport compared to distributing the fillers throughout the membrane matrix. In a series of studies conducted by Murali *et al.* [111–113], different inorganic particles (carbon nanotube (CNT), SiO_2, and zeolite) were incorporated into the PEBAX selective layer of membrane. The findings from the works indicated that the presence of inorganic fillers were prominent in boosting the membrane performances. Using 2 wt% of CNT could lead to 490% improvement in CO_2 permeability followed by 96% increase in CO_2/N_2 selectivity. The promising results can be attributed to the characteristics of CNT, which provide specific affinity toward CO_2. For zeolite, an increase of 22% of CO_2 permeability and 79% of CO_2/CH_4 selectivity was achieved when incorporating 5 wt% of zeolite 4A. This is due to the additional molecular sieving effect created by zeolite. The embodiment of silica nanoparticles on the other hand altered the diffusivity by creating more cavities for the diffusing molecules. Incorporating 0.3 wt% of SiO_2

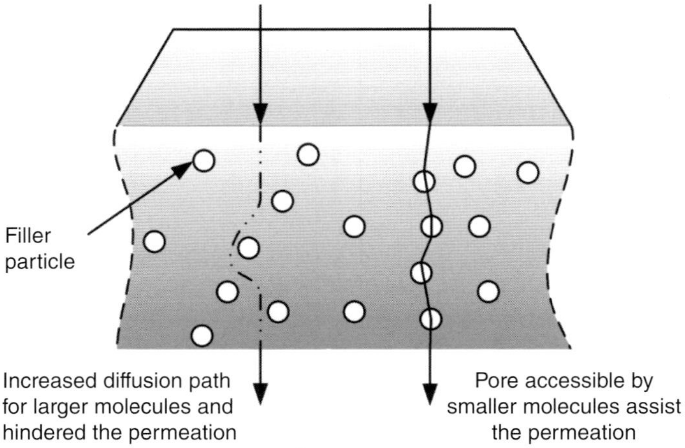

Figure 4.25 Illustration of gas permeation across membrane embedded with fillers.

was reported to enhance CO_2 permeability and CO_2/CH_4 selectivity by 710% and 88%, respectively, compared to the control membrane.

Although embedding fillers into the selective layer of a membrane seems to be more effective for enhancing gas separation performance, many scientists prefer to introduce fillers into the membrane matrix by blending method (during dope preparation). This is due to the simplicity of the method. Figure 4.25 illustrates how nano-scale non-porous fillers would disrupt polymer chain packing and create free volume for better gas diffusivity. Some examples of nonporous fillers used in a polymer matrix are graphene oxide, SiO_2, and clay minerals.

Zeolites as inorganic fillers have been in prominence due to their high intrinsic separation properties and high thermal and mechanical strength as well as high chemical stability. While there are 218 different types of zeolites, only a few have shown to be promising fillers for CO_2/CH_4 separation. Among them, Zeolite 4A, SAPO-34, Faujasite (FAU), and ZSM-5 were found to have positive features for improving membrane gas separation properties. Chen *et al.* [114] reported 140% improvement in CO_2 permeability and 60% in CO_2/CH_4 selectivity when the PI membrane was blended with 25 wt% (total composition) FAU/EMT-type zeolite. Similarly, Junaidi *et al.* [115] reported that CO_2 permeance increased from 105 to 706 GPU, while CO_2/CH_4 selectivity increased from 15 to 30.7 when PSF membrane was incorporated with SAPO-34. The presence of SAPO-34 restricts the diffusion of large gases (e.g., CH_4, N_2) through its porous channel whereas diffusion of smaller gases (e.g., CO_2) is unaffected. As pore diameter of zeolite is relatively larger than gas molecules, the separation-based molecular sieving mechanism is not the dominant factor. It is the preferential adsorption site of zeolite that determines good selectivity. Zeolite preference toward quadrupole moment of CO_2 has hindered CH_4 permeation, leading to increased gas pair selectivity.

Incorporation of carbon molecular sieves (CMSs) into polymer matrices has also been previously reported. CMS is considered because it has better affinity with glassy polymers compared to other classes of fillers, allowing good adhesion at the interface without leading to interfacial defects. In a series of works conducted by Vu *et al.* [116–118], CMS was prepared before incorporation into the PI (Matrimid® 5218) membrane matrix. With respect to the pure Matrimid membrane, the CO_2 permeability was improved by 26%, while CO_2/CH_4 was increased by 47% upon CMS incorporation. Similar observation was also experienced when CMS was embodied into PEI (Ultem®) membrane where CO_2 permeability and CO_2/CH_4 selectivity was improved by 200% and 38%, respectively. The increases in separation properties are attributed to the highly selective and permeable of the CMS-incorporated membrane.

As for MOFs, their application for gas separation is not well developed compared to other applications such as gas storage and gas adsorbent. Ordoñez *et al.* [119] reported the potential of using ZIF-8 as filler to improve membrane performance for CO_2/CH_4 separation. It is interesting to know that the membrane selectivity was not compromised by the ZIF-8 agglomeration even though high loading of ZIF-8 was utilized within the PI matrix (Figure 4.26). The results showed that enormous increment in CO_2/CH_4 selectivity was achieved when approximately 3 wt% (total composition) of ZIF-8 was incorporated. The ZIF-8-modified membrane exhibited CO_2/CH_4 selectivity of around 125 compared to 31.5 shown by the control membrane, that is, almost 300% improvement in selectivity. Similarly, Zornoza *et al.* [120] reported remarkable improvement in CO_2/CH_4 selectivity (150%) upon incorporation of 2.5 wt% (total composition) NH_2-MIL-53(Al) into membrane made of PSF. The role of MOFs in enhancing membrane performance can be generally summarized as follows; (i) enhancement of CO_2 diffusivity coefficient through polymer chain disruption exceeding CH_4 [121], (ii) interaction between quadrupole moment of CO_2 with the weak electrostatic field of ZIF-8 [122], and/or (iii) molecular sieving induced by MOF pores [119].

As reported in the work of Ahn *et al.* [123, 124], the incorporated SiO_2 tended to improve the CO_2/CH_4 selectivity by restricting polymer chain mobility. The presence of silica in the membrane could hinder the permeability of CO_2 but to a lesser extent compared to CH_4 and N_2. In certain cases, incorporation of SiO_2 could increase the solubility coefficient of CO_2 in the membrane by creating good interaction with —OH group of silica particle. A study by Sadeghi *et al.* [125] showed that gas permeability across a polybenzimidazole (PBI) membrane was dominant by diffusivity over solubility coefficient. Incorporation of silica subsequently rigidified polymer chains and decreased the diffusivity coefficient of the penetrant; where larger molecules were highly affected. In addition, the existence of —OH group on the membrane surface upon SiO_2 incorporation tended to enhance gas dissolution in the membrane, which subsequently increased the solubility coefficient of all gases. By incorporating 4 wt% (total composition) SiO_2 into membrane, CO_2 permeability was reported to increase by 340% with CO_2/CH_4 selectivity improved by 120%.

Recent research has focused on the use of nano-sized inorganic fillers because it offers higher interfacial area and better interaction with a polymer matrix. Nordin *et al.* [126] evaluated the effect of ZIF-8 with different sizes (134, 288, and 493 nm) on the PSF membrane. It was found that the incorporation of the smallest size of

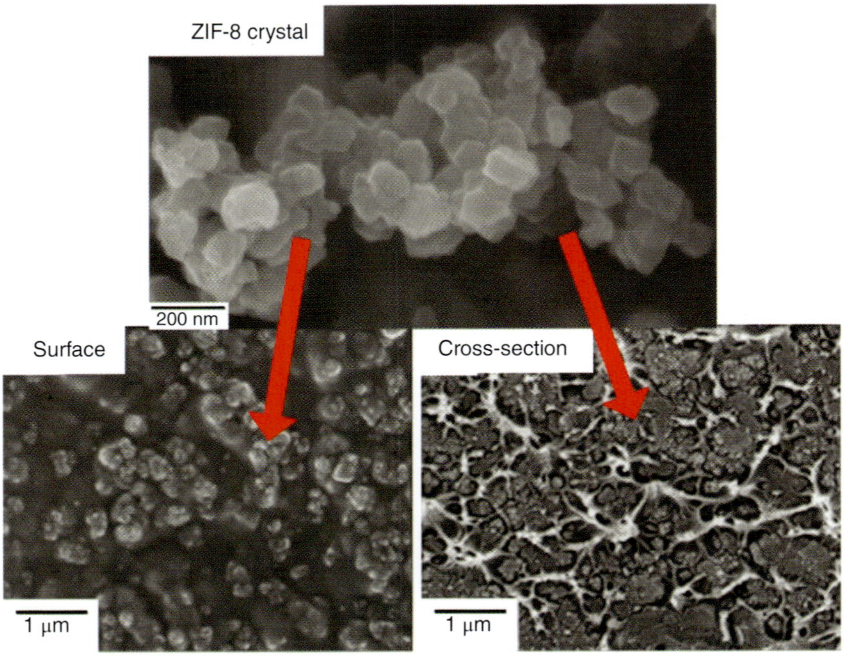

Figure 4.26 MMM consisting of ZIF-8 crystal embedded into polyimide matrix. Well dispersed of ZIF-8 crystal was observed on the surface and throughout membrane morphology [119].

ZIF-8 (134 nm) could enhance the CO_2/CH_4 selectivity of the control PSF membrane from 19.43 to 28.5 compared to the larger fillers that negatively affected the selectivity. As the gas separation takes place on the top surface layer, it is hypothesized that a larger filler would hardly reside on the selective layer and is likely to pierce through the selective layer. It might, therefore, result in the formation of macrovoids that makes the larger molecules to permeate easily. Compared to the large particles, small particles tend to provide a large number of active sites for gas/filler particle interaction with minimum filler piercing through the selective layer as illustrated in Figure 4.27a. Nordin *et al.* [127] have further reported that 0.0013 wt% (total composition) of ZIF-8 is the ideal loading to improve the membrane CO_2 permeance and CO_2/CH_4 selectivity. It is because excessive use of filler (3 wt% of total composition) would provoke the filler to agglomerate on the selective layer, creating interfacial voids as illustrated in Figure 4.27b. The consequence of the poor structural integrity was reported to affect separation characteristics of the resultant membranes.

4.3.4 Solvent Resistant Nanofiltration Membrane for Organic Solvent Application

Solvent resistant nanofiltration (SRNF) is an emerging technology in which membranes are used for organic solvent separation and purifications. This type of membrane technology holds enormous potential as it allows separation of small compounds with MW ranging from 200 to 1400 g mol^{-1} from organic solvents.

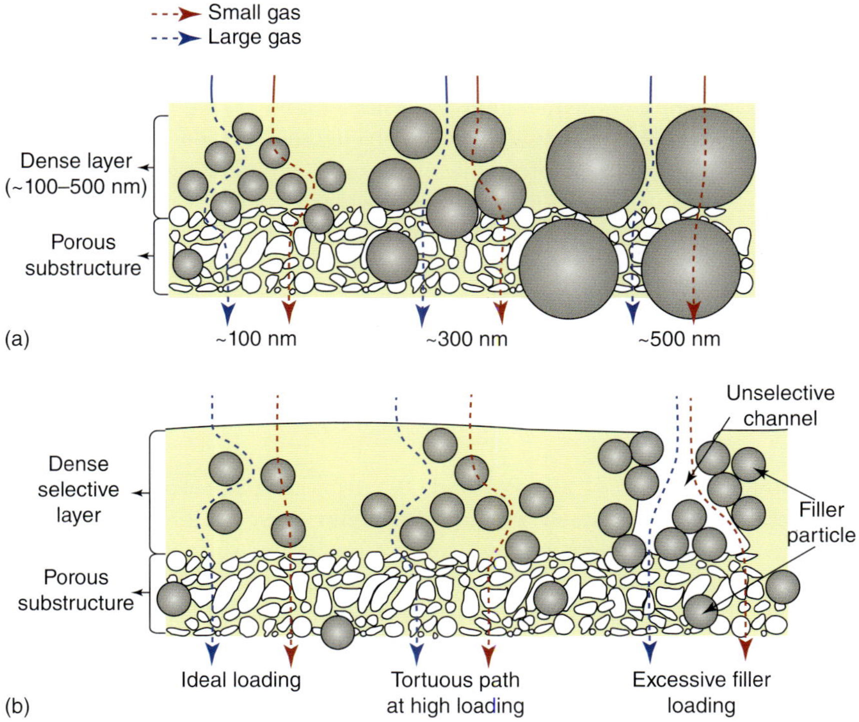

Figure 4.27 Illustration of gas permeation across polymeric membrane embedded with inorganic particles, (a) the use of particle with different size [126] and (b) the use of same particle but at different loading [127].

It has been proven to be significant in expanding the spectrum of membrane applications from aqueous systems primarily for water purification and other water-related treatments to filtration and concentration of organic solutions. The possible industrial application of these technologies include recovery of solvent in lube oil dewaxing processes, degumming of vegetable oil, reuse of extraction solvent in the food industry, and purification of active pharmaceutical ingredients (APIs) [128].

Most commonly used membranes for SRNF applications are asymmetric polymeric membranes that typically consist of a dense selective layer on the top of a microporous structure. A major challenge in the further development of SRNF membranes is the compatibility of the membrane under extreme operating conditions like harsh and aggressive media, elevated pH, and high temperatures, while maintaining a reasonable long-term separation performance and reproducibility. In addition, exploring for approaches that could be used to synthesize thermally and chemically stable SRNF membranes is another challenging research area. A number of methods have been employed to modify and improve the separation properties of SRNF membranes (mainly polymeric membranes) without sacrificing their permeation properties. Modification methods that have been proposed and applied for tuning the characteristics of SRNF membranes can be divided into two categories: (i) either only the surface is modified or (ii) the surface is modified along with the bulk membrane material. Incorporation of nanoparticles into the polymer matrix is another method that influences both the surface and the bulk structure of the membrane simultaneously.

Photo irradiation-induced modification is an effective method used for tuning the SRNF membrane surface properties due to its low operation cost and mild reaction conditions. This method has been recognized as a well-known technique for the preparation of polymeric membranes with improved surface properties from precursors, initiating polymerization, and cross-linking reactions. As an example, Li *et al.* [129] prepared asymmetric membranes via blending a block copolymer of polystyrene-*b*-poly(ethylene oxide) with poly(acrylic acid) homopolymer. The membranes were then cross-linked by UV irradiation in air at a wavelength of 254 nm. Upon modification using a simple UV irradiation method, the membranes were not only stable to chlorine solutions, but also to a variety of organic solvents.

Struzynska-Piron *et al.* [130] employed UV curing after a phase inversion process to fabricate solvent-stable PSF and PI membranes. The cross-linked PSF and PI membranes showed good performances in organic solvents such as ethyl acetate (EA) and isopropanol (IPA). Cross-linked PSF membrane for instance showed 91% rejection and a permeance of $8.1\,l\,m^{-2}\,h^{-1}\,bar^{-1}$ for Rose Bengal (RB)/EA mixture and 94% rejection and a permeance of $0.2\,l\,m^{-2}\,h^{-1}\,bar^{-1}$ for RB/IPA mixture. UV cross-linked PI membrane on the other hand showed 96% rejection and a permeance of $1.4\,l\,m^{-2}\,h^{-1}\,bar^{-1}$ for RB/IPA mixture. Other work conducted by the same research group showed that the use of a combination of acyl phosphine oxide-based photo-initiators with a penta-acrylate cross-linker could modify the surface properties of PSF-based membrane, extending its solvent resistances to acetone, *n*-butyl acetate, toluene, and xylene [131]. The acyl phosphines are known to undergo photobleaching, thus allowing UV irradiation to penetrate progressively deeper into the membrane.

Although IP technique is mainly used to synthesize TFC membranes for aqueous NF and RO application, it has also been used to develop membranes for filtration in organic solvent media. Since the top layer of the TFC membranes plays the main role in the separation of organic solvents, it is important to obtain a very stable top layer with high separation abilities and permeation properties. A PA top-layer prepared from MPD and TMC was synthesized on top of a cross-linked PI support layer. To increase the hydrophobic character of the membranes and stability in aprotic solvents, a mixture of triacyl and monoacyl chlorides was used, while free acyl chloride groups, left on the membrane surface after IP, were reacted with hydrophobic molecules. This way, the membranes were reported to have significantly higher permeance without sacrificing rejection for aprotic solvents, including toluene and EA [132].

Li *et al.* [133] synthesized a polypyrrole (PPy)-modified top layer by *in situ* polymerization on different microporous support membranes (Figure 4.28). A solution of pyrrole monomer and ethanol was cast on the porous support and polymerized to form a PPy selective layer. The membranes showed a very high retention of the negatively charged RB in different solvent systems, comparable to commercial membranes, but at much higher permeance. The permeance and

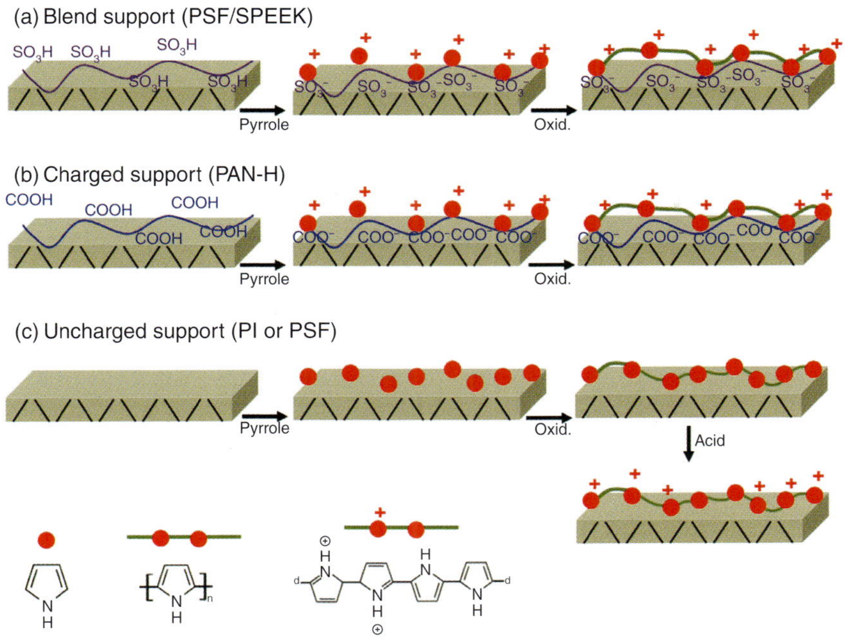

Figure 4.28 Preparation of PPy-modified SRNF membranes on three different porous supports [133].

the rejection of IPA/RB mixture for instance were found to be 1.11 $l\,m^{-2}\,h^{-1}\,bar^{-1}$ and 98%, respectively for PPy-modified PSF/SPEEK membrane. PPy-modified PAN-H membranes on the other hand showed 98% rejection and a permeance of 67.1 $l\,m^{-2}\,h^{-1}\,bar^{-1}$ for THF/RB. For Pyy-modified PI and PSF membranes tested using IPA/RB mixtures, the permeance and the rejection of 0.04 and 3.84 $l\,m^{-2}\,h^{-1}$ and 90% and 85% were found, respectively. They claimed that the high selectivity of the negatively charged RB is due to the introduction of PPy top layer that makes the porous surface of support denser. Besides, the interaction between the positively charged PPy and negatively charged RB also plays a role during the filtration process.

Very recently, Karan *et al.* [134] prepared PA nanofilms with outstanding rejections to separate negatively charged solutes from methanol. They controlled the rate of interfacial reaction between MPD and TMC on a sacrificial layer of cadmium hydroxide nanostrands to prepare a sub-10 nm PA film on P84 and alumina support. The formed nanostrand layers were removed after the PA formation by acid dissolution. It was reported that the DMF-activated PA nanofilm on alumina support, which was prepared with a 1-min interfacial reaction of 3 wt% of MPD in aqueous solution, resulted in an excellent permeance performance with a high rejection. For the mixtures of methanol and 6-hydroxy-2-naphthalenesulfonic acid sodium salt, methyl orange, naphthalene brown and acid fuchsine, permeance, and rejection performances of 52.2, 52.1, 52.0, and 51.8 $l\,m^{-2}\,h^{-1}\,bar^{-1}$ and 98.5%, 98.9%, 99.9%, and 99.9% were found, respectively.

The incorporation of inorganic particles into a polymeric matrix is another promising approach to improve the chemical, thermal, and mechanical stability of membranes for solvent applications. SRNF membrane incorporated with inorganic particles can be prepared in three different ways: (i) dispersing the inorganic filler in the polymer solution, (ii) *in situ* polymerization, that is, dispersion of nanoparticles in solutions before polymerization, and (iii) self-assembling and pre-assembling of nanoparticles, that is, the synthesis of inorganic nanoparticles using precursors during the membrane formation and nanoparticle functionalization before polymerization. Up to now, a variety of inorganic fillers have been attempted in the fabrication of SRNF membranes. These include zeolite, TiO_2, mesoporous SiO_2, CNT, MOF, activated carbon, and so on.

The use of inorganic materials to enhance the performance of SRNF membranes was first reported by Gevers *et al.* [135] in 2005 using SiO_2, carbon, and zeolites as fillers for PDMS-based membranes. They found that the addition of inorganic fillers to the PDMS improved the surface characteristics of the membrane for the separation of nonpolar solvents by reducing the swelling of the PDMS network. Zeolite-filled PDMS showed much stronger antiswelling properties than PDMS films filled with silica and carbon. This is attributed to an improved dispersion of zeolite in the polymer solution, which produces stronger cross-linking effects. Soroko and Livingston [136] worked on chemically cross-linked PI membranes by dispersing TiO_2 nanoparticles in the polymer dope solution. They observed a change in morphology for membranes containing TiO_2 compared to the membrane without addition of nanoparticles. Macrovoid formation was suppressed by the addition of TiO_2 particles and completely disappeared at higher loadings of particles. These membranes were then tested in pure solvents (DMF and ethanol) and styrene oligomer–solvent mixtures. The membrane compaction resistance was significantly improved with separation performance remaining unchanged upon addition of 10 wt% TiO_2. TiO_2 was claimed to have the capability of improving membrane mechanical properties by preventing membrane porous structure from collapsing.

Vanherck *et al.* [137] reported the performance of SRNF membranes containing gold nanoparticles in a pre-made CA membrane in water, ethanol, IPA, and dye (MW: 632 g mol^{-1})-ethanol solution. The membrane was heated by light irradiation during the separation process. They observed significant increase in flux up to 15% and 400% for water and solvents, respectively, when using the membranes containing 2 wt% nanoparticles. However, the photo-thermal heating showed no significant effect on dye rejection. Siddique *et al.* [138] in recent years have developed a new generation of compaction-free SRNF using an aminopropyltrimethoxysilane (APTMS) as a precursor to generate an inorganic organosiloxane network in the PI membrane matrix for APIs purification. The membrane showed enhanced stability in harsh solvents such as acetone, DMF, and dichloromethane and at temperatures up to 100 °C. Stability in the solvents was achieved due to the cross-linking of polymer with APTMS. In addition, the inorganic network increased the strength of the membrane by preventing the membrane pores from collapsing, making it resistant against pressure and heat treatment. The membrane performance was also compared with the commercially available SRNF membrane (Duramem™ 300, Evonik Membrane Technology Ltd) and the results showed that the in-house made membrane was more resistant against pressure compaction (tested at pressure up to 30 bar), although a lower solvent flux was recorded.

The incorporation of MOFs as a filler for the preparation of SRNF membranes with improved surface properties was explored by Basu *et al.* [139]. MOFs are a relatively new class of porous crystalline materials constructed from metal ions/clusters and multidentate organic linkers. The newly made membranes prepared by the incorporation of Cu-BTC, MIL-47, MIL-53(Al), and ZIF-8 into the PDMS matrix showed enhanced RB rejection from 87% (in the control PDMS) to 95–98% in IPA, owing to the reduced polymer swelling and improved surface size exclusion effect upon filler incorporation. Campbell *et al.* [140] prepared SRNF membrane by adding MOF Cu-BTC in the PI dope solutions. The performance of the membrane was later tested in PS-acetone mixtures and compared with the control PI membrane. Results showed that the Cu-BTC-modified membrane showed higher PS rejections and lower flux decline than the control membrane, suggesting that the Cu-BTC could change the transport properties of the membrane and provide a rigid support to the entire membrane structure. Very recently, a PPSU-based SRNF membrane incorporated with Cu-BTC was fabricated by dispersing Cu-BTC particles in PPSU dope solution prior to the casting process [141]. The presence of Cu-BTC particles successfully enhanced the thermal and chemical stabilities of the membranes by hindering the polymer chain mobility. The results indicated that when 0.8 wt% Cu-BTC was incorporated into the PPSU membrane, methanol flux increased by 43% while MWCO of the membrane decreased by 18% in comparison with the neat PPSU membrane, when both were tested using 10 ppm of methanol/dye solutions at 6 bar. The newly developed membrane also showed a great improvement in terms of resistance to compaction, suggesting that Cu-BTC particles are of importance in improving membrane rigidity and strength.

The potential of thin film nanocomposite (TFN) membranes that are composed of a very thin film containing nanoparticles on a porous support has also been demonstrated for SRNF applications. Namvar-Mahboub *et al.* [142] fabricated TFN membrane by the incorporation of amine-functionalized UZM-5 nanoparticles into the PA thin selective layer synthesized via IP of MPD with TMC (Figure 4.29). The membranes were then studied in a dewaxing solvent recovery process. The presence of UZM-5 in the PA layer was found to improve both the oil rejection and the permeate flux under an optimal concentration of 0.02% w/v of UZM-5. The increase in the permeate flux upon the addition of UZM-5 was attributed to the existence of pore channels of the particles used. The average pore diameter of 16.8 Å in the modified UZM-5 tended to provide a pathway for toluene (6.1 Å) and MEK (5.2 Å) to pass through while rejecting bigger lube oil macromolecules. The cross-linking degree of PA layer however was negatively affected by the excessive use of UZM-5. More surface defects were detected, leading to significant decline in oil rejection.

Sorribas *et al.* [143] conducted the same IP reaction and synthesized TFN membranes containing 50–150 nm size MOF nanoparticles (ZIF-8, MIL-53(Al),

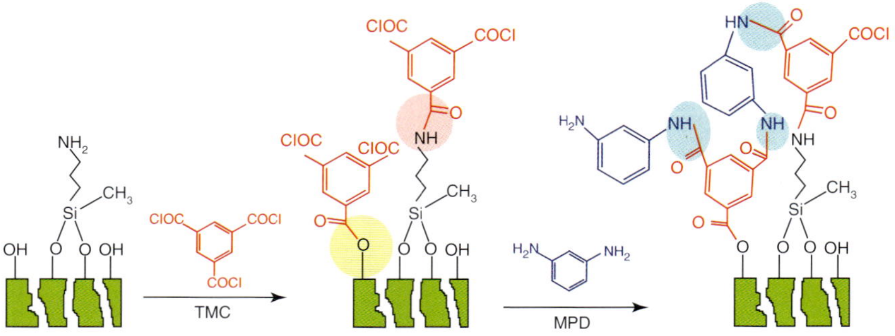

Figure 4.29 Schematic of chemical reactions between an amine-functionalized-UZM-5, TMC, and MPD during the PA synthesis [142].

NH_2-MIL-53(Al), and MIL-101(Cr)) on top of cross-linked PI porous supports. When compared to the filler-free membrane, the organic solvent (methanol and THF) fluxes of the TFN membranes were reported to increase without sacrificing styrene oligomers rejections (>90%). Peyravi *et al.* [144] also synthesized the TFN membranes but with the use of surface-modified TiO_2 nanoparticles. The TFN membranes represented higher methanol flux than filler-free TFC membranes with a slight sacrifice in dye rejection during the NF of a methanol/dye solution.

PEM membranes made of LbL assembly method are also considered as one of the potential membrane structures for SRNF applications since the thin selective layer properties could be easily manipulated to achieve high permeability and selectivity. Furthermore, the insolubility of polyelectrolytes in common organic solvents renders them highly suitable for SRNF. One of the examples of PE combination to produce multilayer membrane for solvent application is poly(diallyldimethylammonium chloride) (PDDA)/SPEEK. This membrane with a PDDA/SPEEK on a hydrolyzed PAN support was potentially applied for the separation of charged organic compounds [145]. The dyes that were utilized for the separation were charged aromatic derivatives such as RB, crystal violet, methyl orange, and acid fuchsine. NaCl was added in the PE solutions as the presence of NaCl tended to cause the conformation of multilayer membrane to be more loopy and coiled, which in turn resulted in thicker and looser membrane structures [146]. With respect to performance, PDDA/SPEEK multilayer membrane exhibited promising solvent stability and separation characteristics in organic solvents, including aprotic solvents, like THF and DMF.

Other polyelectrolyte combinations applied in SRNF are PDDA/polyacrylic acid (PAA), PDDA/PSS, and PDDA/PVS [147, 148]. For PDDA/PAA, the membranes were prepared under different conditions of NaCl and pH. The top-layer thickness was decreased by increasing the pH from 2 to 4, while a thicker top layer was formed at higher pH values. These membranes showed comparable selectivity but the permeance was lower than for previous PDDA/SPEEK membranes. The separation characteristics of membranes synthesized from PDDA/PSS and PDDA/PVS were clearly higher when the polyanions were used in the H-form compared to the Na-form. PDDA/SPEEK has also been successfully deposited on a H-PAN/Si support and applied for the filtration of organic solvents [149]. This developed membrane also showed excellent potential in the challenging polar aprotic solvents and found to be a good candidate for SRNF.

4.4 Conclusions

In this chapter, the surface modification of polymeric membranes by different methods, namely blending, surface coating (IP, LbL, sol–gel coating, and spin coating), and photo-initiated polymerization is described and discussed in detail by experimental examples and illustrations. This is a review of the surface modification of polymeric membranes used in different fields (UF, (SR)NF, RO, FO, gas separation, etc.), covering aqueous, organic solvent, and gas phases. In most of the cases, the surface-modified membranes exhibited better performance than those of unmodified membranes, in particular permeability and selectivity. Improvements in membrane antifouling properties (water and wastewater process), organic solvent stability (SRNF applications) as well as plasticization resistance (gas separation) could also be achieved, by choosing the right types of functionality and modification methods. The future direction of *R* and *D* will be focused more on the evaluation of long-term durability of surface-modified membranes under harsh conditions.

Abbreviations

6FAPBS	2,5-bis(4-Amino-2-trifluoromethyl-phenoxy) benzenesulfonic acid
6FBABDS	4,4′-bis(4-Amino-2-trifluoromethyl-phenoxy)biphenyl-4,4′-disulfonic acid
API	Active pharmaceutical ingredient
APTMS	Aminopropyltrimethoxysilane
BP	Benzophenone
BTEC	Biphenyl tetraacyl chloride
CA	Cellulose acetate
CAB	Cellulose acetate butyrate
CD	Cyclodextrin
CHBA	Cyclohexanebis(methylamine)
CMCNa	Sodium carboxymethyl cellulose
CMS	Carbon molecular sieve
CNT	Carbon nanotube
CTC	Charge transfer complex
DABA	3,5-Diamino-*N*-(4-aminophenyl)benzamide
DADMAC	Diallyl dimethyl ammonium chloride
DMF	Dimethylformamide
EA	Ethyl acetate
EDA	Ethylenediamine

EG	Ethylene glycol
FFV	Fractional free volume
FMBO	Fe–Mn binary oxide
FO	Forward osmosis
HFO	Hydrous ferric oxide
HMO	Hydrous manganese dioxide
HNT	Halloysite nanotube
HPEI	Hyperbranched polyethyleneimine
IP	Interfacial polymerization
IPA	Isopropanol
IPC	Isophthaloyl chloride
LbL	Layer-by-layer
MD	Membrane distillation
MIT	Molecular imprinting technology
MMM	Mixed matrix membrane
MOF	Metal–organic framework
MPD	*m*-Phenylenediamine
MW	Molecular weight
MWCNT	Multiwalled carbon nanotube
MWCO	Molecular weight cut-off
NaSS	Sodium *p*-styrene sulfonate
NF	Nanofiltration
NMP	*n*-Methyl-2-pyrrolidone
PA	Polyamide
PAA	Polyacrylic acid
PAH	Poly(allylamine hydrochloride)
PAI	Polyamide-imide
PAN	Polyacrylonitrile
PAN-*b*-PEG	Polyacrylonitrile-*block*-polyethylene glycol
PAN-*g*-PEO	Polyacrylonitrile-*graft*-poly(ethylene oxide)
PAN-*r*-DMMSA	polyacrylonitrile-*random*-*N*,*N*-dimethyl-*N*-methacryloxyethyl-*N*-(3-sulfopropyl)
PAS	Poly(arylsulfone)
PBI	Polybenzimidazole
PDA	Polydopamine
PDDA	Poly(diallyldimethylammonium chloride)
PDMAEMA	Poly(*N*,*N*-dimethylaminoethyl methacrylate)
PDMS	Polydimethylsiloxane
PE	Polyelectrolyte
PEG	Poly(ethylene glycol)
PEI	Polyetherimide
PEM	Polyelectrolyte multilayer
PEMM	Polyelectrolyte multilayer membrane
PEO	Poly(ethylene oxide)
PES	Polyethersulfone
PET	Polyethylene terephthalate
PI	Polyimide

PIM-1	Polymer of intrinsic microporosity
PIP	Piperazine
PMAPS	Poly(3-(*N*-2-methacryloxyethyl-*N*,*N*-dimethyl) ammonatopropanesultone)
PS-*b*-PEG	Polystyrene-*block*-polyethylene glycol
PRO	Pressure retarded osmosis
PSF	Polysulfone
PSS	Poly(sodium 4-styrene-sulfonate)
PTS	Phosphorylated TiO_2-SiO_2
PP	Polypropylene
PPESK	Poly(phthalazinone ether sulfone ketone)
PPSU	Polyphenylsulfone
PPy	Polypyrrole
PU	Polyurethane
PVA	Poly(vinyl) alcohol
PVDF	Polyvinylidene fluoride
PVP	Polyvinylpyrrolidone
PVS	Poly(vinyl sulfate)
RB5	Reactive Black 5
RO	Reverse osmosis
SiO_2	Silicon dioxide
SMPR	Submerged membrane photocatalytic reactor
SPEEK	Sulfonated poly(ether ether ketone)
SPES-NH_2	Sulfonated cardo poly(arylene ether sulfone)
SRNF	Solvent resistant nanofiltration
sPPSU	Sulfonated polyphenylenesulfone
SWNT	Single-walled nanotube
SZP	Zr-doped hybrid silica
TETA	Trimethylenetetramine
TFC	Thin film composite
TFN	Thin film nanocomposite
TiO_2	Titanium dioxide
TNT	Titania nanotube
TOC	Total organic carbon
UF	Ultrafiltration
XDC	Xylylene dichloride

References

1 Zheng, Y.M., Zou, S.W., Nanayakkara, K.G.N., Matsuura, T., and Chen, J.P. (2011) *J. Membr. Sci.*, **374**, 1–11.

2 Morgan, P.W. (1965) *Polym. Rev.*, **10**, 19–64.

3 Joseph, N., Ahmadiannamini, P., Hoogenboom, R., and Vankelecom, I.F. (2014) *J. Polym. Chem.*, **5**, 1817–1831.

4 Ichinose, I., Mizuki, S., Ohno, S., Shiraishi, H., and Kunitake, T. (1999) *Polym. J.*, **31**, 1065–1070.

5 Xu, G.R., Wang, S.H., Zhao, H.L., Wu, S.B., Xu, J.M., Li, L., and Liu, X.Y. (2015) *J. Membr. Sci.*, **493**, 428–443.
6 Brinker, C.J. and Scherer, G.W. (1990) *Sol–Gel Science: The Physics and Chemistry of Sol–Gel Processing*, Academic Press, San Diego, CA.
7 Rao, C.N.R., Muller, A., and Cheetham, A.K. (2004) *The Chemistry of Nanomaterials: Synthesis, Properties and Applications*, vol. **2**, Wiley-VCH Verlag GmbH & Co. KGaA, Weinheim.
8 Centexbel The Sol Gel Principle, http://www.centexbel.be/solgel-treatment (accessed 15 April 2016).
9 Gordon England Cold Gas Dynamic Spray Coating Process, http://www.gordonengland.co.uk/xcoldspray.htm (accessed 15 April 2016).
10 Yavaş, H. (2012) Development of indium tin oxide (ITO) nanoparticle incorporated transparent conductive oxide thin films, in *Department of Metallurgical and Materials Engineering*, Middle East Technical University.
11 Pham, N.P., Boellard, E., Sarro, P.M., and Burghartz, J.N. (2002) Spin, spray coating and electrodeposition of photoresist for MEMS structures – A comparison. *Eurosensors 2002*, 81–86.
12 Sahu, N., Parija, B., and Panigrahi, S. (2009) *Indian J. Phys.*, **83**, 493–502.
13 Ma, N., Quan, X., Zhang, Y., Chen, S., and Zhao, H. (2009) *J. Membr. Sci.*, **335**, 58–67.
14 Choi, H., Stathatos, E., and Dionysiou, D.D. (2006) *Appl. Catal., B*, **63**, 60–67.
15 Norrman, K., Ghanbari-Siahkali, A., and Larsen, N.B. (2005) *Annu. Rep. Prog. Chem. Sect. C: Phys. Chem.*, **101**, 174–201.
16 Hall, D.B., Underhill, P., and Torkelson, J.M. (1998) *Polym. Eng. Sci.*, **38**, 2039–2045.
17 Wang, Z., Ma, J., and Wang, P. (2011) *Desalin. Water Treat.*, **34**, 197–203.
18 Nady, N., Franssen, M.C., Zuilhof, H., Eldin, M.S.M., Boom, R., and Schroën, K. (2011) *Desalination*, **275**(1), 1–9.
19 Kochkodan, V., Johnson, D.J., and Hilal, N. (2014) *Adv. Colloid Interface Sci.*, **206**, 116–140.
20 He, D., Susanto, H., and Ulbricht, M. (2009) *Prog. Polym. Sci.*, **34**(1), 62–98.
21 Revanur, R., McCloskey, B., Breitenkamp, K., Freeman, B.D., and Emrick, T. (2007) *Macromolecules*, **40**(10), 3624–3630.
22 Yang, B. and Yang, W. (2003) *J. Macromol. Sci. Part A*, **40**(3), 309–320.
23 Béquet, S., Remigy, J.C., Rouch, J.C., Espenan, J.M., Clifton, M., and Aptel, P. (2002) *Desalination*, **144**, 9–14.
24 Koros, W.J. and Woods, D.G. (2001) *J. Membr. Sci.*, **181**, 157–166.
25 Tan, X., Tan, S.P., Teo, W.K., and Li, K. (2006) *J. Membr. Sci.*, **271**, 59–68.
26 Mukherjee, P., Jones, K.L., and Abitoye, J.O. (2005) *J. Membr. Sci.*, **254**, 303–310.
27 Arthur, J.D., Langhus, B.G., and Patel, C. (2005) *Technical Summary of Oil & Gas Produced Water Treatment Technologies*, All Consulting, LLC.
28 Fakhru'l-Razi, A., Pendashteh, A., Abdullah, L.C., Biak, D.R.A., Madaeni, S.S., and Abidin, Z.Z. (2009) *J. Hazard. Mater.*, **170**, 530–551.
29 Yin, J. and Deng, B. (2015) *J. Membr. Sci.*, **479**, 256–275.

30 Wu, C., Li, A., Li, L., Zhang, L., Wang, H., Qi, X., and Zhang, Q. (2008) *Desalination*, **225**, 312–321.
31 Asatekin, A. and Mayes, A.M. (2009) *Environ. Sci. Technol.*, **43**, 4487–4492.
32 Salahi, A., Mohammadi, T., Behbahani, R.M., and Hemmati, M. (2015) *Korean J. Chem. Eng.*, **32**, 1101–1118.
33 Chakrabarty, B., Ghoshal, A.K., and Purkait, M.K. (2010) *Chem. Eng. J.*, **165**, 447–456.
34 Zhang, Y., Shan, X., Jin, Z., and Wang, Y. (2011) *J. Hazard. Mater.*, **192**, 559–567.
35 Qiu, Y.R., Zhong, H., and Zhang, Q.X. (2009) *Trans. Nonferrous Met. Soc. China*, **19**, 773–777.
36 Yuliwati, E., Ismail, A.F., Matsuura, T., Kassim, M.A., and Abdullah, M.S. (2011) *Desalination*, **283**, 214–220.
37 Li, Y.S., Yan, L., Xiang, C.B., and Hong, L.J. (2006) *Desalination*, **196**, 76–83.
38 Yi, X.S., Yu, S.L., Shi, W.X., Sun, N., Jin, L.M., Wang, S., Zhang, B., Ma, C., and Sun, L.P. (2011) *Desalination*, **281**, 179–184.
39 Jamshidi Gohari, R., Halakoo, E., Lau, W.J., Kassim, M.A., Matsuura, T., and Ismail, A.F. (2014) *RSC Adv.*, **4**, 17587–17596.
40 Zhao, X., Su, Y., Chen, W., Peng, J., and Jiang, Z. (2012) *J. Membr. Sci.*, **415–416**, 824–834.
41 Zhu, Y., Zhang, F., Wang, D., Pei, X.F., Zhang, W., and Jin, J. (2013) *J. Mater. Chem. A*, **1**, 5758–5765.
42 Xu, Z.L., Chung, T.S., and Huang, Y. (1999) *J. Appl. Polym. Sci.*, **74**, 2220–2233.
43 Ju, H., McCloskey, B.D., Sagle, A.C., Wu, Y.H., Kusuma, V.A., and Freeman, B.D. (2008) *J. Membr. Sci.*, **307**, 260–267.
44 Ong, C.S., Lau, W.J., Goh, P.S., Ng, B.C., and Ismail, A.F. (2014) *Desalination*, **353**, 48–56.
45 Moslehyani, A., Ismail, A.F., Othman, M.H.D., and Matsuura, T. (2015) *Desalination*, **363**, 99–111.
46 Duong, P.H.H., Chung, T.-S., Wei, S., and Irish, L. (2014) *Environmental Science & Technology*, **48**, 4537–4545.
47 Han, G., De Wit, J.S., and Chung, T.-S. (2015) *Water Res.*, **81**, 54–63.
48 Firmino, P.I.M., da Silva, M.E.R., Cervantes, F.J., and dos Santos, A.B. (2010) *Bioresour. Technol.*, **101**, 7773–7779.
49 Dasgupta, J., Sikder, J., Chakraborty, S., Curcio, S., and Drioli, E. (2015) *J. Environ. Manage.*, **147**, 55–72.
50 Shao, L., Cheng, X.Q., Liu, Y., Quan, S., Ma, J., Zhao, S.Z., and Wang, K.Y. (2013) *J. Membr. Sci.*, **430**, 96–105.
51 Liu, Y., Zhang, S., Zhou, Z., Ren, J., Geng, Z., Luan, J., and Wang, G. (2012) *J. Membr. Sci.*, **394**, 218–229.
52 Mondal, M. and De, S. (2016) *Chem. Eng. J.*, **285**, 304–318.
53 Yu, S., Chen, Z., Cheng, Q., Lü, Z., Liu, M., and Gao, C. (2012) *Sep. Purif. Technol.*, **88**, 121–129.
54 Ismail, A.F. and Lau, W.J. (2009) *Desalin. Water Treat.*, **6**, 281–288.

55 Yu, S., Liu, M., Ma, M., Qi, M., Lü, Z., and Gao, C. (2010) *J. Membr. Sci.*, **350**, 83–91.

56 Sun, S.P., Hatton, T.A., Chan, S.Y., and Chung, T.S. (2012) *J. Membr. Sci.*, **401**, 152–162.

57 Zhong, P.S., Widjojo, N., Chung, T.S., Weber, M., and Maletzko, C. (2012) *J. Membr. Sci.*, **417**, 52–60.

58 Zhang, R., Su, Y., Zhao, X., Li, Y., Zhao, J., and Jiang, Z. (2014) *J. Membr. Sci.*, **470**, 9–17.

59 Liu, F., Ma, B.R., Zhou, D., Zhu, L.J., Fu, Y.Y., and Xue, L.X. (2015) *React. Funct. Polym.*, **86**, 191–198.

60 Akbari, A., Desclaux, S., Rouch, J.C., Aptel, P., and Remigy, J.C. (2006) *J. Membr. Sci.*, **286**, 342–350.

61 Han, R., Zhang, S., Zhao, W., Li, X., and Jian, X. (2009) *Sep. Purif. Technol.*, **67**, 26–30.

62 Wei, X., Kong, X., Sun, C., and Chen, J. (2013) *Chem. Eng. J.*, **223**, 172–182.

63 Daraei, P., Madaeni, S.S., Salehi, E., Ghaemi, N., Ghari, H.S., Khadivi, M.A., and Rostami, E. (2013) *J. Membr. Sci.*, **436**, 97–108.

64 Mokhtar, N.M., Lau, W.J., Ng, B.C., Ismail, A.F., and Veerasamy, D. (2015) *Desalin. Water Treat.*, **56**, 1999–2012.

65 Mokhtar, N.M., Lau, W.J., Ismail, A.F., and Ng, B.C. (2014) *RSC Adv.*, **4**, 63367–63379.

66 Ghosh, A.K. and Hoek, E. (2009) *J. Membr. Sci.*, **336**(1), 140–148.

67 Misdan, N., Lau, W.J., Ismail, A.F., and Matsuura, T. (2013) *Desalination*, **329**, 9–18.

68 Li, L., Zhang, S., Zhang, X., and Zheng, G. (2008) *J. Membr. Sci.*, **315**, 20–27.

69 Widjojo, N., Chung, T.-S., Weber, M., Maletzko, C., and Warzelhan, V. (2011) *J. Membr. Sci.*, **383**, 214–223.

70 Bui, N.N., Lind, M.L., Hoek, E.M.V., and McCutcheon, J.R. (2011) *J. Membr. Sci.*, **385**, 10–19.

71 Hoover, L.A., Schiffman, J.D., and Elimelech, M. (2013) *Desalination*, **308**, 73–81.

72 Wei, J., Liu, X., Qiu, C., Wang, R., and Tang, C.Y. (2011) *J. Membr. Sci.*, **381**, 110–117.

73 Li, X., Wang, K.Y., Helmer, B.J., and Chung, T.S. (2012) *Ind. Eng. Chem. Res.*, **51**, 10039–10050.

74 Chen, G., Li, S., Zhang, X., and Zhang, S. (2008) *J. Membr. Sci.*, **310**, 102–109.

75 Wang, H., Li, L., Zhang, X., and Zhang, S. (2010) *J. Membr. Sci.*, **353**, 78–84.

76 Lau, W.J., Gray, S., Matsuura, T., Emadzadeh, D., Paul Chen, J., and Ismail, A.F. (2015) *Water Res.*, **80**, 306–324.

77 Jeong, B.-H., Hoek, E.M.V., Yan, Y., Subramani, A., Huang, X., Hurwitz, G., Ghosh, A.K., and Jawor, A. (2007) *J. Membr. Sci.*, **294**, 1–7.

78 Lind, M.L., Ghosh, A.K., Jawor, A., Huang, X., Hou, W., Yang, Y., and Hoek, E.M.V. (2009) *Langmuir*, **25**, 10139–10145.

79 Baroña, G.N.B., Lim, J., Choi, M., and Jung, B. (2013) *Desalination*, **325**, 138–147.

80 Yin, J., Yang, Y., Hu, Z., and Deng, B. (2013) *J. Membr. Sci.*, **441**, 73–82.

81 Ma, N., Wei, J., Liao, R., and Tang, C.Y. (2012) *J. Membr. Sci.*, **405–406**, 149–157.

82 Saren, Q., Qiu, C.Q., and Tang, C.Y. (2011) *Environ. Sci. Technol.*, **45**, 5201–5208.

83 Jin, W., Toutianoush, A., and Tieke, B. (2003) *Langmuir*, **19**, 2550–2553.

84 Liu, C., Shi, L., and Wang, R. (2015) *React. Funct. Polym.*, **86**, 154–160.

85 Setiawan, L., Wang, R., Shi, L., Li, K., and Fane, A.G. (2012) *J. Membr. Sci.*, **421–422**, 238–246.

86 Chen, L., Therien-Aubin, H., Wong, M.C.Y., Hoek, E.M.V., and Ober, C.K. (2013) *J. Mater. Chem. B*, **1**, 5651–5658.

87 Jamshidi Gohari, R., Lau, W.J., Matsuura, T., and Ismail, A.F. (2013) *Sep. Purif. Technol.*, **118**, 64–72.

88 Chatterjee, S. and De, S. (2015) *Environ. Sci. Water Res. Technol.*, **1**, 227–243.

89 Gohari, R.J., Lau, W.J., Halakoo, E., Ismail, A.F., Korminouri, F., Matsuura, T., Jamshidi Gohari, M.S., and Chowdhury, M.N.K. (2015) *New J. Chem.*, **39**, 8263–8272.

90 He, J., Matsuura, T., and Chen, J.P. (2014) *J. Membr. Sci.*, **452**, 433–445.

91 Bhaumik, M., Noubactep, C., Gupta, V.K., McCrindle, R.I., and Maity, A. (2015) *Chem. Eng. J.*, **271**, 135–146.

92 Jamshidi Gohari, R., Lau, W.J., Matsuura, T., Halakoo, E., and Ismail, A.F. (2013) *Sep. Purif. Technol.*, **120**, 59–68.

93 Abdullah, N., Gohari, R.J., Yusof, N., Ismail, A.F., Juhana, J., Lau, W.J., and Matsuura, T. (2016) *Chem. Eng. J.*, **289**, 28–37.

94 Putri, S.A., Humairo, F.Y., Ong, C.S., Widiastuti, N., Jaafar, J., and Ismail, A.F. (2016) *IPTEK J. Proc. Ser.*, **2**, 1.

95 Business Wire (2015) Markets and Markets. Membranes Market by Type, by Technology, by Region, and by Application – Global Forecast to 2020.

96 Baker, R.W. (2012) *Gas Separation. Membrane Technology and Applications*, John Wiley & Sons, Ltd.

97 Robeson, L.M. (2008) *J. Membr. Sci.*, **320**, 390–400.

98 Visser, T., Masetto, N., and Wessling, M. (2007) *J. Membr. Sci.*, **306**, 16–28.

99 Yong, W.F., Li, F.Y., Xiao, Y.C., Li, P., Pramoda, K.P., Tong, Y.W., and Chung, T.S. (2012) *J. Membr. Sci.*, **407–408**, 47–57.

100 Yong, W.F., Li, F.Y., Xiao, Y.C., Chung, T.S., and Tong, Y.W. (2013) *J. Membr. Sci.*, **443**, 156–169.

101 Car, A., Stropnik, C., Yave, W., and Peinemann, K.V. (2008) *Sep. Purif. Technol.*, **62**, 110–117.

102 Shao, L., Chung, T.-S., Goh, S.H., and Pramoda, K.P. (2005) *J. Membr. Sci.*, **267**, 78–89.

103 Shao, L., Chung, T.-S., Goh, S.H., and Pramoda, K.P. (2005) *J. Membr. Sci.*, **256**, 46–56.

104 Khan, M.M., Bengtson, G., Shishatskiy, S., Gacal, B.N., Mushfequr Rahman, M., Neumann, S., Filiz, V., and Abetz, V. (2013) *Eur. Polym. J.*, **49**, 4157–4166.

105 Askari, M., Xiao, Y., Li, P., and Chung, T.-S. (2012) *J. Membr. Sci.*, **390–391**, 141–151.

106 Zhao, J., Wang, Z., Wang, J., and Wang, S. (2006) *J. Membr. Sci.*, **283**, 346–356.

107 Du, R., Chakma, A., and Feng, X. (2007) *J. Membr. Sci.*, **290**, 19–28.

108 Henis, J.M. and Tripodi, M.K. (1980) Multicomponent membranes for gas separations. US Patent 05/832,481.

109 Zulhairun, A., Fachrurrazi, Z., Izwanne, M.N., and Ismail, A.F. (2015) *Sep. Purif. Technol.*, **146**, 85–93.

110 Blume, I. and Pinnau, I. (1990) Composite membrane, method of preparation and use. US Patent 07/295,686.

111 Murali, R.S., Sridhar, S., Sankarshana, T., and Ravikumar, Y. (2010) *Ind. Eng. Chem. Res.*, **49**, 6530–6538.

112 Murali, R.S., Ismail, A.F., Rahman, M.A., and Sridhar, S. (2014) *Sep. Purif. Technol.*, **129**, 1–8.

113 Murali, R.S., Praveen, K., Ismail, A.F., and Sridhar, S. (2014) *Microporous Mesoporous Mater.*, **197**, 291–298.

114 Chen, X.Y., Nik, O.G., Rodrigue, D., and Kaliaguine, S. (2012) *Polymer*, **53**, 3269–3280.

115 Junaidi, M.U.M., Leo, C.P., Ahmad, A.L., Kamal, S.N.M., and Chew, T.L. (2014) *Fuel Process. Technol.*, **118**, 125–132.

116 Vu, D.Q., Koros, W.J., and Miller, S.J. (2003) *J. Membr. Sci.*, **221**, 233–239.

117 Vu, D.Q., Koros, W.J., and Miller, S.J. (2003) *J. Membr. Sci.*, **211**, 311–334.

118 Vu, D.Q., Koros, W.J., and Miller, S.J. (2003) *J. Membr. Sci.*, **211**, 335–348.

119 Ordoñez, M.J.C., Balkus, K.J., Ferraris, J.P., and Musselman, I.H. (2010) *J. Membr. Sci.*, **361**, 28–37.

120 Zornoza, B., Martinez-Joaristi, A., Serra-Crespo, P., Tellez, C., Coronas, J., Gascon, J., and Kapteijn, F. (2011) *Chem. Commun. (Camb.)*, **47**, 9522–9524.

121 Song, Q., Nataraj, S.K., Roussenova, M.V., Tan, J.C., Hughes, D.J., Li, W., Bourgoin, P., Alam, M.A., Cheetham, A.K., Al-Muhtaseb, S.A., and Sivaniah, E. (2015) *Energy Environ. Sci.*, **5**, 8359–8369.

122 Basu, S., Cano-Odena, A., and Vankelecom, I.F.J. (2011) *Sep. Purif. Technol.*, **81**, 31–40.

123 Ahn, J., Chung, W.-J., Pinnau, I., and Guiver, M.D. (2008) *J. Membr. Sci.*, **314**, 123–133.

124 Ahn, J., Chung, W.-J., Pinnau, I., Song, J., Du, N., Robertson, G.P., and Guiver, M.D. (2010) *J. Membr. Sci.*, **346**, 280–287.

125 Sadeghi, M., Semsarzadeh, M.A., and Moadel, H. (2009) *J. Membr. Sci.*, **331**, 21–30.

126 Nordin, N.A.H.M., Ismail, A.F., Mustafa, A., Murali, R.S., and Matsuura, T. (2014) *RSC Adv.*, **4**, 52530–52541.

127 Nordin, N.A.H.M., Ismail, A.F., Mustafa, A., Murali, R.S., and Matsuura, T. (2015) *RSC Adv.*, **5**, 30206–30215.

128 Vandezande, P., Gevers, L.E.M., and Vankelecom, I.F.J. (2008) *Chem. Soc. Rev.*, **37**, 365–405.
129 Li, X., Fustin, C.-A., Lefevre, N., Gohy, J.-F., Feyter, S.D., Baerdemaeker, J.D., Egger, W., and Vankelecom, I.F.J. (2010) *J. Mater. Chem.*, **20**, 4333–4339.
130 Struzynska-Piron, I., Loccufier, J., Vanmaele, L., and Vankelecom, I.F.J. (2013) *Chem. Commun.*, **49**, 11494–11496.
131 Struzynska-Piron, I., Loccufier, J., Vanmaele, L., and Vankelecom, I.F.J. (2014) *Macromol. Chem. Phys.*, **215**, 614–623.
132 Jimenez Solomon, M.F., Bhole, Y., and Livingston, A.G. (2013) *J. Membr. Sci.*, **434**, 193–203.
133 Li, X., Vandezande, P., and Vankelecom, I.F.J. (2008) *J. Membr. Sci.*, **320**, 143–150.
134 Karan, S., Jiang, Z., and Livingston, A.G. (2015) *Science*, **348**, 1347–1351.
135 Gevers, L.E.M., Vankelecom, I.F.J., and Jacobs, P.A. (2005) *Chem. Commun.*, **19**, 2500–2502.
136 Soroko, I. and Livingston, A. (2009) *J. Membr. Sci.*, **343**, 189–198.
137 Vanherck, K., Hermans, S., Verbiest, T., and Vankelecom, I. (2011) *J. Mater. Chem.*, **21**, 6079–6087.
138 Siddique, H., Rundquist, E., Bhole, Y., Peeva, L.G., and Livingston, A.G. (2014) *J. Membr. Sci.*, **452**, 354–366.
139 Basu, S., Maes, M., Cano-Odena, A., Alaerts, L., De Vos, D.E., and Vankelecom, I.F.J. (2009) *J. Membr. Sci.*, **344**, 190–198.
140 Campbell, J., Szekely, G., Davies, R.P., Braddock, D.C., and Livingston, A.G. (2014) *J. Mater. Chem. A*, **2**, 9260–9271.
141 Sani, N.A.A., Lau, W.J., and Ismail, A.F. (2015) *RSC Adv.*, **5**, 13000–13010.
142 Namvar-Mahboub, M., Pakizeh, M., and Davari, S. (2014) *J. Membr. Sci.*, **459**, 22–32.
143 Sorribas, S., Gorgojo, P., Téllez, C., Coronas, J., and Livingston, A.G. (2013) *J. Am. Chem. Soc.*, **135**, 15201–15208.
144 Peyravi, M., Jahanshahi, M., Rahimpour, A., Javadi, A., and Hajavi, S. (2014) *Chem. Eng. J.*, **241**, 155–166.
145 Li, X., De Feyter, S., Chen, D., Aldea, S., Vandezande, P., Prez, F.D., and Vankelecom, I.F.J. (2008) *Chem. Mater.*, **20**, 3876–3883.
146 Li, X., Goyens, W., Ahmadiannamini, P., Vanderlinden, W., De Feyter, S., and Vankelecom, I. (2010) *J. Membr. Sci.*, **358**, 150–157.
147 Ahmadiannamini, P., Li, X., Goyens, W., Joseph, N., Meesschaert, B., and Vankelecom, I.F.J. (2012) *J. Membr. Sci.*, **394–395**, 98–106.
148 Ahmadiannamini, P., Li, X., Goyens, W., Meesschaert, B., Vanderlinden, W., De Feyter, S., and Vankelecom, I.F.J. (2012) *J. Membr. Sci.*, **403–404**, 216–226.
149 Chen, D. (2013) *J. Appl. Polym. Sci.*, **129**, 3156–3161.

5

Langmuir–Blodgett Films: Sensor and Biomedical Applications and Comparisons with the Layer-by-Layer Method

Epameinondas Leontidis

5.1 Introduction

Thin organic (molecular) films on surfaces have attracted strong technological interest, especially in applications where (i) the long-life of the coating is not an issue and (ii) molecular organization provides properties at the assembly level that do not exist at the molecular (bulk solution) level. Whenever fragility, lack of perfect order, and defect density are not deleterious for the application at hand, organic thin films provide good solutions, sometimes with unique advantages [1]. The main types of thin organic films are summarized in Figure 5.1. These are spin-coated films, organic vapor-phase deposited films, inkjet-printed films, Langmuir–Blodgett (LB) films, self-assembled monolayers (SAMs), and Layer-by-layer (LbL) films.

Many of the methods for the production of thin organic films have been adopted by nanotechnology for the production of hybrid organic–inorganic films, containing molecular layers, nanoparticles, sheet-like inorganic materials, and biopolymers. The present chapter focuses mainly on LB films and compares them to LbL films; these two types of films share the following structural property: In both cases, components of solutions are deposited on solids in the form of sequential molecular or nanoparticle layers. In both methods, the films are usually deposited by dipping solid substrates in appropriate precursor solutions. In the LbL case, molecules from the bulk solution phase are deposited on the substrate, while, in the LB case, molecules from the surface of the solution are deposited.

Langmuir monolayers and LB films have been an active area of research for more than a century. The possibility to study phenomena in two dimensions or to examine simplified model systems of biological membranes have always created a fascination for these structures. The following extract from a classic review by Irving Langmuir is typical: "To-day, however, I propose to tell you of a real two-dimensional world in which phenomena occur that are analogous to those described in "Flatland." I plan to tell you of the behavior of molecules and atoms that are held at the surfaces of three-dimensional solids and liquids. The chemist has long described molecules that are held in this way on surfaces as adsorbed molecules. I will show you that we can have adsorbed

Surface Treatments for Biological, Chemical and Physical Applications, First Edition.
Edited by Mehmet Gürsoy and Mustafa Karaman.

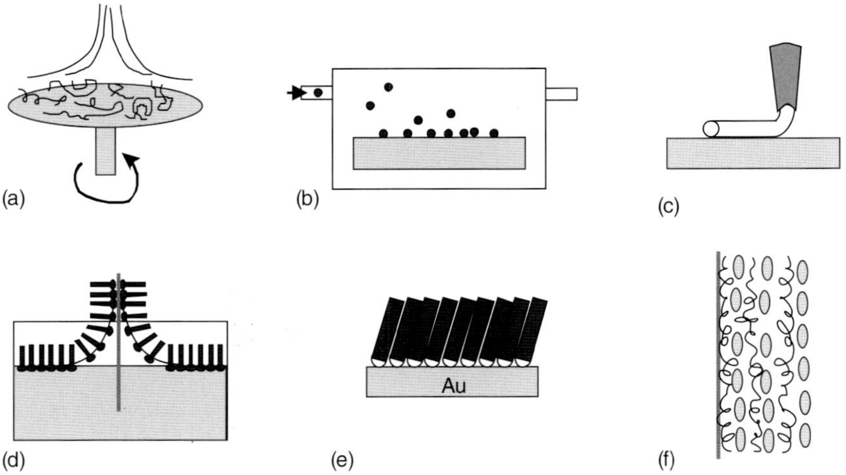

Figure 5.1 Various types of technologically important thin organic films: (a) spin-coated films, (b) organic vapor-phase deposited films, (c) inkjet-printed films, (d) Langmuir–Blodgett (LB) films, (e) self-assembled monolayers (SAMs), and (f) layer-by-layer (LbL) films.

films which really constitute two-dimensional gases, two-dimensional liquids, and two-dimensional solids" [2]. LB films are traditionally deposited on solid substrates from condensed, insoluble molecular monolayers at the air–water surface using a vertical dipping deposition technique. From the 1960s to the 1980s, following improvements in the design and control of LB troughs and the rapid developments in the electronics industry, LB films acquired great significance and the related research grew explosively, because of the promise of producing layered molecular structures with well-defined optical and electronic properties and a minimal defect density [3]. LB films are also significant biological membrane mimics, through which one may examine the interactions of molecules, ions, drugs, and biopolymers with cell membranes [4–7] as well as the behavior of pulmonary surfactants [5, 8]. The rapid developments during those decades in the study of Langmuir monolayers and LB films led to a number of important monographs in the 1990s [3, 5–7]. The seminal book of Ulman on "Thin Organic Films" discussed essentially two types of systems: LB films and SAMs [3].

The versatile LbL method for building molecular films appeared in the 1990s [9, 10]. In the original method, layers of cationic and anionic polyelectrolytes are deposited one at a time by alternate dipping of a solid substrate in their respective solutions (Figure 5.2).

The method developed quickly, and soon made use of attractive interactions between the polymer layers other than pure electrostatics [11, 12]. A large number of molecules may be solubilized in the polymer layers, making the structures suitable for biomedical [12, 13], drug delivery [12, 13], or sensor [12, 14] applications. One or both of the polyelectrolytes can be replaced by biopolymers [15] or nanoparticles [16], or even exfoliated minerals [17], and so the method is thought to be suitable for all types of materials applications, as it does not

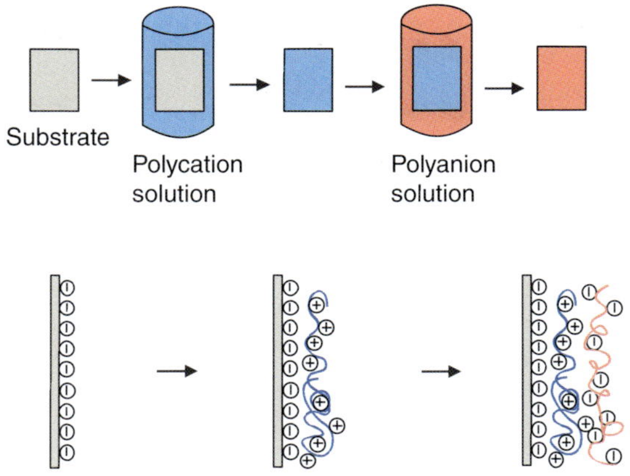

Figure 5.2 Layer-by-layer deposition of layers of cationic and anionic polyelectrolyte.

require expensive specialized equipment and is very flexible. More recently it was proved that polyelectrolyte multilayers can be coated on nanoparticles [18], internal pores of mesoporous materials [19], and even emulsion droplets [20], opening up a large range of applications of LbL technology in drug delivery [21] and food science [22].

The development and rapid proliferation of LbL methods had an unexpected negative impact on the image of the LB method, as it enhanced the false impression that LB films are of limited use in materials and energy applications. LB technology is also wrongly thought by some to be applicable only to relatively simple molecular monolayers; if this were true, the LB method would certainly be doomed to be replaced by the LbL methods for most technological applications of thin organic and hybrid films. A point has been reached when the research community needs to be reminded of the power, breadth, and versatility of the LB method, and of its complementarity with the LbL method. The recent excellent review of Ariga *et al.* [23] is addressing exactly this problem, highlighting applications of LB methods in materials chemistry.

Many excellent monographs and reviews on LB films exist [3–7, 23–26]. No monographs have appeared however after the 1990s, because the relevant literature is so huge as to be practically beyond the reach of a single volume of work. It is characteristic that the chapter on LB films in the 1991 monograph by Ulman [3] contained already more than 600 references. At the same time, LbL literature has literally exploded and several reviews have covered the structure and applications of LbL films [11–14, 20–22, 27–30]. In this chapter, we do not try to review the literature on LB films in a comprehensive way. Instead, we focus on the LB method and explain its various adaptations. We will also compare LB to LbL and discuss the comparative strengths and weaknesses of the two methods. In particular, following on the steps of Ariga *et al.* [23] we will point out that both LB and LbL films should no more be thought of as simply "thin organic films." We will finally try to show that LB and LbL methods are to a much larger extent complementary rather than alternate methods to solve the same problems. The outline of the chapter is as follows: Following this introduction we discuss LB films in Section 5.2. We focus on the materials used for the production of LB films, on their structure and properties. In Section 5.3, we discuss the formation of LB films using nanoparticles. In Sections 5.4 and 5.5, we present the applications of LB films in sensor technology and in biomedicine. In Section 5.6, we will attempt the comparison between the LB and LbL methods.

5.2 Langmuir–Blodgett Films: General Discussion

5.2.1 Deposition Methods, Film Materials, and Substrates

LB films are organic films deposited on solid substrates from insoluble monolayers at the air–water surface. There are two major ways of depositing components from the air–water surface to a solid substrate. These are (i) the most often used original dipping method of Blodgett and (ii) the horizontal or "Schaeffer" deposition, which can be performed by contact from the top, or by lifting from the subphase of the monolayer. More elaborate deposition methods, such as the alternate layer deposition, are essentially variants of the classical dipping method or of the horizontal method (Figure 5.3).

The film materials must be necessarily hydrophobic and essentially water-insoluble. This restriction is not as severe as it seems. The film materials need not be molecular. Hydrophobic molecules and polymers were deposited for almost a century, before the method was being applied to a broad range of insoluble clusters and particles. Figure 5.4 shows the major classes of materials that can be deposited from the air–water surface to form LB films. These are hydrophobic molecules, including supermolecules like fullerenes, nanoparticles of all shapes with or without hydrophobic surface coatings, water-insoluble proteins, hydrophobic polymers, and even exfoliated minerals such as graphene and graphite oxide.

The usual requirement for these materials is that they should form stable, compact layers at the air–water surface. Monolayer stability is usually ascertained by

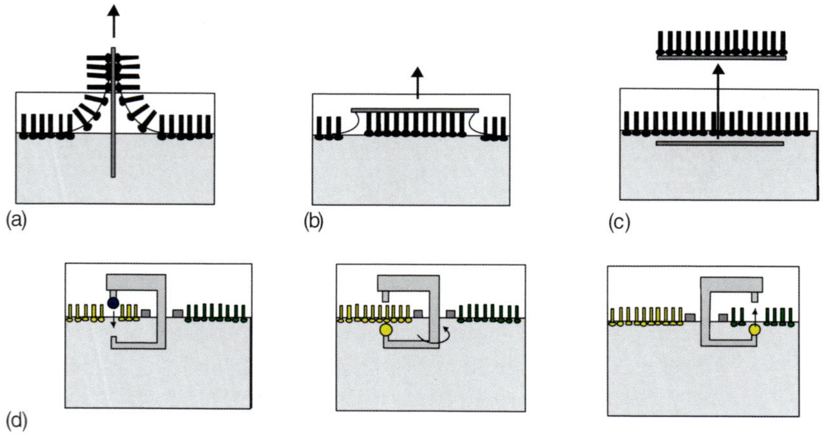

Figure 5.3 Methods to deposit monolayers from the air–water surface to solid substrates. (a) Vertical deposition. (b) Horizontal "contact" deposition. (c) Horizontal "lifting" deposition. (d) Alternate deposition.

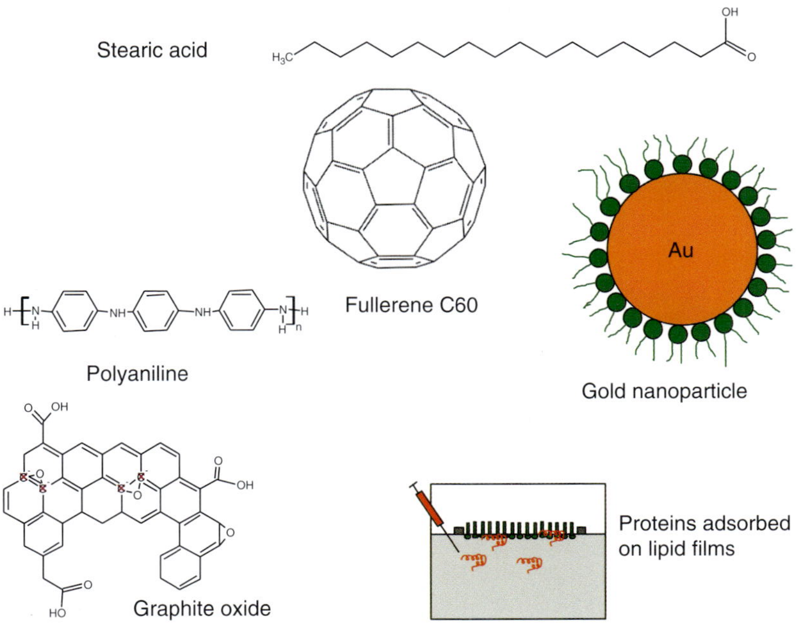

Figure 5.4 Film materials appropriate for LB deposition.

Figure 5.5 How to measure the surface pressure–area isotherm of a Langmuir monolayer.

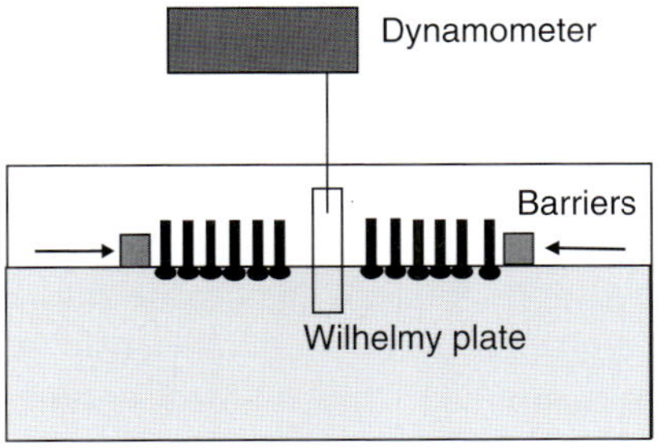

measuring the surface pressure–area isotherms on the water trough of the LB system. The monolayer area is automatically varied between two moving barriers and the surface pressure of the monolayer is measured using the Wilhelmy plate method or its variants. The principle of the method is illustrated in Figure 5.5.

The surface pressure (π) is defined as the difference of the surface tension of the pure water surface (γ_0) minus the surface tension of the surface containing the insoluble monolayer (γ).

$$\pi = \gamma_0 - \gamma \tag{5.1}$$

Monolayer stability is ascertained by measuring surface pressure relaxation (surface pressure vs time at a specific area or area vs time at a specific surface pressure), and by performing compression–decompression cycles to check hysteresis phenomena. Stable monolayers are those that are characterized by negligible pressure relaxation and hysteresis phenomena. However, it often

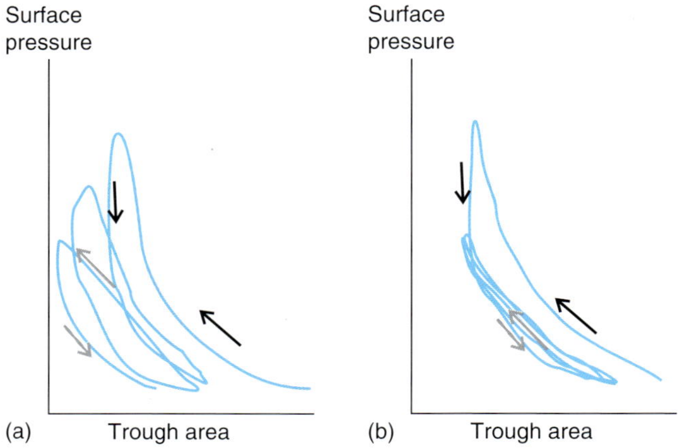

Figure 5.6 Two types of isotherm hysteresis. (a) Shifting hysteresis loops indicate unstable monolayers. (b) Converging hysteresis loops indicating monolayers that relax to a stable structure and can be used for deposition.

suffices to achieve a steady state to have a monolayer that can be deposited on solid substrates. Thus, even when there is significant isotherm hysteresis, the monolayer can be successfully deposited provided that the hysteresis loops disappear after a few cycles (Figure 5.6). This is sometimes the case for polymer monolayers, the interfacial structure of which relaxes slowly and eventually converges to a stable structure. Monolayer stability and successful deposition may sometimes depend on the solvent used to spread the hydrophobic component on the air–water surface, as, for example, in the case of metal-free phthalocyanine [31]. In addition, it often happens that molecules or components that do not form stable monolayers by themselves, either because they are very hydrophobic [32], or because they dissolve in the aqueous subphase [33], may still be deposited quite efficiently as LB films when mixed with other molecules or components that do form stable monolayers, such as stearic or arachidic acid. This increases considerably the generality of the LB method.

In all cases, the quality of the deposition for each upward or downward dipping "stroke" for vertical deposition is simply assessed by measuring the deposition ratio θ, given by the ratio of the area change of the monolayer at the air–water surface (ΔA_{mon}) to the area of the solid substrate coated (A_{c}):

$$\theta = \frac{\Delta A_{\mathrm{mon}}}{A_{\mathrm{c}}} \tag{5.2}$$

High-quality depositions have θ values close to unity. Besides the nature of the film materials, deposition quality depends on a large number of parameters, including the surface pressure at which the deposition is carried out, temperature, environmental humidity and impurities, and the speeds of immersion and removal of the solid substrate in the subphase. The effects of all these parameters have been examined in detail in early LB literature, and are discussed in great detail in the older monographs on LB methods [3, 5–7]. Here we will only discuss the nature of the solid substrate on which the monolayers are deposited.

Since the deposition of the films is made at the molecular layer level and the film properties are very sensitive to impurities, the substrate must be as clean as possible, and, if possible, *molecularly* smooth. The substrate is chosen depending on the application of the film. Normal glass is often used to establish the optimal deposition conditions, quartz glass is used for optical applications, and ITO or FTO glass is used for electrical applications. Mica, silicon, and various metal substrates have been used depending on the application. For example, gold electrodes used in Quartz Crystal Microbalance (QCM) resonators have often been used as substrates [34]. In recent years, the advances of patterning techniques have increased the choice of LB substrates. Although the LB method itself may be used to produce molecularly controlled patterned surfaces, as has been reviewed by Ariga *et al.* [23] and will be briefly discussed below, it is also possible to use patterned substrates to achieve deposition on particular patterns on surfaces. These substrates may be metallic electrodes for conductivity measurements [7, 35], metal-oxide-semiconductor field effect transistors (MOSFETs) [36], polymeric films with patterns generated by lithography [37], gold plates with patterned deposition of SAMs [38], and silicon with a designed array of holes [39]. LB films have been deposited even on holey grids used for TEM work [40]! Monodisperse SiO_2 nanoparticles have been deposited on a Si:H solar cell [41, 42] or a GaAs solar cell covered with an Al grid [43]. The range and variability of substrates used nowadays for LB deposition is as broad as the choice of deposition material!

5.2.2 Applications of LB Films

The use of LB films has been proposed for a very broad range of applications, but the emphasis on various application fronts has shifted over the years. It is instructive to juxtapose the main LB film applications described in the 1991 and 1996 books by Ulman [3] and Petty [7] and those in the 2013 review articles by Cataldo and Pignataro [44] and Ariga *et al.* [23] to get a clear idea of the evolution of ideas in the area of thin organic films and understand why some of the promises expected of LB films have not materialized. This juxtaposition is carried out in Table 5.1:

Table 5.1 Main applications of LB films from the viewpoints of 1991 and 2013.

Main LB applications in the 1990s [3, 7]	Main LB applications in 2013 [23, 44]
Nonlinear optics	Nanofabrication and nanopatterning
Piezoelectric and pyroelectric devices	Nanostructured materials (q-dots, nanowires, nanosheets)
Resist applications	Nanoporous and mesoporous films
Electrically conducting films	Electrical conducting films for solar cells
Gas penetration	Humidity barrier films for solar cells
Electrochromism and photochromism	Photoelectrochromism
Photoconductivity	Biomedicine
Optical sensors and biosensors	Sensors

Comparing the applications in Table 5.1 one can observe a rather significant emphasis shift from precisely tuned molecular films for optoelectronics applications to nanoparticle films and structured materials applications. It is interesting that a method originally developed for the precise placement of molecules on surfaces is currently included in the palette of nanotechnology and used to deposit much larger-scale components on surfaces. The reasons why interest in LB methods for applications in molecular electronics has decreased in recent years is discussed by Motschmann and Möhwald in their review [24]. Two problems have plagued the extensive applications of LB films in nanoelectronics. First, even at the best possible conditions LB films are never perfectly ordered but contain a significant density of defects, which have deleterious consequences for such applications as molecular conductance and switching. Second, given the uncertainty principle, it may be unrealistic to expect that a molecule will remain in a specific quantum state until forced by a nanocircuit or by optical means to switch to another state. However, LB films are still extensively used as components in solar cell designs [26, 44]. Given the somewhat reduced interest for applications of LB films in molecular electronics and nonlinear optics, the fact that these applications have been reviewed expertly and extensively in the past [3, 7, 26, 44] and the relative inexperience of the author in this area, the present chapter will not discuss optoelectronic properties and applications of LB films. Nanopatterning using LB films is a very new and exciting direction of research, but will not be reviewed here, as it was recently covered quite extensively by Ariga *et al.* [23]. In the next three sections we will examine instead three other applications of LB films and coatings. Applications of LB films as chemical sensors and biosensors appear to have retained their importance over the years, and their current status will be examined in some detail. Uses of LB coatings in biomedical devices is not extensive, but it holds some promise and will be also discussed. However, before discussing these applications of LB films we will present in some detail the production of simple and mixed nanoparticle films using the LB method, since this enables the coupling of LB technology to nanotechnology and changes the emphasis of LB research from the molecular to the nanoparticulate scale and from the organic to the inorganic domain.

5.3 LB Films of Nanoparticles

Nanoparticles can be processed in many ways, which have been recently reviewed by Yang *et al.* [45]. Some of the available methods, like spin coating, or drop coating are simple and do not require expensive equipment, while inkjet printing is well suited for larger scale production of nanoparticulate films. It is, therefore, surprising that the LB method is considered a very useful tool for nanoparticle manipulation and LB deposition of nanoparticles has been repeatedly reviewed in recent years [46–49]. Here, we will present only some key developments since little time has elapsed since these excellent reviews. Suffice to say that LB films provide a bottom-up approach for the production of nanoparticle devices, allowing good to excellent control at the individual nanoparticle scale. They are ideal for the production of very well-ordered films.

In the first place, LB films have been used for the *in situ* production of nanoparticles. Sulfide semiconductor nanoparticles were produced by reacting LB films of fatty acid salts of Cd^{2+}, Pb^{2+}, Cu^{2+}, and so on, with H_2S gas. This approach was rather popular for a few years in the late 1980s and early 1990s [50, 51]. It was hoped that the confinement and order of fatty acid LB film would produce an array of narrow-size-distribution nanoparticles, and even a nanoparticle supercrystal. However, it soon became clear that the difficulty to control the gas diffusion and reaction in the monolayer and the very softness of the template would not allow the needed control of size distribution of the nanoparticles. Another key idea at that time was to produce nanoparticles under Langmuir monolayers in the condensed phase [52]. The idea was to use the organized crystal monolayer as a template, by reacting S^{2-} ions in the solution with metal cations attached to the monolayer surface. This was also a rather limited approach. However, in the early 1990s excellent methods were found for preparation of monodisperse semiconductor nanoparticles in organic phases [53]. The surfaces of these nanoparticles were functionalized with organic stabilizers, rendering them hydrophobic and eminently suitable to form stable LB monolayers and LB films. There was a large activity in this direction in the 1990s, and soon the parameters were found for the formation of stable nanoparticle monolayers and their transfer to solid substrates [54–56]. Since that time the LB method has been popular for the production of organized assemblies of quantum dots, but has been generalized to nanowires (1-D nanostructures) of semiconductors [57] and metals [58], and nanosheets of various materials [59, 60] as well. Graphite oxide is one of these 2-D materials that is currently of great scientific interest and has also been deposited as LB films [61–63]. As further proof of the involvement of the LB method with nanotechnology, it has been possible to successfully create films of carbon nanotubes [64, 65], zeolites [66], and metal–organic frameworks (MOFs) [67]. It is practically impossible to describe all the types of nanomaterials that are currently transferred to solid substrates with the LB method, and it is beyond the scope of this review to describe the properties of these structures and the many applications of these films.

5.4 LB Films as Sensors

5.4.1 Types of Sensors

Sensors based on LB films have been fabricated for many decades, and constitute one of the earliest technological applications of LB technology. The original applications of LB films as chemical sensors are reviewed by Moriizumi [68], Tieke [69], and Ulman [3]. There are several different categories of sensors, based on the physicochemical principle behind the signal production. These are summarized in Table 5.2.

LB films have been associated with almost all types of sensors listed in Table 5.2. Sensors may also be classified according to the nature of the analyte (e.g., biosensors), or the nature of the phase in which the measurement takes place (gas, liquid, solid, interface).

Table 5.2 Types of sensors and related measurement principles.

Sensor type	Quantity measured
Thermal	Thermal conductivity, heat capacity
Dielectric	Dielectric constant, refractive index
Electrical	Electrical conductivity/resistivity
Optical	Fluorescence signal, absorbance (UV–vis, SPR), reflectance
Acoustic	Pressure, strain, torque (but also temperature and mass)
Mass	Oscillator frequency (QCM)
Electrochemical	Potential, current, impedance

5.4.2 Gas Sensors

From the beginning it has been clear that LB films are more useful as gas sensors, because of their relative instability and easier degradation when in contact with liquid environments. Gas sensors based on porphyrins and phthalocyanines as well as on derivatives of perylene, tetracene, and squarylium dyes, were developed for a large range of gases, including NO_x, SO_2, CO, NH_3, H_2S, HCl, HNO_3, Cl_2, I_2, O_2, and some volatile organic chemicals (VOCs). Porphyrins and phthalocyanines were preferred as sensing molecules for a variety of reasons [70]: They are thermally and environmentally stable molecules, and they bind a large range of metal atoms at their centers (e.g., Zn, Pd, Pt, Fe, Co, Ni, Cu, Lu, and other lanthanides), which tune their electronic, catalytic, and optical properties. They do not form very stable or transferable monolayers by themselves, hence they are usually co-deposited with film-forming molecules, such as stearic or arachidic acid [71, 72]. There exist however specially modified porphyrin compounds, which can be transferred to solid substrates without the aid of film-forming lipids or fatty acids [36, 70]. It is sometimes found that the surface dilution of phthalocyanines with fatty acids leads to better sensor performance of the derived films [70]. Over the years there have been several attempts to replace phthalocyanines with molecules that are better LB film-formers. These are usually double-chain metallosurfactants with headgroups that can efficiently create complexes with many different metal atoms [73, 74]. Despite these efforts, substituted phthalocyanines and porphyrins have retained their strong position (Figure 5.7).

For VOCs, cavitand molecules are more suitable. These include calixarenes [75–78], the cavity size, the rim, or the interior of which may be tuned to sharpen the selectivity toward particular organic compounds. MOF particles deposited using the LB method may also serve as an alternative gas-sensing element, since they also contain tunable cavities that provide selectivity [79]. Effective gas sensing is also achieved using LB films of conductive polymers [80], which are often coupled to other polymers or inorganic components or even graphite oxide layers [81] to improve film-forming behavior, response, or selectivity.

Regarding the sensing principle, LB films can sense gaseous analytes with various mechanisms. Charge-transfer reactions with molecules in the gas may change the film conductivity [82], or the adsorption may change film

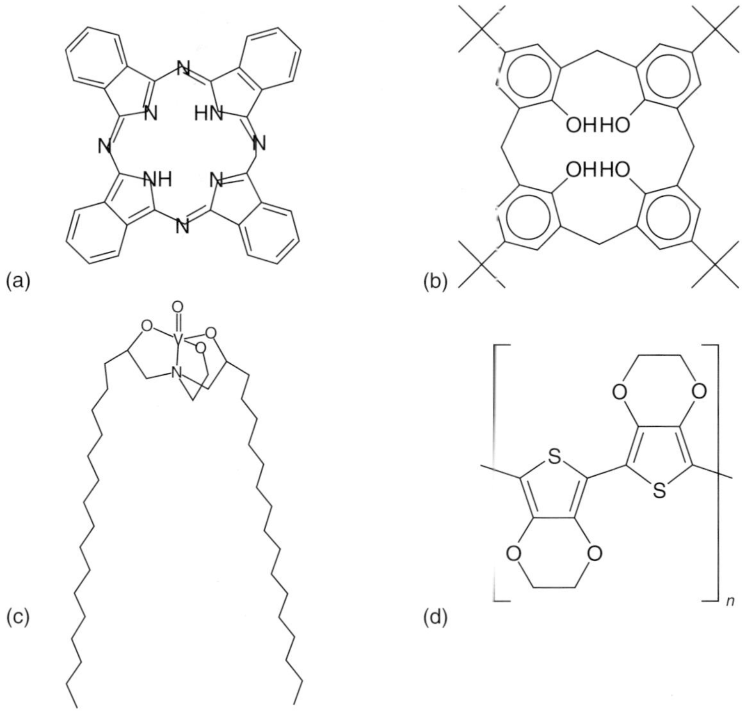

Figure 5.7 Molecules used in LB-based gas-sensors. (a) Phthalocyanine, (b) calix-4-arene, (c) metallosurfactant from [73], and (d) PEDOT (conductive polymers).

fluorescence [71], and it will certainly change film mass [83]. Since the 1980s, the use of LB films in surface plasmon resonance (SPR) sensors [84], and with field-effect transistors (FETs) [85] has started gaining ground, and has become popular. SPR sensors depend on the shift of the SPR of a gold film in the visible spectrum, when a dielectric layer is adsorbed on it. The principle behind a SPR sensor based on LB film technology is presented in Figure 5.8. Briefly, the presence of the LB film already shifts the gold resonance, but when analytes are adsorbed on the LB film, its dielectric constant changes further producing a further change of the gold resonance wavelength, which should ideally be proportional to the bound analyte quantity. SPR instruments are still rather expensive, but the versatility of the method for sensing applications is great, since the sensing principle is almost as general as that of mass sensors.

Already the first LB-based gas sensors that were based on FETs could boast of very low detection limits (e.g., 0.1 ppm for NO_2), good stability and selectivity, and fast response and recovery [85]. Response and, especially, recovery times are among the most important criteria to decide if a sensor is useful or not. Improved performance of electronics and reduced signal-to-noise ratio, plus coupling to nanofabrication methods have boosted FET sensor use in recent years [36, 80]. Some principles of operation of gas-sensors based on LB films and FETs are presented in Figure 5.9. The SOI-MOSFET from [36] is very similar to the simple TFT in Figure 5.9a.

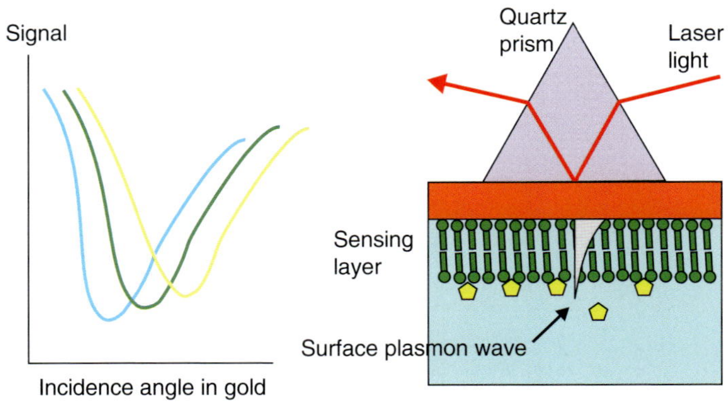

Figure 5.8 An SPR sensor based on LB film response.

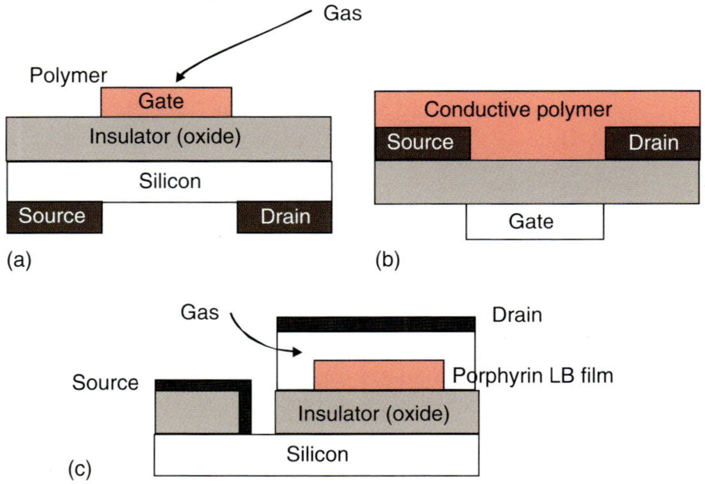

Figure 5.9 Some FET principles for gas sensors based on LB films. (a) TFT [80], (b) IGFET [80], and (c) SOI-MOSFET [36].

The SOI-MOSFET depicted in Figure 5.9c was in fact used in the important article of Takulapalli *et al.* [36], who deposited a zinc porphyrin layer on a silicon-on-insulator (SOI) FET. They demonstrated that pyridine vapor adsorbs on the zinc center and the signal is generated through a charge-transfer mechanism to the oxide layer in such an efficient way that amine concentrations in the range of attomoles can be detected. In general it can be said that the principles behind gas sensors are fairly well established for the last 30–40 years and that current effort is to improve the sensitivity and selectivity of the devices through appropriate coupling of components at the nanoscale. At the same time there is always a quest either for completely new molecular materials for sensing [73, 74] or for new molecular combinations of known materials, as in the work of Sizov *et al.* who combined silicate and thiophene chemistry to obtain conductive molecules with good sensing properties and good behavior under LB processing

Table 5.3 Selected recent examples of LB films used as gas sensors.

References	Analyte	Sensing method	Type of sensing film
[79]	Water, acetone, chloroform, ethanol	Optical (absorbance shift)	Metal–organic framework
[81]	NO_2, SO_2, H_2S, HCl	Electrical (resistance)	PEDOT conductive polymer coupled to graphene oxide
[87]	H_2	Electrical (resistance)	$W_{18}O_{49}$ nanowires on nanopatterned substrate
[88]	H_2	Electrical (conductance)	Pd nanosheets
[89]	H_2	Electrical (resistance and conductance)	Carbon nanotubes
[78]	Chloroform, acetone, *n*-propanol, methanol, hexane, cyclohexane	Acoustic	Microfabricated resonator array of cyclodextrin, porphyrin, calixarene, and cucurbituril

conditions [86]. The area of gas sensing using LB technology has clearly advanced in recent years by taking advantage of the progress in nanotechnology. Table 5.3 summarizes a few recent papers to highlight this point:

In all cases the sensing film has changed to a nanostructure that is deposited on suitable electrodes by the LB method. Nanowires and nanosheets are now regularly deposited as LB films, as will be discussed in Section 5.4. These types of sensing structures often offer excellent sensitivity and selectivity toward specific analytes. Of the beautiful work contained in these papers it is useful to highlight the *electronic nose* sensor array proposed by Lu *et al.* [78]. These authors deposited by LB independent monolayers of a calixarene, a porphyrine, a cyclodextrin, and a cucurbituril on different sensing substrates, which were then connected to form an acoustic resonator nanofilm acting as an electronic nose with sensitivity toward a range of different VOCs. This type of multiple sensor demonstrates how LB technology can be integrated with nanotechnology to provide considerable advances in chemical sensors.

5.4.3 Sensors for Ions and Other Solution Components

Sensor mechanisms for these types of sensors tend to be different from those for gas sensors and depend more strongly on recognition processes. This area is, therefore, strongly coupled to developments in supramolecular chemistry as the films are composed of such components. Otherwise, the sensing device principles are the same as for gas sensors, and there exist electrical, optical, mass, and so on, sensors for these applications as well, although complex electrical sensors (FETs, etc.) are much more common. Since the sensing film is in contact with solutions, the requirements placed on it for long-term structure stability and stability toward oxidation are more stringent.

Early work for ion sensors involved mixed LB films of phospholipids with compounds known to bind ions or small molecules, such as valinomycin for K^+ binding [90], crown ethers, and cyclodextrins for various alkali cations and for molecular anions [91], and even complicated phthalocyaninato polysiloxane polymers, which provided excellent H^+ sensitivity [92]. Work on such sensors already at that stage dealt with some quite complex physicochemical phenomena, as can be seen in the work of Sugawara *et al.* [91], where "stimulant" ions are used to connect to ionophores, generating a structure change in the LB film and facilitating the sensing of the "marker" ion by the underlying electrode. In this interesting paper the authors discuss selectivities of the LB sensors for many groups of inorganic and organic anions and cations, including ATP, ADP, and AMP, or phthalate, isophthalate, and terephthalate.

Instead of reviewing the large area of LB sensors for analytes in solutions, we choose again to tabulate a number of important recent papers, which clearly show the types of films used, the measurement methods, and the occasional coupling with nanotechnology (Table 5.4).

This table summarizes a very broad range of chemical analyses from solution. Besides simple ions, such as H^+ and K^+, one finds various alcohols and sugars, drugs and pesticides, and small molecules of biological significance. Much of the analysis work is done with electrochemistry, as the films are often components of

Table 5.4 Selected examples of recent solution chemical sensors based on LB films.

References	Analyte	Sensing method	Type of sensing film
[93]	Tl^+, Pb^{2+}	Voltammetry	Calixarene
[94]	Catechol	Voltammetry	$LuPc_2$ (double-decker phthalocyaninato complex)
[95]	H^+	Electrical (ISOFET)	Hexyl-thiophene
[96]	K^+	Electrical (OFET)	Valinomycin on PMMA
[97]	Phenol	Mass (QCM)	Cu-phthalocyanine
[98]	Cu^{2+}, methylene blue, saccharose	Electrochemical (impedance)	Azo-substituted perylene
[34]	Norepinephrine	Mass (QCM)	Thiophene deposited on polystyrene nanoparticle monolayer
[99]	Salbutamol	Voltammetry	DNA coupled to poly(aniline) (PANI)
[100]	Various metal cations	Voltammetry	PANI and organophilic montmorillonite
[101]	Gold nanoparticles	Voltammetry	Nanoparticle-imprinted polymer
[102]	Dopamine, Uric acid	Voltammetry	Calixarene
[103]	Methyl paraben	Voltammetry	Multiwall carbon nanotubes (MW-CNTs)
[104]	Daidzein	Voltammetry	MW-CNT combine with PANI

specific electrodes. The impact of nanotechnology is visible in this table as well, as the films often contain carbon nanotubes or nanoparticles. In [98] an e-tongue is created. All sensing units are based on the same azo-derivative of pyridine deposited on Pt interdigitated electrodes, but interestingly some are deposited with the LB method and some by physical vapor deposition (PVD). The resulting sensor can sense a broad range of analytes and the authors discuss how the sensitivity and selectivity depend on sensor geometry and thickness and type of deposition. In [100] it was found that co-deposition of PANI and montmorillonite nanoparticles can provide much better response to cations than that provided by the individual components of the sensor. In [101] the concept of a sensor film is coupled to the idea of nanoparticle-imprinting. Films of cellulose acetate that complex gold nanoparticles are deposited using LB technology. The nanoparticles are then removed electrochemically and the resulting surface can act as an efficient sensor for gold nanoparticles of the same average size. All these exciting developments show that this is a very active area of research where LB technology can be a valuable tool.

5.4.4 Biosensors

The area of biosensors is a science in itself. A quick search reveals that books on biosensors number in the hundreds! LB technology has been occasionally used for biosensor construction, probably less frequently than expected, because of the difficulty of immobilizing and stabilizing proteins in LB films, while at the same time retaining film integrity and structure as the sensor works. The two main applications of LB films in this area evolved in the 1980s [68, 69]. Typical glucose sensors involved glucose oxidase (GOD) mixed layers with fatty acids, lecithin/cholesterol mixtures, or synthetic double-chain lipids. Conductive polymer layers were often added to improve the signal [105]. Such sensors were proved at the time to provide faster response times and good signal stability. The most important issue in sensor preparation was the way to introduce the protein within the LB film, as the protein can be adsorbed on the film from the subphase or be co-spread as a mixture with the lipids. The first choice is apparently the best for hydrophilic proteins. Typical immunosensors also appeared involving antigen–antibody recognition reactions, and were originally based on immunoglobulin G [106]. The same principles are still used for the design of biosensors, as can be seen by more recent works, and the emphasis is on the molecular design of the sensing monolayer. The structure is seldom as simple as in the work of Caseli *et al.* who immobilized alcohol dehydrogenase in a dimyristoyl phosphatidic acid (DMPA) monolayer to make an impedance sensor that detects ethanol [107]. To detect pyrogallol with tyrosinase Pavinatto *et al.* [72] had to create mixed monolayers of arachidic acid and lutetium bis-phthalocyanine and adsorb tyrosinase from the subphase prior to LB deposition that creates a voltammetric sensor. Cabaj *et al.* have made much more elaborate biosensor designs. In one case linoleic acid, octadecyl-trimethylamine, and a specialized copolymer containing phenoxazine and benzothiadiazole were combined to form a LB film on which tyrosinase and laccase could be immobilized by adsorption [108]. The result was a fairly

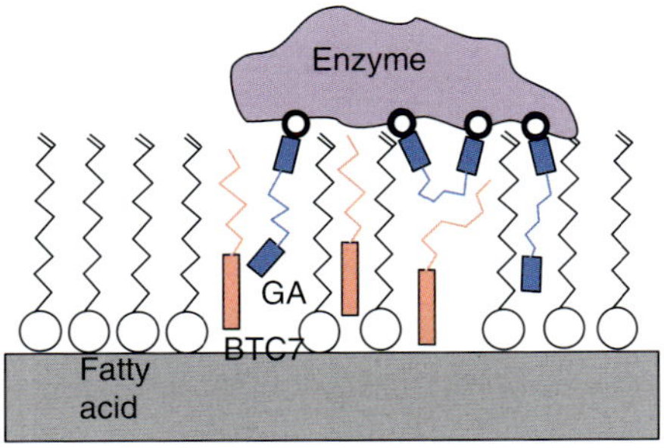

Figure 5.10 Laccase immobilization by glutaraldehyde in the LB film of [109]. The glutaraldehyde molecules (GA) serve to anchor laccase on the LB film.

robust sensor, which could be used for the detection of aminophenol, catechol, tyrosine, L-DOPA, and even syringaldazine. Even more extreme is the design by Cabaj *et al.* of a biosensor which uses a film of 22-tricoseionic acid, the signaling molecule heptyl-bis-thiophene-carbazole (BTC-7) and glutaraldehyde. Laccase was attached to this film covalently through the glutaraldehyde action and retained its activity for a considerable length of time [109]. The principle of protein immobilization and stabilization of this work is presented in Figure 5.10.

An analogous complex design of an LB immunosensor is presented by Prabhakar *et al.* [110]. In their work a DNA aptamer is attached to poly(aniline) via glutaraldehyde cross-linking. Mixed LB films of this compound with stearic acid are prepared and used to capture ochratoxin A. The film impedance changes upon reaction. Khan and Park used LB technology as part of a process to prepare a glucose biosensor [111]. They prepared a new block copolymer of poly(acrylic acid) and poly(acrylic acid-*b*-4-cyanobiphenyl-4′-undecylacrylate). This monolayer was deposited on a specially prepared TEM grid, the surface of which contained the liquid crystalline material 4-cyano-4′-pentylbiphenyl (5CB). This served for intimate contact with the LB film. Horseradish peroxidase and GOD were then adsorbed on the LB film and used to sense glucose. This elaborate work serves to show that LB can be an effective component of broader sensor designs. As a final example of novel biosensor/immunosensor designs we mention the work of Sharma *et al.* [112]. These authors formed an LN film of CdSe nanoparticles mixed with stearic acid. Thiol-terminated probe DNA was adsorbed on the Cd atoms of the CdSe nanoparticles by directly contacting the DNA solution with the film. The resulting bioelectrode was used as a voltammetric sensor for the detection of complementary target DNA associated with chronic myelogenous leukemia. The few selected examples presented here demonstrate the broad range of applicability of the LB method in the biosensor area.

5.5 LB Films in Biomedicine

Biomedical applications of LB films, other than in biosensor formation are not very numerous, because there exist other technologies that can produce similar results at a smaller cost. LB methods have advantages when precise confinement of biologically active material is required. When structural order of a device is not required for an application, LbL films are usually preferred

[13], even though they are degradable in aqueous environments. In this section we will present a few recent examples of biomedical applications of LB films to illustrate the point that LB technology *does* have a place in biomedicine and *should* be considered as a principal or complementary method in the biomedical palette. LB films can be used as supports for drugs in devices that deliver drugs from surfaces. Pierstorff and Ho provide such an example [113]. In their work dexamethasone and doxorubicin were added to a Langmuir monolayer of a copolymer. The resulting equilibrium film was transferred to glass slides and brought to contact with a murine macrophage cells. It was found that the film is suitable for anti-inflammatory chemotherapy. In this work, the LbL method was also used in a complementary fashion for a production of multilayers, showing that a combination of LB and LbL methods can be advantageous. de Souza *et al.* simulated hydroxyapatite formation on medical implants used as bone scaffolds, by depositing monolayers of dihexadecyl phosphate on stainless steel and titanium substrates [114]. These LB films were then exposed to a simulated blood plasma and the growth of carbonated hydroxyapatite in the presence of phosphate buffer was monitored. It was found that osteoblasts could proliferate on the films grown on Ti-substrates, thus proving the suitability of the material and the appropriate conditions for bone growth. In this example we see a complex biomedical application of LB films, where LB technology is essential as it allows the careful control of interfacial structure and processes. Mangadlao *et al.* proved that graphene oxide sheets deposited by LB on poly(ethylene terephthalate) substrates exhibited good antibacterial action, which was not related to the edges of the sheets [115]. This is an application for which LB is probably not the only possible solution, but it is a viable alternative to other methods. Turina *et al.* created LB films of synaptosomal membranes from bovine brain cortex [116]. The monolayers of the biological material, which contained the $GABA_A$ receptor, were transferred on glass beads, and then brought in contact with flunitrazepam, which is a benzodiazepine drug acting on the receptor. In this complex biomedical application it is imperative to know and control the structure of the biological film as much as possible, hence LB technology is essential. The final example is more *futuristic.* Venugopalan *et al.* deposited polystyrene beads on a silicon substrate, further modified them by plasma etching and finally coated them with metallic, magnetic and biocompatible coatings [117]. The particles were removed from the films and used as nanopropellers (magnetic *nanovoyagers*) in blood plasma. It was possible to manipulate the motion of the particles and observe their biocompatibility. This is an advanced materials application, in which LB technology provides an important first step.

The above examples demonstrate that LB technology can be an efficient component in complex biomedical applications, often used in conjunction with many other methods to provide unique innovative materials.

5.6 LB and LbL Methods: a Brief Comparison

As mentioned earlier, in recent years the LbL method appears to be favored by more researchers over the older LB method. LbL technology is very simple and does not require complex instrumentation. The layers can be formed by dipping, spin-coating [118] and spraying [119], and automatic machines have been developed for coating a large number of alternating layers on a substrate [120]. The method is very versatile as it relies on a large range of intermolecular interactions (electrostatics, H-bonding, charge-transfer, biospecific interactions, metal coordination, and others) [11, 12], and can use many different structural elements, such as polyelectrolytes, block copolymers, inorganic nanoparticles, nanowires and nanosheets, proteins, polyoxometalates, and other large clusters, proteins and other biopolymers, and so on. Their manipulation with external triggers, such as pH, light, or biospecific interactions is quite easy.

There is no doubt that the two methods can be used as alternatives for similar applications, and in many cases the LbL method may produce better results. This was, for example, illustrated by Onda *et al.* [121], who dissolved GOD in LB and LbL films that did not interact with the protein and found that its activity was much higher in the LbL film, because the mobility of the enzyme there is not severely hindered. But there are cases where LbL films cannot be used at all, while LB films are working perfectly, and this is clearly the case of nonlinear optical molecular arrays, where the signal intensity depends strongly on the molecular order and orientation [3, 7]. In general, LB films are to be preferred when order, orientation and tight molecular packing are advantages. There are cases where the precise tuning of intermolecular or interparticle distances may have a profound effect on the film performance. Such a beautiful example is provided by Ariga *et al.*, who demonstrated piezoluminescence of some special molecules, which depended on the surface pressure of the molecular film at the air–water surface [122]. As a simpler example it is easy to see that tunneling processes between nanoparticles will depend on the interparticle distance, which can be much more easily controlled in LB films [123].

The two methods differ in another fundamental way. LbL films are deposited most of the time from aqueous solutions, hence the materials deposited are water-soluble and the resulting film can, therefore, be degraded rather easily when brought in contact with aqueous phases. Cross-linking the film components may alleviate this problem [124], but has a rather random effect on the structure and might affect active molecules dissolved in the film. In contrast, LB films are usually deposited from the water solution surface and are composed of hydrophobic components. While these may not dissolve easily in aqueous phases, molecular components are usually rather weakly held to the substrates and the films may crack or peel off. Cross-linking strategies have been employed for LB films, especially when applied to protein immobilization [125]. Several clever strategies have also been employed to fix the component molecules on surfaces, without using real covalent bonding. Such is, for example, the case of the glued layers proposed by the Regen group [126, 127]. These issues may become less serious for nanoparticulate films, which are more rigid and can be confined under pressure more easily.

The LB and LbL methodologies are in the end complementary, rather than competing technologies, and a laboratory developing coating applications in the long term should consider acquiring or developing both. The richness of components that can be deposited with these two methodologies clearly demonstrates that LB and LbL films can no more be classified as thin organic films, but must be viewed as extremely versatile nanotechnology tools.

References

1 Swalen, J.D., Allara, D.L., Andrade, J.D., Chandross, E.A., Garoff, S., Israelachvili, J., McCarthy, T.J., Murray, R., Pease, R.F., Rabolt, J.F., Wynne, K.J., and Yu, H. (1987) Molecular monolayers and films. *Langmuir*, **3**, 932–950.

2 Langmuir, I. (1936) Two-dimensional gases, liquids, and solids. *Science*, **84**(2183), 379–383.

3 Ulman, A. (1991) *An Introduction to Ultrathin Organic Films. From Langmuir–Blodgett to Self-Assembly*, Academic Press Inc., San Diego, CA.

4 Gaines, G.L. (1966) *Insoluble Monolayers at Liquid–Gas Interfaces*, Wiley-Interscience, New York.

5 MacRitchie, F. (1990) *Chemistry at Interfaces*, Academic Press, San Diego, CA.

6 Roberts, G. (ed.) (1990) *Langmuir–Blodgett Films*, Plenum Publishing Co., New York.

7 Petty, M.C. (1996) *Langmuir–Blodgett Films. An Introduction*, Cambridge University Press, Cambridge.

8 Pérez-Gil, J. (2008) Structure of pulmonary surfactant membranes and films: the role of proteins and lipid–protein interactions. *Biochim. Biophys. Acta*, **1778**(7–8), 1676–1695.

9 Decher, G. and Hong, J.-D. (1991) Buildup of ultrathin multilayer films by a self-assembly process: I. Consecutive adsorption of anionic and cationic bipolar amphiphiles. *Makromol. Chem. Macromol. Symp.*, **46**, 321–327.

10 Decher, G. (1997) Fuzzy nanoassemblies: toward layered polymeric multicomposites. *Science*, **277**, 1232–1237.

11 Quinn, J.F., Johnston, A.P.R., Such, G.K., Zelikin, A.N., and Caruso, F. (2007) Next generation, sequentially assembled ultrathin films: beyond electrostatics. *Chem. Soc. Rev.*, **36**, 707–718.

12 Ariga, K., Hill, J.P., and Ji, Q. (2007) Layer-by-layer assembly as a versatile bottom-up nanofabrication technique for exploratory research and realistic application. *Phys. Chem. Chem. Phys.*, **9**, 2319–2340.

13 Tang, Z., Wang, Y., Podsiadlo, P., and Kotov, N.A. (2006) Biomedical applications of layer-by-layer assembly: from biomimetics to tissue engineering. *Adv. Mater.*, **18**, 3203–3224.

14 Crespilho, F.N., Zucolotto, V., Oliveira, O.N., and Nart, F.C. (2006) Electrochemistry of layer-by-layer films: a review. *Int. J. Electrochem. Sci.*, **1**, 194–214.

15 Caruso, F., Niikura, K., Furlong, D.N., and Okahata, Y. (1997) 2. Assembly of alternating polyelectrolyte and protein multilayer films for immunosensing. *Langmuir*, **13**, 3427–3433.

16 Kotov, N.A., Dékány, I., and Fendler, J.H. (1995) Layer-by-layer self-assembly of polyelectrolyte-semiconductor nanoparticle composite films. *J. Phys. Chem.*, **99**, 13065–13069.
17 Fang, M., Kim, C.H., Saupe, G.B., Kim, H.-N., Waraksa, C.C., Miwa, T., Fujishima, A., and Mallouk, T.E. (1999) Layer-by-layer growth and condensation reactions of niobate and titanoniobate thin films. *Chem. Mater.*, **11**, 1526–1532.
18 Sukhorukov, G.B., Donath, E., Davis, S., Lichtenfeld, H., Caruso, F., Popov, V.I., and Möhwald, H. (1998) Stepwise polyelectrolyte assembly on particle surfaces: a novel approach to colloid design. *Polym. Adv. Technol.*, **9**, 759–767.
19 Ai, S., Lu, G., He, Q., and Li, J. (2003) Highly flexible polyelectrolyte nanotubes. *J. Am. Chem. Soc.*, **125**, 11140–11141.
20 Guzey, D. and McClements, D.J. (2006) Formation, stability and properties of multilayer emulsions for application in the food industry. *Adv. Colloid Interface Sci.*, **128–130**, 227–248.
21 Ariga, K., Lvov, Y.M., Kawakami, K., Ji, Q., and Hill, J.P. (2011) Layer-by-layer self-assembled shells for drug delivery. *Adv. Drug Delivery Rev.*, **63**, 762–771.
22 Shchukina, E.M. and Shchukin, D.G. (2012) Layer-by-layer coated emulsion microparticles as storage and delivery tool. *Curr. Opin. Colloid Interface Sci.*, **17**, 281–289.
23 Ariga, K., Yamauchi, Y., Mori, T., and Hill, J.P. (2013) 25th anniversary article: what can be done with the Langmuir–Blodgett method? Recent developments and its critical role in materials science. *Adv. Mater.*, **25**, 6477–6512.
24 Motschmann, H. and Möhwald, H. (2001) in *Handbook of Applied Surface and Colloid Chemistry* (ed. K. Holmberg), John Wiley & Sons, Ltd., Chichester, pp. 629–648.
25 Talham, D.R. (2004) Conducting and magnetic Langmuir–Blodgett films. *Chem. Rev.*, **104**, 5479–5501.
26 Talham, D.R., Yamamoto, T., and Meisel, M.W. (2008) Langmuir–Blodgett films of molecular organic materials. *J. Phys. Condens. Matter*, **20**, 184006.
27 Decher, G. (2012) in *Layer-by-Layer Assembly (Putting Molecules to Work) in Multilayer Thin Films: Sequential Assembly of Nanocomposite Materials*, 2nd edn (eds G. Decher and J.B. Schlennof), Wiley-VCH Verlag GmbH & Co., pp. 1–21.
28 Li, Y., Wang, X., and Sun, J. (2012) Layer-by-layer assembly for rapid fabrication of thick polymeric films. *Chem. Soc. Rev.*, **41**, 5998–6009.
29 Borges, J. and Mano, J.F. (2014) Molecular interactions driving the layer-by-layer assembly of multilayers. *Chem. Rev.*, **114**(18), 8883–8942.
30 Xiao, F.-X., Pagliaro, M., Xu, Y.-J., and Liu, B. (2016) Layer-by-layer assembly of versatile nanoarchitectures with diverse dimensionality: a new perspective for rational construction of multilayer assemblies. *Chem. Soc. Rev.*, **45**(11), 3088–3121. doi: 10.1039/C5CS00781j

31 Baker, S., Petty, M.C., Roberts, G.G., and Twigg, M.V. (1983) The preparation and properties of stable metal-free phthalocyanine Langmuir–Blodgett films. *Thin Solid Films*, **99**, 53–59.

32 Gayathri, S.S. and Patnaik, A. (2007) Interfacial behaviour of brominated fullerene (C60Br24) and stearic acid mixed Langmuir films at air–water interface. *Chem. Phys. Lett.*, **433**, 317–322.

33 Berg, J.M. and Eriksson, L.G.T. (1994) Mixed monolayers and Langmuir–Blodgett films consisting of a fatty amine and a bipolar substance. *Langmuir*, **10**, 1213–1224.

34 Pernites, R.B., Venkata, S.K., Tiu, B.D.B., Yago, A.C.C., and Advincula, R.C. (2012) Nanostructured, molecularly imprinted, and template-patterned polythiophenes for Chiral sensing and differentiation. *Small*, **8**, 1669–1674.

35 Sakai, K., Kawada, H., Takamatsu, O., Matsuda, H., Eguchi, K., and Nakagiri, T. (1989) Electrical memory switching in Langmuir–Blodgett films. *Thin Solid Films*, **179**(1–2), 137–142.

36 Takulapalli, B.R., Laws, G.M., Liddell, P.A., Andréasson, J., Erno, Z., Gust, D., and Thornton, T.J. (2008) Electrical detection of amine ligation to a metalloporphyrin via a hybrid SOI-MOSFET. *J. Am. Chem. Soc.*, **130**, 2226–2233.

37 Harirchian-Saei, S., Wang, M.C.P., Gates, B.D., and Moffitt, M.G. (2010) Patterning block copolymer aggregates via Langmuir–Blodgett transfer to microcontact-printed substrates. *Langmuir*, **26**(8), 5998–6008.

38 Bourgoin, J.-P. and Palacin, S. (1998) Selective deposition of Langmuir–Blodgett films of a phthalocyanine onto patterned substrates. *Langmuir*, **14**, 3967–3970.

39 Hur, J. and Won, Y.-Y. (2008) Fabrication of high-quality non-close-packed 2D colloid crystals by template-guided Langmuir–Blodgett particle deposition. *Soft Matter*, **4**, 1261–1269.

40 Sachan, A.K. and Galla, H.-J. (2013) Bidirectional surface analysis of monomolecular membrane harboring nanoscale reversible collapse structures. *Nano Lett.*, **13**, 961–966.

41 Zhu, J., Yu, Z., Burkhard, G.F., Hsu, C.-M., Connor, S.T., Xu, Y., Wang, Q., McGehee, M., Fan, S., and Cui, Y. (2009) Optical absorption enhancement in amorphous silicon nanowire and nanocone arrays. *Nano Lett.*, **9**(1), 279–282.

42 Zhu, J., Hsu, C.-M., Yu, Z., Fan, S., and Cui, Y. (2010) Nanodome solar cells with efficient light management and self-cleaning. *Nano Lett.*, **10**, 1979–1984.

43 Liang, D., Kang, Y., Huo, Y., Chen, Y., Cui, Y., and Harris, J.S. (2013) High-efficiency nanostructured window GaAs solar cells. *Nano Lett.*, **13**, 4850–4856.

44 Cataldo, S. and Pignataro, B. (2013) Polymeric thin films for organic electronics: properties and adaptive structure. *Materials*, **6**, 1159–1190.

45 Yang, J., Choi, M.K., Kim, D.-H., and Hyeon, T. (2016) Designed assembly and integration of colloidal nanocrystals for device applications. *Adv. Mater.*, **28**, 1176–1207.

46 Tao, A.R., Huang, J., and Yang, P. (2008) Langmuir–Blodgettry of nanocrystals and nanowires. *Acc. Chem. Res.*, **41**(12), 1662–1673.
47 Acharya, S., Hill, J.P., and Ariga, K. (2009) Soft Langmuir–Blodgett technique for hard nanomaterials. *Adv. Mater.*, **21**, 2959–2981.
48 Bardosova, M., Pemble, M.E., Povey, I.M., and Tredgold, R.H. (2010) The Langmuir–Blodgett approach to making colloidal photonic crystals from silica spheres. *Adv. Mater.*, **22**, 3104–3124.
49 Park, J.Y. and Advincula, R.C. (2011) Nanostructuring polymers, colloids, and nanomaterials at the air–water interface through Langmuir and Langmuir–Blodgett techniques. *Soft Matter*, **7**, 9829–9843.
50 Smotkin, E.S., Lee, C., Bard, A.B., Campion, A., Fox, M.A., Mallouk, T.E., Webber, S.E., and White, J.M. (1988) Size quantization effects in cadmium sulfide layers formed by a Langmuir–Blodgett technique. *Chem. Phys. Lett.*, **152**(2–3), 265–268.
51 Geddes, N.J., Urquhart, R.S., Furlong, D.N., Lawrence, C.R., Tanaka, K., and Okahata, Y. (1993) Growth of CdS particles in cadmium arachidate films: monitoring by surface plasmon resonance, UV-visible absorption spectroscopy, and quartz crystal microgravimetry. *J. Phys. Chem.*, **97**, 13767–13772.
52 Zhao, X.K., Yang, J., McCormick, L.D., and Fendler, J.H. (1992) Epitaxial formation of lead sulfide crystals under arachidic acid monolayers. *J. Phys. Chem.*, **96**(24), 9933–9939.
53 Murray, C.B., Kagan, C.R., and Bawendi, M.G. (1995) Self-organization of CdSe nanocrystallites into three-dimensional quantum dot superlattices. *Science*, **270**, 1335–1338.
54 Dabbousi, B.O., Murray, C.B., Ruber, M.F., and Bawendi, M.G. (1994) Langmuir–Blodgett manipulation of size-selected CdSe nanocrystallites. *Chem. Mater.*, **6**, 216–219.
55 Tian, Y. and Fendler, J.H. (1996) Langmuir–Blodgett film formation from fluorescence-activated, surfactant-capped, size-selected CdS nanoparticles spread on water surfaces. *Chem. Mater.*, **8**, 969–974.
56 Heath, J.R., Knobler, C.M., and Leff, D.V. (1997) Pressure/temperature phase diagrams and superlattices of organically functionalized metal nanocrystal monolayers: the influence of particle size, size distribution, and surface passivant. *J. Phys. Chem. B*, **101**, 189–197.
57 Acharya, S., Das, B., Thupakula, U., Ariga, K., Sarma, D.D., Israelachvili, J., and Golan, Y. (2013) A bottom-up approach toward fabrication of ultrathin PbS sheets. *Nano Lett.*, **13**, 409–415.
58 Liu, J.-W., Wang, J.-L., Wang, Z.-H., Huang, W.-R., and Yu, S.-H. (2014) Manipulating nanowire assembly for flexible transparent electrodes. *Angew. Chem. Int. Ed.*, **53**, 13477–13482.
59 Akatsuka, K., Haga, M., Ebina, Y., Osada, M., Fukuda, K., and Sasaki, T. (2009) Construction of highly ordered lamellar nanostructures through Langmuir–Blodgett deposition of molecularly thin titania nanosheets tens of micrometers wide and their excellent dielectric properties. *ACS Nano*, **3**(5), 1097–1106.

60 Yuan, H., Lubbers, R., Besselink, R., Nijland, M., and ten Elshof, J.E. (2014) Improved Langmuir–Blodgett titanate films via in situ exfoliation study and optimization of deposition parameters. *ACS Appl. Mater. Interfaces*, **6**, 8567–8574.

61 Cote, L.J., Kim, F., and Huang, J. (2009) Langmuir–Blodgett assembly of graphite oxide single layers. *J. Am. Chem. Soc.*, **131**, 1043–1049.

62 Harrison, K.L., Biedermann, L.B., and Zavadil, K.R. (2015) Mechanical properties of water-assembled graphene oxide Langmuir monolayers: guiding controlled transfer. *Langmuir*, **31**, 9825–9832.

63 Hidalgo, R.S., López-Diaz, D., and Velázquez, M.M. (2015) Graphene oxide thin films: influence of chemical structure and deposition methodology. *Langmuir*, **31**, 2697–2705.

64 Caseli, L., Tiburcio, V.L.B., Vargas, F.F.R., Marangoni, S., and Siqueira, J.R. (2012) Enhanced architecture of lipid-carbon nanotubes as Langmuir–Blodgett films to investigate the enzyme activity of phospholipases from snake venom. *J. Phys. Chem. B*, **116**, 13424–13429.

65 Massey, M.K., Rosamond, M.C., Pearson, C., Zeze, D.A., and Petty, M.C. (2012) Electrical behavior of Langmuir–Blodgett networks of sorted metallic and semiconducting single-walled carbon nanotubes. *Langmuir*, **28**, 15385–15391.

66 Tosheva, L., Valtchev, V.P., Mihailova, B., and Doyle, A.M. (2007) Zeolite beta films prepared via the Langmuir–Blodgett technique. *J. Phys. Chem. C*, **111**, 12052–12057.

67 Makiura, R., Motoyama, S., Umemura, Y., Yamanaka, H., Sakata, O., and Kitagawa, H. (2010) Surface nano-architecture of a metal–organic framework. *Nat. Mater.*, **9**, 565–571.

68 Moriizumi, T. (1988) Langmuir–Blodgett films as chemical sensors. *Thin Solid Films*, **160**, 413–429.

69 Tieke, B. (1991) Langmuir–Blodgett membranes for separation and sensing. *Adv. Mater.*, **3**, 532–541.

70 Valli, L. (2005) Phthalocyanine-based Langmuir–Blodgett films as chemical sensors. *Adv. Colloid Interface Sci.*, **116**, 13–44.

71 Beswick, R.B. and Pitt, C.W. (1988) Optical detection of toxic gases using fluorescent porphyrin Langmuir–Blodgett films. *J. Colloid Interface Sci.*, **124** (1), 146–155.

72 Pavinatto, F.J., Fernandes, E.G.R., Alessio, P., Constantino, C.J.L., de Saja, J.A., Zucolotto, V., Apetrei, C., Oliviera, O.N., and Rodriguez-Mendez, M.L. (2011) Optimized architecture for Tyrosinase-containing Langmuir–Blodgett films to detect pyrogallol. *J. Mater. Chem.*, **21**, 4995–5003.

73 Mallouri, R., Keramidas, A.D., Brezesinski, G., and Leontidis, E. (2008) Monolayer properties of surface-active metalorganic complexes with a tunable headgroup. *J. Colloid Interface Sci.*, **317**, 544–555.

74 Santafé, A.A.-M., Blum, L.J., Marquette, C.A., and Girard-Egrot, A.P. (2010) Chelating Langmuir–Blodgett film: a new versatile chemiluminescent sensing layer for biosensor applications. *Langmuir*, **26**(3), 2160–2166.

75 Daly, S.M., Grassi, M., Shenoy, D.K., Ugozzoli, F., and Dalcanale, E. (2007) Supramolecular surface plasmon resonance (SPR) sensors for organophosphorus vapor detection. *J. Mater. Chem.*, **17**, 1809–1818.

76 Çapan, R., Özbek, Z., Göktaş, H., Şen, S., Ince, F.G., Özel, M.E., Stanciu, G.A., and Davis, F. (2010) Characterization of Langmuir–Blodgett films of a calix[8]arene and sensing properties towards volatile organic vapors. *Sens. Actuators, B*, **148**, 358–365.

77 Ozmen, M., Ozbek, Z., Bayrakci, M., Ertul, S., Ersoz, M., and Capan, R. (2014) Preparation and gas sensing properties of Langmuir–Blodgett thin-films of calix[n]arenes: investigation of cavity effect. *Sens. Actuators, B*, **195**, 156–164.

78 Lu, Y., Chang, Y., Tang, N., Qu, H., Liu, J., Pang, W., Zhang, H., Zhang, D., and Duan, X. (2015) Detection of volatile organic compounds using microfabricated resonator array functionalized with supramolecular monolayers. *ACS Appl. Mater. Interfaces*, **7**, 17893–17903.

79 Cui, C., Liu, Y., Xu, H., Li, S., Zhang, W., Cui, P., and Huo, F. (2014) Self-assembled metal–organic frameworks crystals for chemical vapor sensing. *Small*, **10**(18), 3672–3676.

80 Bai, H. and Shi, G. (2007) Gas sensors based on conducting polymers. *Sensors*, **7**, 267–307.

81 Yang, Y., Li, S., Yang, W., Yuan, W., Xu, J., and Jiang, Y. (2014) In situ polymerization deposition of porous conducting polymer on reduced graphene oxide for gas sensor. *ACS Appl. Mater. Interfaces*, **6**, 13807–13814.

82 Tieke, B. and Wegmann, A. (1989) Electrical properties of new donor amphiphiles in Langmuir–Blodgett films. *Thin Solid Films*, **179**, 109–120.

83 Snow, A.W., Barger, W.R., Klusty, M., Wohltjen, H., and Jarvis, N.L. (1986) Simultaneous electrical conductivity and piezoelectric mass measurements on iodine-doped phthalocyanine Langmuir–Blodgett films. *Langmuir*, **2**, 513–519.

84 Lloyd, J.P., Pearson, C., and Petty, M.C. (1988) Surface plasmon resonance studies of gas effects in phthalocyanine Langmuir–Blodgett films. *Thin Solid Films*, **160**, 431–443.

85 Gu, C., Sun, L., Zhang, T., Li, T., and Zhang, X. (1998) High-sensitivity phthalocyanine LB film gas sensor based on field effect transistors. *Thin Solid Films*, **327–329**, 383–386.

86 Sizov, A.S., Anisimov, D.S., Agina, E.V., Borschev, O.V., Bakirov, A.V., Shcherbina, M.A., Grigorian, S., Bruevich, V.V., Chvalun, S.N., Paraschuk, D.Y., and Ponomarenko, S.A. (2014) Easily processable highly ordered Langmuir–Blodgett films of quaterthiophene disiloxane dimer for monolayer organic field-effect transistor. *Langmuir*, **30**, 15327–15334.

87 Cheng, W., Ju, Y., Payamyar, P., Primc, D., Rao, J., Willa, C., Koziej, D., and Niederberger, M. (2015) Large-area alignment of tungsten oxide nanowires over flat and patterned substrates for room-temperature gas sensing. *Angew. Chem. Int. Ed.*, **54**, 340–344.

88 Pan, Y.-T., Yin, X., Kwok, K.S., and Yang, H. (2014) Higher-order nanostructures of two-dimensional palladium nanosheets for fast hydrogen sensing. *Nano Lett.*, **14**, 5953–5959.

89 Lee, J.-H., Kang, W.-S., Najeeb, C.K., Choi, B.-S., Choi, S.-W., Lee, H.J., Lee, S.S., and Kim, J.-H. (2013) A hydrogen gas sensor using single-walled carbon nanotube Langmuir–Blodgett films decorated with palladium nanoparticles. *Sens. Actuators, B*, **188**, 169–175.

90 Howarth, V.A., Cui, D.F., Petty, M.C., Ancelin, H., and Yarwood, J. (1989) Phospholipid-based potassium-selective Langmuir–Blodgett films. *Thin Solid Films*, **180**(1-2), 111–115.

91 Sugawara, M., Kataoka, M., Odashima, K., and Umezawa, Y. (1989) Biomimetic ion-channel sensors based on host-guest molecular recognition in Langmuir–Blodgett membrane assemblies. *Thin Solid Films*, **180**, 129–133.

92 Vogel, A., Hoffmann, B., Schwiegk, S., and Wegner, G. (1991) Novel Langmuir–Blodgett membranes for silicon-based ion sensors. *Sens. Actuators, B*, **4**, 65–71.

93 Zou, L., Zhang, Y., Qin, H., and Ye, B. (2009) Simultaneous determination of thallium and lead on a chemically modified electrode with Langmuir–Blodgett film of a *p*-tert-butylcalix[4]arene derivative. *Electroanalysis*, **21**, 2563–2568.

94 Alessio, P., Pavinatto, F.J., Oliveira, O.N., De Saja Saez, J.A., Constantino, C.J.L., and Rodríguez-Méndez, M.L. (2010) Detection of catechol using mixed Langmuir–Blodgett films of a phospholipid and phthalocyanines as voltammetric sensors. *Analyst*, **135**, 2591–2599.

95 Ritjareonwattu, S., Yun, Y., Pearson, C., and Petty, M.C. (2010) Enhanced sensitivity of an organic field-effect transistor pH sensor using a fatty acid Langmuir–Blodgett film. *Org. Electron.*, **11**, 1792–1795.

96 Ritjareonwattu, S., Yun, Y., Pearson, C., and Petty, M.C. (2012) An ion sensitive organic field-effect transistor incorporating the ionophore valinomycin. *IEEE Sens. J.*, **12**(5), 1181–1186.

97 Giancane, G., Basova, T., Hassan, A., Gümüş, G., Gürek, A.G., Ahsen, V., and Valli, L. (2012) Investigations and application in piezoelectric phenol sensor of Langmuir–Schäfer films of a copper phthalocyanine derivative functionalized with bulky substituents. *J. Colloid Interface Sci.*, **377**, 176–183.

98 Volpati, D., Aoki, P.H.B., Dantas, C.A.R., Paulovich, F.V., de Oliveira, M.C.F., Oliveira, O.N., Riul, A., Aroca, R.F., and Constantino, J.L. (2012) Toward the optimization of an e-tongue system using information visualization: a case study with perylene tetracarboxylic derivative films in the sensing units. *Langmuir*, **28**, 1029–1040.

99 Zou, L., Li, Y., Cao, S., and Ye, B. (2014) A glassy carbon electrode modified with Langmuir–Blodgett film composed of DNA and polyaniline for the sensitive determination of salbutamol. *Electroanalysis*, **26**, 1051–1058.

100 de Barros, A., Ferreira, M., Constantino, C.J.L., Bortoleto, J.R.R., and Ferreira, M. (2015) Synergy between polyaniline and OMt clay mineral in Langmuir–Blodgett films for the simultaneous detection of traces of metal ions. *ACS Appl. Mater. Interfaces*, **7**, 6828–6834.

101 Bruchiel-Spanier, N. and Mandler, D. (2015) Nanoparticle-imprinted polymers: shell-selective recognition of Au nanoparticles by imprinting using the Langmuir–Blodgett method. *ChemElectroChem*, **2**, 795–802.

102 Wang, F., Chi, C., Yu, B., and Ye, B. (2015) Simultaneous voltammetric determination of dopamine and uric acid based on Langmuir–Blodgett film of calixarene modified glassy carbon electrode. *Sens. Actuators, B*, **221**, 1586–1593.

103 Wang, L., Li, Y., Li, G., and Ye, B. (2015) A new strategy for enhancing electrochemical sensing from MWCNTs modified electrode with Langmuir–Blodgett film and used in determination of methylparaben. *Sens. Actuators, B*, **211**, 332–338.

104 Wang, L., Li, Y., Wang, Q., Zou, L., and Ye, B. (2016) The construction of well-aligned MWCNTs-PANI Langmuir–Blodgett film modified glassy carbon electrode and its analytical application. *Sens. Actuators, B*, **228**, 214–220.

105 Li, J.R., Cai, M., Chen, T.F., and Jiang, L. (1989) Enzyme electrodes with conductive polymer membranes and Langmuir–Blodgett films. *Thin Solid Films*, **180**, 205–210.

106 Erokhin, V.V., Kayushina, R.L., Lvov, Y., and Feigin, L.A. (1990) Langmuir–Blodgett films of immunoglobulins as sensing elements. *II Nuovo Cimento*, **12D**(9), 1254–1258.

107 Caseli, L., Perinotto, A.C., Viitala, T., Zucolotto, V., and Oliveira, O.N. (2009) Immobilization of alcohol dehydrogenase in phospholipid Langmuir–Blodgett films to detect ethanol. *Langmuir*, **2009**, 3057–3061.

108 Cabaj, J., Soloducho, J., and Nowakowska-Oleksy, A. (2010) Langmuir–Blodgett film based biosensor for estimation of phenol derivatives. *Sens. Actuators, B*, **143**, 508–515.

109 Cabaj, J., Idzik, K., Soloducho, J., Chyla, A., Bryjak, J., and Doskocz, J. (2008) Well-ordered thin films as practical components of biosensors. *Thin Solid Films*, **516**, 1171–1174.

110 Prabhakar, N., Matharu, Z., and Malhotra, B.D. (2011) Polyaniline Langmuir–Blodgett film based aptasensor for ochratoxin A detection. *Biosens. Bioelectron.*, **26**, 4006–4011.

111 Khan, M. and Park, S.-Y. (2015) Glucose biosensor based on GOx/HRP bienzyme at liquid–crystal/aqueous interface. *J. Colloid Interface Sci.*, **457**, 281–288.

112 Sharma, A., Pandey, C.M., Matharu, Z., Soni, U., Sapra, S., Sumana, G., Pandey, M.K., Chaterjee, T., and Malhotra, B.D. (2012) Nanopatterned cadmium selenide Langmuir–Blodgett platform for leukemia detection. *Anal. Chem.*, **84**, 3082–3089.

113 Pierstorff, E. and Ho, D. (2008) Nanomembrane-driven co-elution and integration of active chemotherapeutic and anti-inflammatory agents. *Int. J. Nanomed.*, **3**(4), 425–433.

114 de Souza, I.D., Cruz, M.A.E., de Faria, A.N., Zancanela, D.C., Simão, A.M.S., Ciancaglini, P., and Ramos, A.P. (2014) Formation of carbonated hydroxyapatite films on metallic surfaces using dihexadecyl phosphate–LB film as template. *Colloids Surf., B*, **118**, 31–40.

115 Mangadlao, J.D., Santos, C.M., de Leon, A.C.C., and Rodrigues, D.F. (2015) On the antibacterial mechanism of graphene oxide (GO) Langmuir–Blodgett films. *Chem. Commun.*, **51**, 2886–2889.

116 Turina, A.V., Clop, P.D., and Perillo, M.A. (2015) Synaptosomal membrane-based Langmuir–Blodgett films: a platform for studies on γ-aminobutyric acid type a receptor binding properties. *Langmuir*, **31**, 1792–1801.

117 Venugopalan, P.L., Sai, R., Chandorkar, Y., Basu, B., Shivashankar, S., and Ghosh, A. (2014) Conformal cytocompatible ferrite coatings facilitate the realization of a nanovoyager in human blood. *Nano Lett.*, **14**, 1968–1975.

118 Lee, S.-S., Hong, J.-D., Kim, C.H., Kim, K., Koo, J.P., and Lee, K.-B. (2001) Layer-by-layer deposited multilayer assemblies of ionene-type polyelectrolytes based on the spin-coating method. *Macromolecules*, **34**, 5358–5360.

119 Izquierdo, A., Ono, S.S., Vögel, J.-C., Schaaf, P., and Decher, G. (2005) Dipping versus spraying: exploring the deposition conditions for speeding up layer-by-layer assembly. *Langmuir*, **21**(16), 7558–7567.

120 Clark, S.L. and Hammond, P.T. (1999) Engineering the microfabrication of layer-by-layer thin films. *Adv. Mater.*, **10**(18), 1515–1519.

121 Onda, M., Ariga, K., and Kunitake, T. (1999) Activity and stability of glucose oxidase in molecular films assembled alternately with polyions. *J. Biosci. Bioeng.*, **87**(1), 69–75.

122 Ariga, K., Nakanishi, T., Terasaka, Y., Tsuji, H., Sakai, D., and Kikuchi, J.-I. (2005) Piezoluminescence at the air-water interface through dynamic molecular recognition driven by lateral pressure application. *Langmuir*, **21**, 976–981.

123 Chen, S. (2003) Langmuir monolayers of gold nanoparticles: from ohmic to rectifying charge transfer. *Anal. Chim. Acta*, **496**, 29–37.

124 Zhou, Y., Cheng, M., Zhu, X., An, Q., and Shi, F. (2013) A facile method to prepare molecularly imprinted layer-by-layer nanostructured multilayers using postinfiltration and a subsequent photo-cross-linking strategy. *ACS Appl. Mater. Interfaces*, **5**, 8308–8313.

125 Aizawa, M., Owaku, K., Matsuzawa, M., Shinohara, H., and Ikariyama, Y. (1989) Molecular film technology for biosensors. *Thin Solid Films*, **180**, 227–233.

126 McCullough, D.H., Janout, V., Li, J., Hsu, J.T., Truong, Q., Wilusz, E., and Regen, S.L. (2004) Glued Langmuir–Blodgett bilayers from porous versus nonporous surfactants. *J. Am. Chem. Soc.*, **126**, 9916–9917.

127 Wang, M., Janout, V., and Regen, S.L. (2012) Minimizing defects in polymer-based Langmuir–Blodgett monolayers and bilayers via gluing. *Langmuir*, **28**, 4614–4617.

6

Surface Modification of Biopolymer-Based Nanoforms and Their Biological Applications

Susana C.M. Fernandes

6.1 Introduction

Natural products have inspired man for millennia. Their rich biological, biochemical, and structural diversity has prompted scientists to isolate, understand, and use them in different areas, including in biological applications. This is the case of the biopolymers that are originally extracted from vegetal (cellulose) and animal waste resources (e.g., chitin, chitosan, and collagen) or by microorganisms (e.g., bacterial cellulose (BC) and pullulan) [1–3]. These natural polymers have found large-scale use in high added value sectors (including biomedical) and represent a multibillion-euro market worldwide (e.g., chitosan €19 billion in 2015 [4]; collagen €2.7 billion in 2020 [5]; hyaluronic acid €2.3 billion in 2017 [6]).

The importance of biopolymers is linked to their abundance, renewable and recyclable nature, biodegradable character, as well as to their easy functionalization capabilities that help in turning them into valuable materials.

This chapter addresses chemical surface modifications of biopolymers-based nanoforms, namely, nanocellulose, including BC and nanochitin, for biological applications. These biopolymers provide a sustainable solution to the need for new (bio)materials in biological applications due to their unique biological properties like low toxicity, biocompatibility, biodegradability, and bioactivity. Nonetheless, further chemical functionalization has been advanced in order to optimize their intrinsic properties and/or generate novel functions to develop new materials, in particular, functional bionanocomposites. These modifications are crucial in determining the potential of these materials in biological applications such as in food technology, pharmaceutical and biomedical industry, and in the biotechnology field.

6.2 Nanocellulose and Nanochitin

Cellulose and chitin are the most widespread natural polysaccharides, which perform structure-forming and protective functions in flora and fauna, respectively. Cellulose and chitin either used as such or modified have often been assessed as alternative for petrol-based counterparts, not only as sustainable resources but

Surface Treatments for Biological, Chemical and Physical Applications, First Edition.
Edited by Mehmet Gürsoy and Mustafa Karaman.

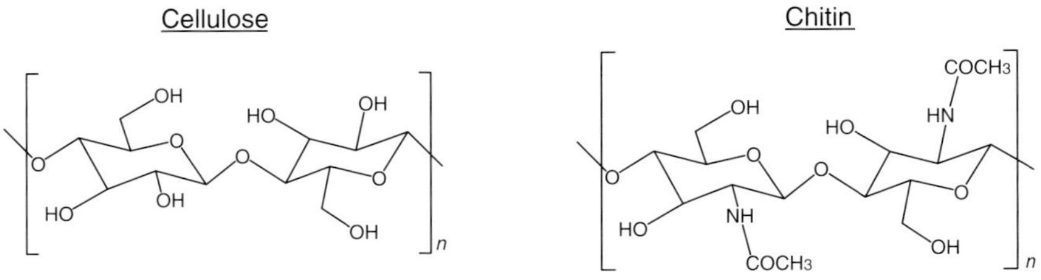

Figure 6.1 Chemical structures of cellulose and chitin.

also as attractive materials with specific properties and functionalities [1–3, 7]. Despite the structural similarity of these polysaccharides (Figure 6.1), their properties (e.g., crystallinity, solubility, and aptitude to chemical modification) are quite distinct, because of the only structural difference that exists in the replacement of an OH group at position C-2 in each saccharide unit of cellulose by an acetamido group or/and a residual amino counterpart in chitin, resulting in different functionalities that could be exploited for the development of new sophisticated materials.

Cellulose is considered the most abundant natural polymer and the oldest used on earth [1].

This polysaccharide is biosynthesized in a relatively pure form, as in cotton, and largely combined with hemicelluloses and lignin in the cell walls of woody plants (e.g., softwoods and hardwoods) and in annual plants [8].

Cellulose is a linear polymer constituted by β-D-glucopyranose molecules linked by β(1 → 4) glycosidic bonds (Figure 6.1). The repetitive unit of this polysaccharide is cellobiose – a dysaccharide that constitute two adjacent glucose units. As each glucose unit stands three OH groups at the positions C-2, C-3, and C-6, hydrogen bonds are an important aspect in this biopolymer's chemistry and ultrastructure: (i) intra-molecular hydrogen bonds are responsible for keeping each structural unit firmly in its place, resulting in a stiff chain while (ii) inter-molecular bonds hold adjacent chains together forming elementary fibrils with high tensile strength, which are the basic units of the cellulose fiber's morphology [8]. Both crystalline (where molecules follow an ordered arrangement) and amorphous (where there is no regular arrangement) regions of cellulose contribute to the final fiber's properties.

During the biosynthesis of cellulose, the molecules pile onto each other forming microfibrils, with typically a diameter of 2–30 nm, which form both crystalline and amorphous regions. Further, the microfibrils aggregate into fibrils (diameters of 30–100 nm and lengths of 100–500 μm), and finally into cellulose fibers (diameters of 100–400 nm and lengths of 0.5–4 mm) [9, 10]. Amorphous regions are more available showing high reactivity toward other chemical species, while crystalline regions are so densely packed and tightly bound that their hydroxyl groups become unavailable [11]. Using effective chemical, mechanical, or a combination of these methods, the fibrils can be disintegrated from the fibers to form a uniform micro- or nano-sized material to obtain cellulose nanocrystals (CNCs) (Figure 6.2) and cellulose nanofibrils (CNFs).

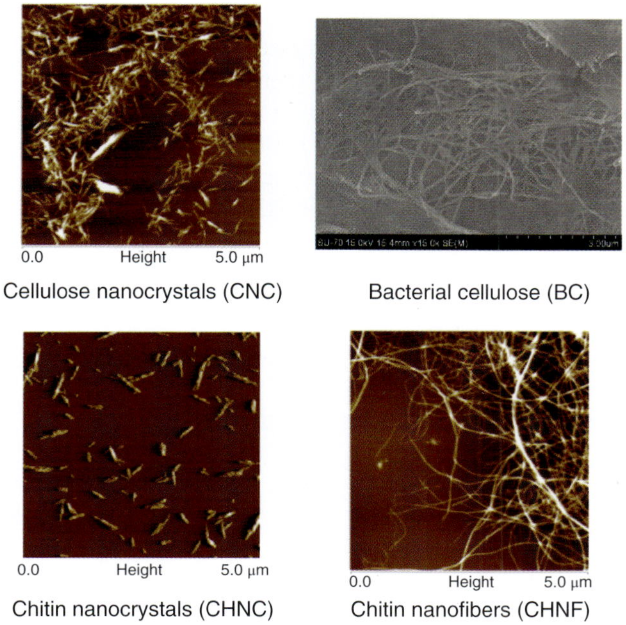

Figure 6.2 Morphology of cellulose nanocrystals, bacterial cellulose, and chitin nanoforms.

Apart from its vegetable origin, cellulose is also biosynthesized by bacteria (e.g., *Gluconacetobacter genus*). BC (also known as *microbial cellulose*) can be produced in the form of a highly swollen gel (~90% water), which forms a 3D network of nano- and microfibrils with 10–100 nm width (Figure 6.2) [12, 13]. BC presents unique properties including high mechanical strength, high crystallinity, high water holding capacity, biocompatibility, and high porosity, which make it a very promising biopolymer. The applications of BC have increased considerably in the last decade, mainly in the biomedical field (e.g., as wound healing [14]), and specific high technology areas (as reinforcing agent for paper [15], nanocomposites [7, 16–18], and nanopapers [19, 20], etc.). BC could be produced in both stationary and agitated cultures. The stationary culture results in the accumulation of a gelatinous membrane of cellulose at the air/liquid interface, while the agitated culture results in fibrous suspensions [21]. The culture media composition and conditions also influence the nanostructure, such as crystallinity and mechanical properties, of BC.

Chitin is considered the second most abundant biopolymer on earth after cellulose, and can be found in different living systems namely in insect cuticles, crustacean shells, and fungi cell walls [22]. Chitin is a linear polymer composed of *N*-acetyl-2-amido-2-deoxy-D-glucose units linked by $\beta(1 \rightarrow 4)$ bonds (Figure 6.1). Isolated chitin is a highly ordered copolymer of *N*-acetyl-D-glucosamine as the major component, and D-glucosamine as a minor constituent. These residual monomers are present in the native chitin or are formed through hydrolysis of some acetamido groups during the isolation and purification processes. In nature, chitin forms part of a well-organized hierarchical structure in the exoskeleton of many invertebrates of amorphous regions where crystallites coexist to form assemblies of strong extended fibers increasing from the nanometer to the millimeter scale. Depending on their biological origin, the cross-sectional width of chitin crystalline fibrils could range from 2.5 to 25 nm, [23]. Like with cellulose, different chemical and mechanical methods have been developed to extract/isolate chitin nanofillers presenting rod-like (nanocrystals, CHNC) or fibrous (nanofibers, CHNF) structures (Figure 6.2) [3, 24–26]. Thanks to their morphology, bioactivity, and biocompatibility, these nanofillers are considered as fascinating fillers for the processing of functional materials and bionanocomposites [3, 27–31].

6.3 The Unique Biological Properties of Nanocellulose and Nanochitin

6.3.1 Nanocellulose

Cellulose nanoforms, that is, nanocrystals, nanofibrils, and BC, have gained much attention not only because of their remarkable physical properties but also due to their biological properties: biodegradability, biocompatibility, and low cytotoxicity [32, 33].

6.3.1.1 Biodegradability

Cellulose degradation occurs via hydrolysis by cellulase enzymes that hydrolyze its β-1,4 D-glucose linkages. Cellulases are made by many types of fungus and bacteria, but are not present in animals. Ruminants like cows and horses host symbiotic cellulase-secreting bacteria, which allow them to degrade cellulose into simple glucose [34].

Because of the lack of cellulases in the human body and in animals, there are no mechanisms for the large-scale breakdown of cellulose; and cellulose may be considered as non-biodegradable *in vivo* or really slowly degradable. Non-enzymatic, spontaneous biodegradability of cellulose chains may explain the slow breakdown of unaltered cellulose within the human body, but this has not been sufficiently studied [35]. Enzymes in the human body such as amylase or hydrolyzing stomach acid are capable of significantly affecting cellulose. This could be ideal for some applications, and problematic for others.

The degradation of cellulose can be affected by its compact structure, hydration, and swelling. Miyamoto *et al.* [36] found that the degradation of cellulose and cellulose derivatives in canine specimens depended significantly on the cellulose crystalline form and chemical modification. More recently, Kümmerer *et al.* [37] showed that *in vitro* (in aqueous environments) CNC were more biodegradable than carbon nanotubes.

Different studies have shown that oxidized cellulose is more susceptible to hydrolysis and, is therefore, potentially degradable by the human body. Consequently, recently, multiple works on improving nanocellulose's biodegradability through oxidation has been developed (discussed in Section 6.4).

6.3.1.2 Biocompatibility

One of the main requirements of any material for biomedical application is that it must be biocompatible. The biocompatibility, that is, the capacity to remain in contact with living tissue without causing any side effects, including cytotoxicity, of cellulose-based materials has been investigated and demonstrated [14, 36, 38–40]. For instance, Helenius *et al.* [39], in an *in vivo* study of subcutaneous BC implantation showed that after 12 weeks, no fibrotic capsule or giant cells were detectable by microscopy, indicating no foreign body reaction in rats. More recently, Jia *et al.* [40] found that the combination of microcrystalline cellulose (MCC) and CNC in electrospun composite scaffolds considerably improved cell viability and morphology on vascular smooth muscle cell. The authors suggested that MCC providing anchors for cells grow within the 3D network of scaffolds and CNCs improve the cell adhesion. This suggests that cellulose-based materials are promising additives to potential scaffolds for tissue engineering applications to improve the biocompatibility of scaffolds.

6.3.1.3 Low Cytotoxicity

The number of studies regarding the cytotoxicity of cellulose nanoforms (i.e., CNF, CNC, and BC) is still limited and is at a very preliminary stage. In general, there is no indication of serious injury (e.g., cytotoxicity, inflammatory effects, etc.) of celloluse nanoforms on *in vitro* cellular or genetic level and on *in vivo* animal tests [41–48].

Alexandrescu *et al.* [41] demonstrated the no toxic phenomena of pure CNF (two types of morphology: thin and dense structures and porous structures) *in vitro* with 3T3 fibroblast cells using the test of cell membrane, cell mitochondrial activity, and DNA proliferation. This work involved direct and indirect contact between the materials and the cells [41]. In another report, Vartiainen *et al.* [42] showed no inflammatory effects or cytotoxicity on mouse macrophage and human monocyte; only low acute environmental toxicity was found [42].

Apart from the evaluation of *in vitro* cytotoxicity, Pereira *et al.* [43], also studied the effect of CNF on gene expression. The authors found that low concentrations of CNF (0.02–100 μg ml^{-1}) have low cytotoxicity, whereas high concentrations of CNF (2000–5000 μg ml^{-1}) reduced cell viability and affection of the expression of stress- and apoptosis-associated molecular markers [43].

The cytotoxicity of CNC was tested by two different tests – 3-(4,5-dimethylthiazol-2-yl)-2,5-diphenyltetrazoliumbromide (MTT) assay and lactate dehydrogenase (LDH) assay – against nine different cell lines. In the concentration range (0–50 μg ml^{-1}) and exposure time studied (48 h), no cytotoxic effects of CNC against any of these cell lines were observed [44]. Nevertheless, it was described that aerosolized CNC may induce some respiratory toxicity and inflammatory effects on 3D human lung cells, especially the risk with inhalatory exposure under high concentrations of released CNC powders [45].

BC is considered to be the most biocompatible material in the family of cellulose nanoforms a property correlated to its purity level. To date, no cytotoxicity of BC has been observed in previous studies *in vitro* on osteoblast, L929 fibroblast cells, and human umbilical vein endothelial cells; and *in vivo* using a mice model (BALB/c) BALB/c mouse [46–48].

6.3.2 Nanochitin

Chitin possesses very interesting biological properties namely biodegradability, non-toxity, biocompatibility, and antimicrobial activity. Therefore, they have been used in many applications in biotechnology, mainly in water treatment, in medical and pharmaceutical fields, and personal-care products [49–51].

The biological properties of these compounds are related to the source of chitin and the conditions of the chitin production process and depend strongly on its structural parameters like weight-averaged molecular weight (MW), crystallinity, and degree of N-acetylation (DA). For instance, biodegradability is directly proportional to the DA and inversely proportional to MW; whereas, biocompatibility and antimicrobial activity are inversely proportional to DA. A review paper presented by Kumirska *et al.* [49] highlights the relationship between structural and other key factors of the chitin/chitosan-based materials and their biomedical activity.

Chitin is an eco-friendly and nontoxic or allergic biopolymer.

The biodegradability of chitin is mainly due to its susceptibility to enzymatic hydrolysis by lysozyme, an enzyme present in all tissues of the human body. This enzyme is present in many tissues and secretions namely tears, saliva, blood, and milk, and is released and utilized by phagocytic cells during an inflammatory response to a foreign implant. The degree of acetylation, the MW, and the pH affect biodegradation [52, 53]. Furthermore, chitin has no antigenic properties and, thus, is perfectly compatible not only with animal tissues but also with plant tissues.

Chitin and chitosan are well-established antimicrobial agents against bacteria, fungi, and viruses. The precise mechanisms of the antimicrobial action of chitin and chitosan are yet to be elucidated, but there is a consensus that the interaction of the cationic amino groups of chitosan and residual groups in chitin probably bind to anionic groups of these microorganisms, resulting in growth inhibition [54].

Chitin nanoforms were found to promote cell attachment and spreading of normal human keratinocytes and fibroblasts and also with better antifungal properties compared to chitin microfibers. This may be a consequence of the high surface area available for cell attachment due to their three-dimensional features and high surface area to volume ratios, which are favorable parameters for cell attachment, growth, and proliferation; the higher surface area encourages better accessibility of the residual amino groups that influence the antimicrobial activity.

6.4 Functional Surface Modification

As discussed before, cellulose and chitin nanoforms have diverse biomedical applications due to their intrinsic properties; nonetheless, their modification before the development of functional materials, for instance, introducing active molecules in their chains *via* chemical functionalization, is also important, as this will determine its potential biological applications. Cellulose and chitin are quite interesting because of the presence of two hydroxyl functionalities (primary C-6 and secondary C-3 hydroxyl groups) along their chains; and also residual-free —NH_2 groups in the case of chitin (chitin nanoform surfaces possess amino groups due to acid hydrolysis-induced deacetylation), which provide the possibility of modification through chemical reaction. The chemical modification of these groups gives rise to various new functional materials described in the next sections.

6.4.1 For Biomedical Application

6.4.1.1 To Improve Nanocellulose's Biodegradability

As already mentioned in "The unique biological properties of nanocellulose and nanochitin" section (6.3), cellulose is not enzymatically degradable *in vivo*, which is a limitation in potential applications. Based on the data showing the improvements of the biodegradability of cellulose through periodate oxidation, several studies have been done *in vitro* [55–57] and *in vivo* [57] conditions.

Li *et al.* [55] modified BC using periodate oxidation to give rise to a biodegradable 2,3-dialdehyde BC. The authors demonstrated that the modified BC nano-network was able to degrade into porous scaffold with micro-sized pores in different conditions: water, phosphate buffered saline (PBS), and simulated body fluid (SBF). The degradation process began from the oxidized amorphous part of the network and, simultaneously, hydroxyapatite was formed on the scaffold surface during the process in SBF.

Recently, Czaja *et al.* [57] demonstrated that controlled oxidation of BC membranes pre-irradiated with γ-radiation results in a resorbable membrane that can be rapidly rehydrated in aqueous fluids. The *in vitro* results showed that the degradation of these membranes occurs in two phases: (i) initial rapid degradation of about 70–80% of the entire sample, followed by (ii) slower degradation of an additional 5–10% which eventually levels off leaving a small amount of nonresorbable material. The *in vivo* degradation of the prototype materials in male New Zealand White rabbits, showed clear degradation at all-time points, with the most rapid degradation occurring in the first 2–4 weeks.

6.4.1.2 To Expand Nanocellulose's Biocompatibility

As the surface properties of the biomaterials, namely, surface charge, chemistry, topography, and the presence of hydrophobic and hydrophilic domains have a vital rule in the cell–biomaterial interactions. Consequently, several studies have been developed aiming to enhance nanocellulose's biocompatibility.

The effect of the surface charge of the CNCs on the cellular uptake and cytotoxicity was evaluated by Mahmoud *et al.* [58] in two synthesized fluorescent CNCs: cellulose nanocrystal-fluorescein isothiocyanate (CNC-FITC) and newly synthesized cellulose nanocrystal-rhodamine B isothiocyanate (CNC-RBITC) (Figure 6.3). The *in vitro* cellular uptake studies indicated that the positively charged CNC-RBITC was taken up by human embryonic kidney 293 and *Spodoptera frugiperda* cells without any noticeable cytotoxic effect (Figure 6.3). Regarding the negatively charged CNC–FITC, no significant internalization was observed at physiological pH [58].

CNCs were also used for the targeted delivery of chemotherapeutic agents to cancer cells after grafting in their surface with folic acid (FA) [44]. The

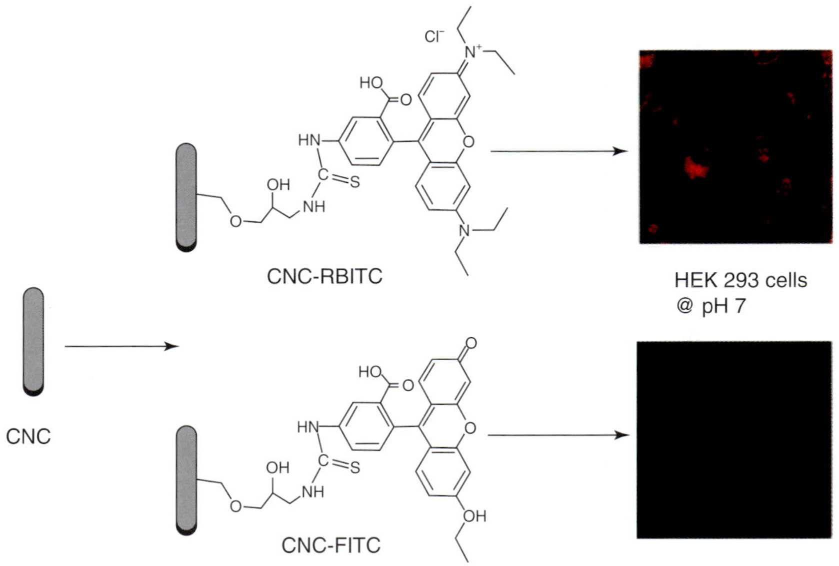

Figure 6.3 Schematic illustration of the CNC-RBITC and CNC–FITC conjugates; and confocal images of the human embryonic kidney (HEK) 293 cells treated with the respective conjugate. (Mahmoud 2010 [58]. Reproduced with permission of American Chemical Society.)

synthesis of FA-grafted CNCs conjugates was made through the following route: first, CNCs were labeled with FITC and were then conjugated with FA. FITC was necessary for the detection of the CNCs in *in vitro* cell-uptake studies. *In vitro* studies showed that the cellular binding/uptake of the fluorescein isothiocyanate–cellulose nanocrystal–folic acid (FITC–CNC–FA) by the folate receptor-mediator, which was overexpressed by human and rat brain tumor cells, was significantly higher than that of the free FA. For instance, cellular binding/uptake of the conjugate by DBTRG-05MG, H4, and C6 cells was 1452, 975, and 46 times higher, respectively, than that of nontargeted CNCs [44]. These studies showed that simple and different surface functionalizations together with observed noncytotoxicity turn modified CNC as a promising material for different biomedical applications like bioimaging and drug delivery systems.

In the past years, we have seen increasing interest in the surface functionalization of BC to advance its biocompatibility and hemocompatibility. Plasma surface modification is an effective surface treatment technique, drawing great interest in biomedical applications. Pertile *et al.* [59] modified BC by nitrogen-containing plasma to improve the cell affinity of the membranes. They found that the nitrogen plasma-treated BC showed an increase in the concentration of functional groups on the BC surface and an improvement in cell adhesion and proliferation of the endothelial and neuroblast cells. The authors also observed that the plasma treatment improved the porosity of the material significantly. This is a significant observation, since the relatively low porosity of BC is the main drawback in the development of tissue engineering applications, because it is the reason for poor cell penetration [59]. To improve blood compatibility, Wang *et al.* [60] grafted zwitterionic carboxybetaine brushes from cellulose membrane via Activator Regenerated by Electron Transfer atom transfer radical polymerization (ARGET-ATRP). The platelet adhesion, hemolytic test, and plasma protein adsorption data demonstrated that cellulose membrane had excellent blood compatibility performed on lower platelet adhesion and protein adsorption without causing hemolysis [60]. Improvement of biocompatibility and hemocompatibility were recently reported regarding the following BC-based materials: peptide (Arg-Gly-Asp)-modified BC membranes [61, 62], BC/polypyrrole [63], and BC/polyvinyl alcohol biocomposites [64]. In another study, the effect of dialdehyde bacterial nanocellulose (DBC) on the cell adhesion and proliferation was also investigated [65]. The DBC were prepared from the selective-oxidation of BC membranes with sodium periodate. The authors found that DBC can mimic extracellular matrix (ECM) 3D structure and could support the epidermal cell adhesion proliferation and proliferation [65].

Alexandrescu *et al.* compared the cytotoxicity on fibroblast cells of pure CNF (see Section 6.3.1.3) and surface modified-CNF with cross-linking agent polyethyleneimine (PEI) and surfactant cetyl trimethylammonium bromide (CTAB). In comparison with pure CNF (no acute toxic phenomen), both modified-CNF samples showed negative effects on biocompatibility. The samples modified with PEI and CTAB caused a significant reduction in cell viability and proliferation [41].

Recently, anionic and cationic microfibrillated cellulose (MFC) samples were prepared via carboxymethylation and glycidyltrimethylammonium chloride (EPTMAC) condensation, respectively to improve the biocompatibility of cellulose nanoforms cytocompatibility [66]. Human dermal fibroblasts were used in the indirect contact cytotoxicity test (the cells are exposed to culture medium extracts of the materials) and in direct contact assay (cells adhesion and viability). The studied materials were in the form of films of unmodified, anionic, and cationic MFC. In the indirect cytotoxicity test no toxicity was detected, showing that the MFC materials are noncytotoxic, independently of the chemical treatment that they have been subjected to. However, the direct contact assays showed that cationic modified-MFC (trimethylammonium-MFC) presents a better cytocompatibility than does unmodified and anionic modified-MFC (carboxymethylated-MFC) [66].

6.4.1.3 To Expand Nanochitin Applications

Surface modification is an effective way to change the surface property and to provide chitin nanoforms surface with other properties. Several modifications to the chitin nanoforms surface have been achieved, including deacetylation, naphthaloylation, chlorination, 2,2,6,6-tetramethylpiperidine-1′-oxy radical (TEMPO)-mediated oxidation, and graft polymerization.

β-Chitin grafted poly(acrylic acid) (PAA) was prepared to obtain hydrogel suitable for wound dressing application [67]. The acrylic acid was first linked to chitin, via ester bonds between the chitin primary alcohol groups and the carboxyl groups of acrylic acid, and then the active grafted moiety was further polymerized upon addition of an initiator to form a network. The cytocompatibility of the optimized chitin-PAA film was tested with L929 mouse

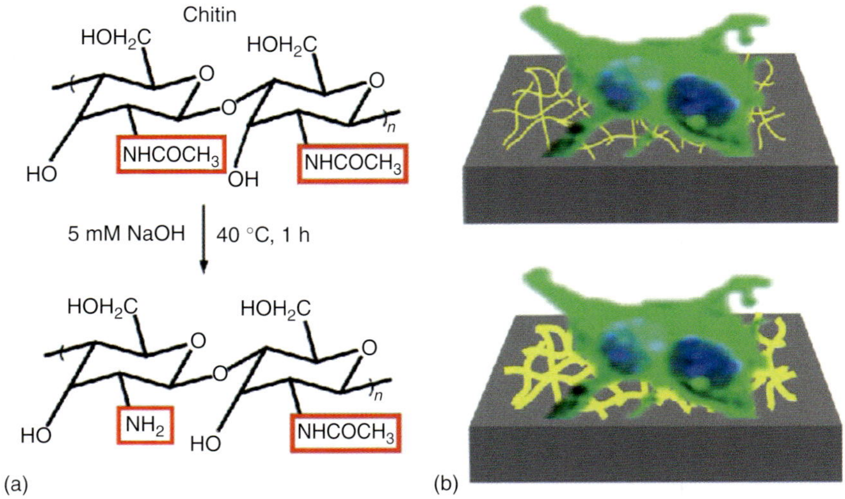

Figure 6.4 (a) Chitin structure and conditions to produce deacetylated chitin (chitosan) and (b) schematics of neurons plated on partially deacetylated chitin. (Cooper 2012 [68]. Reproduced with permission of Royal Society of Chemistry.)

fibroblasts that proliferated and adhered well onto the film. The morphology of the cells on the chitin-PAA film surface was found to be normal after 14 days of culture [67].

Self-assembled chitin nanofibers were applied for the attachment of primary neurons *in vitro* [68]. To develop so, chitin nanofiber surfaces were deacetylated to form 4 to 12 nm diameter chitosan nanofibers that were coupled with poly-D-lysine (PDL) to examine combinatory effects and were structurally analyzed by atomic force microscopy (Figure 6.4). The chitosan substrates were then employed for mouse cortical neuron cultures to examine their capabilities to support cell attachment, neurite coverage, and survival. The 4 nm chitosan nanofibers with PDL supported around 38% neuron viability compared to only 13.5% on traditional PDL surfaces, illustrating significantly improved long-term cell viability. The nanofibrillar chitosan surface could provide an alternative substrate for *in vitro* primary neuron cultures to serve as artificial neural networks for diagnostics and therapeutics.

6.4.2 For Antimicrobial Applications

6.4.2.1 Introduction of Antimicrobial Activity to Cellulose Nanoforms

Contrary to chitin, cellulose does not intrinsically present antimicrobial (i.e., antibacterial and/or antifungal) properties. However, the development of new functional cellulose-based materials (including cellulose nanoforms) with antimicrobial activity is of great interest due to their potential in biological applications. Thus, two main kinds of functionalization have been applied (using inorganic and/or organic agents): (i) chemical grafting of bioactive agents on the surface of the cellulose nanoforms and (ii) incorporation (physical approach) of antibacterial/antimicrobial agents in the cellulosic-based template.

Photobactericidal porphyrin-CNCs materials were prepared by covalently grafting CNC with cationic porphyrin groups, showing excellent efficiency of photodynamic inactivation toward bacteria [69, 70]. The materials showed excellent efficacy toward the photodynamic inactivation of *Acinetobacter baumannii*, multidrug-resistant *A. baumannii*, and methicillin-resistant *Staphylococcus aureus*.

CNC were also conjugated with allicin and lysozyme by a carbodiimide cross-linker to evaluate their antimicrobial properties [71]. Their activity was evaluated by the microdilution method and the results showed that CNC had few antimicrobial activities, but allicin-conjugated nanocellulose and lysozyme-conjugated nanocellulose had good antifungal and antibacterial effects against standard strains of *Candida albicans*, *Aspergillus niger*, *S. aureus*, and *Escherichia coli* [71].

Andresen *et al.* [72] have prepared antimicrobial films based on surface-modified MFC. MFC, was grafted with the quaternary ammonium compound octadecyldimethyl(3-trimethoxysilylpropyl)ammonium chloride (ODDMAC) by a simple adsorption-curing process. Films prepared from the ODDMAC-modified MFC showed substantial antibacterial activity against the gram-positive bacterium *S. aureus* and the gram-negative bacteria *E. coli* and *Pseudomonas aeruginosa* even at very low concentrations of antimicrobial agent immobilized on the surface [72].

Inspired by the intrinsic bactericidal activity of chitosan, which is associated with the presence of free amino groups along the polymeric chain, Fernandes *et al.* [73] have chemically grafted aminoalkyl groups at the surface of BC membranes (Figure 6.5). The aminoalkyl-grafted bacterial nanocellulose (BC-NH_2) membranes were prepared in three steps: (i) hydrolysis of the silane derivative (3-aminopropyltrimethoxysilane, APS), (ii) adsorption of the hydrolyzed species onto BC nanofibrils, and (iii) chemical condensation reaction. Remarkably, the BC-NH_2 membranes showed simultaneously lethal against gram-positive bacterium, *S. aureus* and gram-negative bacterium, *E. coli* and nontoxic to human adipose-derived mesenchymal stem cells [73] (Figure 6.5).

In a different approach, inorganic and/or organic nanoparticles have been incorporated into cellulose nanoforms based-matrices to add antimicrobial activity to those materials. For instance, silver nanoparticles have been used as agents with antibacterial activity to coat cellulosic materials for different applications. Martins *et al.* [74] reported that composites composed of CNF and Ag nanoparticles can be fabricated by electrostatic assembly approach via polyelectrolytes as macromolecular linkers between CNF and Ag nanoparticles. The CNF/Ag nanocomposites were first investigated for their antibacterial properties toward *S. aureus* and *Klebsiella pneumoniae* microorganisms as compared to CNF modified by polyelectrolytes linkers without Ag.

Regarding the organic nanoparticles, Butchosa *et al.* [75] reported the use of partially deacetylated chitin nanocrystals in BC materials (by either *in situ* biosynthesis or post-modification) to develop all polysaccharide antimicrobial composites. It was found that this "green" composite showed strong antibacterial activity against *E. coli* and that this activity was strongly dependent on the deacetilated chitin nanocrystals content.

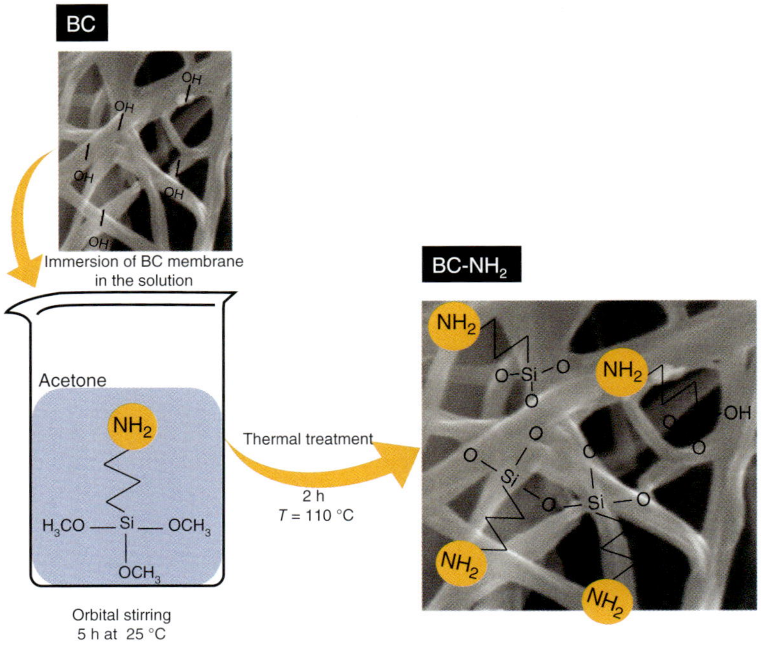

Figure 6.5 Chemical grafting process to develop silane-bacterial cellulose membranes with antimicrobial properties using 3-aminopropyltrimethoxysilane as active agent. (Fernandes 2013 [73]. Reproduced with permission of American Chemical Society.)

6.4.2.2 Expansion of Antimicrobial Activity of Chitin Nanoforms

N-Halamines present efficacy against microorganisms, stability, and nontoxicity to humans [76, 77]. *N*-Halamine chitin nanofibers film was prepared by the reaction of chitin nanofibers film with sodium hypochlorite solution to provide the film with antibacterial and antifungal activities. The N—H bond was substituted to the N—Cl bond at the chitin nanofibers surface. The chlorinated chitin nanofibers film showed strong efficacies against gram-negative and gram-positive bacteria of *E. coli* and *S. aureus*, respectively. Furthermore, the films showed 100% and 80% inhibition of spore germination when faced against fungi: *Alternaria alternata* and *Penicillium digitatum*, respectively.

6.5 Summary and Final Remarks

As renewable, biodegradable, biocompatible, and low-cytotoxic materials, cellulose and chitin nanoforms are attracting and will continue to draw attention due to these unique biological properties. They also present easy surface modification and functionalization, which extend their potential applications in different fields, in particular, biotechnological and biomedical. Different chemical modifications or grafting approaches have been developed to improve the biological properties of these biopolymers. Nonetheless, recent studies have mainly focused on the surface modification of these nanoforms to increase their compatibility with other (bio)polymers for the preparation of nanocomposites application. Less attention has been paid to other application areas. For future investigations, more attention should be focused on developing novel surface modifications/applications of cellulose and chitin nanoforms.

References

1 Gandini, A. (2008) Polymers from renewable resources: a challenge for the future of macromolecular materials. *Macromolecules*, **41**(24), 9491–9504.

2 Fernandes, S.C.M., Freire, C.S.R., Silvestre, A.J.D., Gandini, A., and Neto, C.P. (2011) Novel materials based in chitosan. *Polym. Int.*, **60**, 875–882.

3 Salaberria, A., Labidi, J., and Fernandes, S.C.M. (2015) Different routes to turn chitin into stunning nano-objects. *Eur. Polym. J.*, **68**, 503–515.

4 Wood, L. (2015) Chitin and Chitosan – A Global Strategic Business Report, Global Industry Analysts, Inc.

5 Wood, L. (2015) Collagen Peptide and Gelatin Market- Global Industry Analysis Trends, Size, Share, Growth, Trends & Forecast, 2014–2020.

6 John, J. (2016) Hyaluronic Acid Market poised to surge from USD 6.50 billion in 2014 to USD 10.9 billion, Globally by 2020- Marketresearchstore.com.

7 Freire, C.S.R., Fernandes, S.C.M., Silvestre, A.J., and Neto, C.P. (2013) Novel cellulose-based composites based on nanofibrillated plant and bacterial cellulose: recent advances at the University of Aveiro – A review. *Holzforschung*, **67**, 603–612.

8 Festucci-Buselli, R.A., Otoni, W.C., and Joshi, P.C. (2007) Structure, organization, and functions of cellulose synthase complexes in higher plants. *J. Plant Physiol.*, **19**, 1–13.

9 Klemm, D., Heinze, T., and Wagenknecht, W. (1998) *Comprehensive Cellulose Chemistry*, Wiley-VCH Verlag GmbH, Weinheim.

10 Preston, R.D. (1974) *The Physical Biology of the Plant Cell Wall*, Chapman & Hall, London.

11 Salamone, J.C. (ed.) (1996) *Polymeric Materials Encyclopedia*, vol. **2 (C)**, CRC Press, Florida, p. 1027.

12 Klemm, D., Heublein, B., Fink, H.P., and Bohn, A. (2005) Cellulose: fascinating biopolymer and sustainable raw material. *Angew. Chem. Int. Ed.*, **44**, 3358–3393.

13 Iguchi, M., Yamanaka, S., and Budhiono, A. (2000) Bacterial cellulose – a masterpiece of nature's arts. *J. Mater. Sci.*, **35**, 261–270.

14 Klemm, D., Schumann, D., Udhardt, U., and Marsch, S. (2001) Bacterial synthesized cellulose – artificial blood vessels for microsurgery. *Prog. Polym. Sci.*, **26**, 1561–1603.

15 Mormino, R. and Bungay, H. (2003) Composites of bacterial cellulose and paper made with a rotating disk bioreactor. *Appl. Microbiol. Biotechnol.*, **62**, 503–506.

16 Fernandes, S.C.M., Oliveira, A.L., Freire, C.S.R., Silvestre, A.J.D., Neto, C.P., Gandini, A., and Desbriéres, J. (2009) Novel transparent nanocomposite films based on chitosan and bacterial cellulose. *Green Chem.*, **11**, 2023–2029.

17 Trovatti, E., Fernandes, S.C.M., Rubatat, L., Freire, C.S.R., Silvestre, A.J.D., and Neto, C.P. (2012) Sustainable nanocomposite films prepared entirely from microbial biopolymers. *Cellulose*, **19**, 729–737.
18 Figueiredo, A.G.P.R., Figueiredo, A., Alonso-Varona, A., Fernandes, S.C.M., Palomares, T., Rubio, E., Barros-Timmons, A., Silvestre, A.J.D., Neto, C.P., and Freire, C.S.R. (2013) Biocompatible bacterial cellulose-poly(2-hydroxyethyl methacrylate) nanocomposite films. *Biomed Res. Int.*, **2013**, 1–14, article ID 698141.
19 Gutierrez, J., Fernandes, S.C.M., Mondragon, I., and Tercjak, A. (2013) Multifunctional hybrid nanopapers based on bacterial cellulose and sol-gel synthesized titanium/vanadium oxide nanoparticles. *Cellulose*, **20**, 1301–1311.
20 Gutierrez, J., Fernandes, S.C.M., Mondragon, I., and Tercjak, A. (2012) Conductive photoswitchable green vanadium oxide nanopaper based on bacterial cellulose. *ChemSusChem*, **5**, 2323–2327.
21 Algar, I., Fernandes, S.C.M., Mondragon, G., Castro, C., Gabilondo, N., Retegi, A., and Eceiza, A. (2015) Pineapple agroindustrial residuos for the production of high valeu bacterial cellulose with different morphologies. *J. Appl. Polym. Sci.*, **132**, 41237–41241.
22 Raabe, D., Sachs, C., and Romano, P. (2005) The crustacean exoskeleton as an example of a structurally and mechanically graded biological nanocomposite material. *Acta Mater.*, **53**, 4281–4292.
23 Revol, J.-F. and Marchessault, R.H. (1993) *In vitro* chiral nematic ordering of chitin crystallites. *Int. J. Biol. Macromol.*, **15**, 329–335.
24 Paillet, M. and Dufresne, A. (2001) Chitin whisker reinforced thermoplastic nanocomposites. *Macromolecules*, **34**, 6527–6530.
25 Ifuku, S., Nogi, M., Abe, K., Yoshioka, M., Morimoto, M., Saimoto, H. *et al.* (2011) Simple preparation method of chitin nanofibers with a uniform width of 10–20 nm from prawn shell under neutral conditions. *Carbohydr. Polym.*, **84**, 762–764.
26 Salaberria, A.M., Fernandes, S.C.M., Diaz, R.H., and Labidi, J. (2015) Processing of α-chitin nanofibers by dynamic high pressure homogenization: characterization and antifungal activity against A. niger. *Carbohydr. Polym.*, **116**, 286–291.
27 Saralegi, A., Fernandes, S.C.M., Alonso-Varona, A., Palomares, T., Foster, E.J., Weder, C., Eceiza, A., and Corcuera, M. (2013) Shape-memory bionanocomposites based on chitin nanocrystals and thermoplastic polyurethane with highly crystalline soft segment. *Biomacromolecules*, **14**, 4475–4482.
28 Salaberria, A.M., Labidi, J., and Fernandes, S.C.M. (2014) Chitin nanocrystals and nanofibers as nano-sized fillers into thermoplastic starch-based biocomposites processed by melt-mixing. *Chem. Eng. J.*, **256**, 356–364.
29 Salaberria, A.M., Diaz, R.H., Labidi, J., and Fernandes, S.C.M. (2014) Preparing valuable renewable nanocomposite films based exclusively on oceanic biomass – chitin nanofillers and chitosan. *React. Funct. Polym.*, **89**, 31–39.
30 Herrera, N., Roch, H., Salaberria, A.M., Pino, M.A., Labidi, J., Fernandes, S.C.M., Wodke, T., Radic, D., Leiva, A., and Oksman, K. (2016) Functionalized blown films of plasticized polylactic/chitin nanocomposite: preparation and characterization. *Mater. Des.*, **92**, 846–852.

31 Robles, E., Salaberria, A.M., Diaz, R.H., Fernandes, S.C.M., and Labidi, J. (2016) Self-bonded composite films based on cellulose nanofibers and chitin nanocrystals as antifungal materials. *Carbohydr. Polym.*, **25**, 41–49.
32 Lin, N. and Dufresne, A. (2014) Nanocellulose in biomedicine: current status and future prospect. *Eur. Polym. J.*, **59**, 302–325.
33 Jorfi, M. and Foster, E.J. (2015) Recent advances in nanocellulose for biomedical applications. *J. Appl. Polym. Sci.*, 41719–41738.
34 Garrett, R.H. and Grisham, C.H. (2005) *Biochemistry*, 3rd edn, Garrett & Grisham eds. Brooks/Cole, Belmont, CA, pp. 223–224.
35 Czaja, W.K., Young, D.J., Kawecki, M., and Browm, R.M. (2007) The future prospects of microbial cellulose in biomedical applications. *Biomacromolecules*, **8**, 1–12.
36 Miyamoto, T., Takahashi, S., Ito, H., Inagaki, H., and Noishiki, Y. (1989) Tissue biocompatibility of cellulose and its derivatives. *J. Biomed. Mater. Res.*, **23**, 125–133.
37 Kümmerer, K., Menz, J., Schubert, T., and Thielemans, W. (2011) Biodegradability of organic nanoparticles in the aqueous environment. *Chemosphere*, **82**, 1387–1392.
38 Martson, M., Viljanto, J., Hurme, T., and Saukko, P. (1998) Biocompatibility of cellulose sponge with bone. *Eur. Surg. Res.*, **30**, 426–432.
39 Helenius, G., Bäckdahl, H., Bodin, A., Nannmark, U., Gatenholm, P., and Risberg, B. (2006) In vivo biocompatibility of bacterial cellulose. *J. Biomed. Mater. Res. Part A*, **76A**, 431–438.
40 Jia, B., Li, Y., Yang, B., Xiao, D., Zhang, S., Rajulu, A.V., Kondo, T., Zhang, L., and Zhou, J. (2013) Effect of microcrystal cellulose and cellulose whisker on biocompatibility of cellulose-based electrospun scaffolds. *Cellulose*, **20**, 1911–1923.
41 Alexandrescu, L., Syverud, K., Gatti, A., and Chinga-Carrasco, G. (2013) Cytotoxicity tests of cellulose nanofibril-based structures. *Cellulose*, **20**, 1765–1775.
42 Vartiainen, J., Pöhler, T., Sirola, K., Pylkkänen, L., Alenius, H., Hokkinen, J. *et al.* (2011) Health and environmental safety aspects of friction grinding and spray drying of microfibrillated cellulose. *Cellulose*, **18**, 775–786.
43 Pereira, M.M., Raposo, N.R.B., Brayner, R., Teixeira, E.M., Oliveira, V., Quintão, C.C.R. *et al.* (2013) Cytotoxicity and expression of genes involved in the cellular stress response and apoptosis in mammalian fibroblast exposed to cotton cellulose nanofibers. *Nanotechnology*, **24**, 075103 [p. 8].
44 Dong, S., Hirani, A.A., Colacino, K.R., Lee, Y.W., and Roman, M. (2012) Cytotoxicity and cellular uptake of cellulose nanocrystals. *Nano Life*, **02**, 1241006–1241017.
45 Martin, J.D., Clift, E., Foster, J., Vanhecke, D., Studer, D., Wick, P. *et al.* (2011) Investigating the interaction of cellulose nanofibers derived from cotton with a sophisticated 3D human lung cell co-culture. *Biomacromolecules*, **12**, 3666–3673.
46 Chen, Y.M., Xi, T., Zheng, Y., Guo, T., Hou, J., Wan, Y. *et al.* (2009) *In vitro* cytotoxicity of bacterial cellulose scaffold for tissue engineered bone. *J. Bioact. Compat. Polym.*, **24**, 137–145.

47 Jeong, S.I., Lee, S.E., Yang, H., Jin, Y.H., Park, C.S., and Park, Y.S. (2010) Toxicologic evaluation of bacterial synthesized cellulose in endothelial cells and animals. *Mol. Cell. Toxicol.*, **6**, 373–380.

48 Kim, G.-D., Yang, H., Park, H.R., Park, C.-S., Park, Y.S., and Lee, S.E. (2013) Evaluation of immunoreactivity of *in vitro* and in vivo models against bacterial synthesized cellulose to be used as a prosthetic biomaterial. *Biochip J.*, **7**, 201–209.

49 Kumirska, J., Weinhold, M.X., Thöming, J., and Stepnowski, P. (2011) Biomedical activity of chitin/chitosan based materials – influence of physicochemical properties apart from molecular weight and degree of N-acetylation. *Polymers*, **3**, 1875–1901.

50 Azuma, K., Izumi, R., Osaki, T., Ifuku, S., Morimoto, M., Saimoto, H., Minami, S., and Okamoto, Y. (2015) Chitin, chitosan, and its derivatives for wound healing: old and new materials. *J. Funct. Biomater.*, **6**, 104–142.

51 Rinando, M. (2006) Chitin and chitosan: properties and application. *Prog. Polym. Sci.*, **31**, 603–632.

52 Pantaleone, D., Yalpani, M., and Scollar, M. (1992) Unusual susceptibility of chitosan to enzymic hydrolysis. *Carbohydr. Res.*, **237**, 325–332.

53 Varum, K.M., Myhr, M.M., Hjerde, R.J.N., and Smidsrod, O. (1997) *In vitro* degradation rates of partially N-acetylated chitosans in human serum. *Carbohydr. Res.*, **299**, 99–101.

54 Coma, V., Martial-Gros, A., Garreau, S., Copinet, A., Salin, F., and Deschamps, A. (2002) Edible antimicrobial films based on chitosan matrix. *J. Food Sci.*, **67**, 1162–1169.

55 Li, J., Wan, Y.Z., Li, L.F., Liangm, H., and Wang, J.H. (2009) Preparation and characterization of 2,3-dialdehyde bacterial cellulose for potential biodegradable tissue engineering scaffolds. *Mater. Sci. Eng., C*, **29**, 1635–1642.

56 Luo, H., Xiong, G., Hu, D., Ren, K., Yao, F., Zhu, Y. *et al.* (2013) Characterization of TEMPO-oxidized bacterial cellulose scaffolds for tissue engineering applications. *Mater. Chem. Phys.*, **143**, 373–379.

57 Czaja, W., Kyryliouk, D., DePaula, C.A., and Buechter, D.D. (2014) Oxidation of cirradiated microbial cellulose results in bioresorbable, highly conformable biomaterial. *J. Appl. Polym. Sci.*, **131**, 39995 (1–12).

58 Mahmoud, K.A., Mena, J.A., Male, K.B., Hrapovic, S., Kamen, A., and Luong, J.H.T. (2010) Effect of surface charge on the cellular uptake and cytotoxicity of fluorescent labeled cellulose nanocrystals. *ACS Appl. Mater. Interfaces*, **2**, 2924–2932.

59 Pertile, R.A.N., Andrade, F.K., Alves, C. Jr.,, and Gama, M. (2010) Surface modification of bacterial cellulose by nitrogen-containing plasma for improved interaction with cells. *Carbohydr. Polym.*, **82**, 692–698.

60 Wang, M., Yuan, J., Huang, X., Cai, X., Li, L., and Shen, J. (2013) Grafting of carboxybetaine brush onto cellulose membranes via surface-initiated ARGET-ATRP for improving blood compatibility. *Colloids Surf., B*, **103**, 52–58.

61 Andrade, F.K., Alexandre, N., Amorim, I., Gartner, F., Maurício, A.C., Luís, A.L. *et al.* (2012) Studies on the biocompatibility of bacterial cellulose. *J. Bioact. Compat. Polym.*, **28**, 97–112.

62 Andrade, F.K., Silva, P., Carvalho, M., Castanheira, E.M.S., Soares, R., and Gama, M. (2011) Studies on the hemocompatibility of bacterial cellulose. *J. Biomed. Mater. Res. Part A*, **98**, 554–566.

63 Ferraz, N., Carlsson, D.O., Hong, J., Larsson, R., Fellström, B., Nyholm, L. *et al.* (2012) Haemocompatibility and ion exchange capability of nanocellulose polypyrrole membranes intended for blood purification. *J. R. Soc. Interface*, **9**, 1943–1955.

64 Leitao, A.F., Gupta, S., Silva, J.P., Reviakine, I., and Gama, M. (2013) Hemocompatibility study of a bacterial cellulose/polyvinyl alcohol nanocomposite. *Colloids Surf., B*, **111**, 493–502.

65 Wu, J., Zheng, Y., Yang, Z., Lin, Q., Qiao, K., Chen, X., and Peng, Y. (2014) Influence of dialdehyde bacterial cellulose with the nonlinear elasticity and topology structure of ECM on cell adhesion and proliferation. *RSC Adv.*, **4**, 3998–4009.

66 Hua, K., Carlsson, D.O., Ålander, E., Lindström, T., Strømme, M., Mihranyan, A. *et al.* (2014) Translational study between structure and biological response of nanocellulose from wood and green algae. *RSC Adv.*, **4**, 2892–2903.

67 Tanodekaew, S., Prasitsilp, M., Swasdison, S., Thavornyutikarn, B., Pothsree, T., and Pateepasen, R. (2004) Preparation of acrylic grafted chitin for wound dressing application. *Biomaterials*, **25**, 1453–1460.

68 Cooper, A., Zhong, C., Kinoshita, Y., Morrison, R.S., Rolandi, M., and Zhang, M. (2012) Self-assembled chitin nanofiber templates for artificial neural networks. *J. Mater. Chem.*, **22**, 3105–3109.

69 Feese, E., Sadeghifar, H., Gracz, H.S., Argyropoulos, D.S., and Ghiladi, R.A. (2011) Photobactericidal porphyrin-cellulose nanocrystals: synthesis, characterization, and antimicrobial properties. *Biomacromolecules*, **12**, 3528–3539.

70 Carpenter, B.L., Feese, E., Sadeghifar, H., Argyropoulos, D.S., and Ghiladi, R.A. (2012) Porphyrin-cellulose nanocrystals: a photobactericidal material that exhibits broad spectrum antimicrobial activity. *J. Photochem. Photobiol.*, **88**, 527–536.

71 Jebali, A., Hekmatimoghaddam, S., Behzadi, A., Rezapor, I., Mohammadi, B.H., Jasemizad, T. *et al.* (2013) Antimicrobial activity of nanocellulose conjugated with allicin and lysozyme. *Cellulose*, **20**, 2897–2907.

72 Andresen, M., Stenstad, P., Møretrø, T., Langsrud, S., Syverud, K., Johansson, L.-S. *et al.* (2007) Nonleaching antimicrobial films prepared from surface-modified microfibrillated cellulose. *Biomacromolecules*, **8**, 2149–2155.

73 Fernandes, S.C.M., Sadocco, P., Aonso-Varona, A., Palomares, T., Eceiza, A., Silvestre, A.J.D., Mondragon, I., and Freire, C.S.R. (2013) Bioinspired antimicrobial and biocompatible bacterial cellulose membranes obtained by surface functionalization with aminoalkyl groups. *ACS Appl. Mater. Interfaces*, **5**, 3290.

74 Martins, N.C.T., Freire, C.S.R., Pinto, R.J.B., Fernandes, S.C.M., Neto, C.P., Silvestre, A.J.D. *et al.* (2012) Electrostatic assembly of Ag nanoparticles onto nanofibrillated cellulose for antibacterial paper products. *Cellulose*, **19**, 1425–1436.

75 Butchosa, N., Brown, C., Larsson, P.T., Berglund, L.A., Bulone, V., and Zhou, Q. (2013) Nanocomposites of bacterial cellulose nanofibers and chitin

nanocrystals: fabrication, characterization and bactericidal activity. *Green Chem.*, **15**, 3404–3413.

76 Kocer, H.B., Cerkez, I., Worley, S.D., Broughton, R.M., and Huang, T.S. (2011) Poly-meric antimicrobial *N*-halamine epoxides. *ACS Appl. Mater. Interfaces*, **3**, 2845–2850.

77 Worley, S.D., Li, F., Wu, R., Kim, J., Wei, C., Williams, J.F., Owens, J.R., Wander, J.D., Bargmeyer, A.M., and Shirtiliff, M.E. (2003) A novel N-halamine monomer for preparing biocidal polyurethane coatings. *Surf. Coat. Int. B*, **86**, 273–277.

7

Enzyme-Based Biosensors in Food Industry via Surface Modifications

Nilay Gazel and Huseyin B. Yildiz

7.1 Introduction

Enzyme technology has received great attention with improvements in biological remediation techniques. Enzyme-based biosensors can qualify as a measuring system as they permit the detection of biological events by detecting biochemical changes in biological processes. Advantages of biosensors are identification of even small quantities, cost effectiveness, and requirement of less power and less volume for identifying a wide linear range for usage [1]. Moreover, biosensors have high sensitivity, selectivity, and good response time for different kinds of applications [2].

Biosensor system processing consists of three main parts: analyte detection, signal conversion, and signal reading. Signal conversion can be carried out by optical, thermal, electrical, or electronical components. In conductive polymer-solution part of the system, there will be an electrochemical conversion that occurs at the exact moment of the biochemical reaction and load velocities get higher than the species mass transport speed toward the electrode. This diffusion-controlled electrocatalytic procedure is used for amperometric studies [3]. Biological information is turned into electrical signals in biosensor applications. Biosensors have two main parts: bioreceptors and biorecognition pieces. Bioreceptors can be either biological species such as enzymes or living biological systems that are used as biological mechanisms for recognizing [4].

One of the most important working fields of biosensors is the food industry [5]. Rising demand for high quality and safe food products as well as prevailing consumer trends are important challenges in this industry. In order to meet these requirements, strict auditions for processing and product development are carried out in the food sector [6]. Quality control analyses of manufacturing processes and monitoring the initial and final products are crucial steps in production. These steps have to be performed to prevent enzymatic inactivation, microbial contamination, lack of hygiene, and formation of chemical or biological toxins [7]. There are various different types of analytical methods employed for processing and product development in the food industry. However, most of them are time-consuming, expensive, and nonpractical methods. For this sector, quick, cheap, practical, and reliable methods and equipments

Surface Treatments for Biological, Chemical and Physical Applications, First Edition.
Edited by Mehmet Gürsoy and Mustafa Karaman.

are critical. Due to these reasons biosensors can be desired as alternative means to primitive techniques. To illustrate, levels of hypoxanthine and xanthine are important biomarkers as a sign of meat spoilage, which can be followed by xanthine-oxidase-based biosensors. Alcohol oxidase is used in the measurement of alcohol content in brewing, winemaking, and distilling industries. Glucose oxidase (GOx) is used for determination of glucose in fruit juices or in the fermentation of liquors. Lactose is the main disaccharide present in milk and dairy products and can be hydrolyzed by the enzyme galactosidase. Histamine, which is an important amine for determining fish freshness, can be analyzed by monoamine oxidase enzyme. Oxygen-based biosensors, as their content serve as good biomarkers for determining food quality. Moreover, enzyme-based biosensors can be used to determine food freshness, detection of glucose content in beverages, analysis of cholesterol in butter, food components of sugars, pathogenic organisms test, and so on. In this chapter, detail information will be shared about enzymes, immobilization of enzymes, main characteristics of enzyme-based biosensors, and their usage in food analysis.

7.2 Biosensors

"Sensors" are devices or systems containing control and processing electronics, software, and interconnection networks. Sensors respond to physical or chemical changes in the medium to which they are exposed by generating a product which is a measure of that quantity [8]. Physical sensors, chemical sensors, and biosensors are the types of sensors that are mainly used. Distance, mass, temperature, pH, and so on, can be measured by physical sensors [9]. Chemical sensors can provide responses in the form of analytical signals as a result of chemical reactions. Chemical sensors transform chemical information into an analytically useful and measurable signal [10] (Figure 7.1).

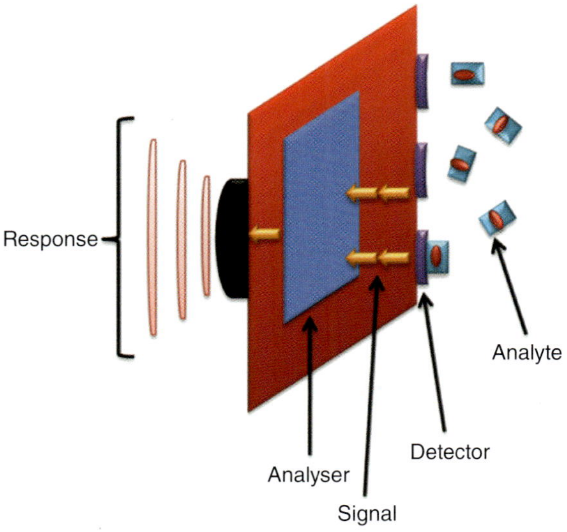

Figure 7.1 Schematic presentation of a sensor. (Kuila 2011 [11]. Reproduced with permission of Elsevier.)

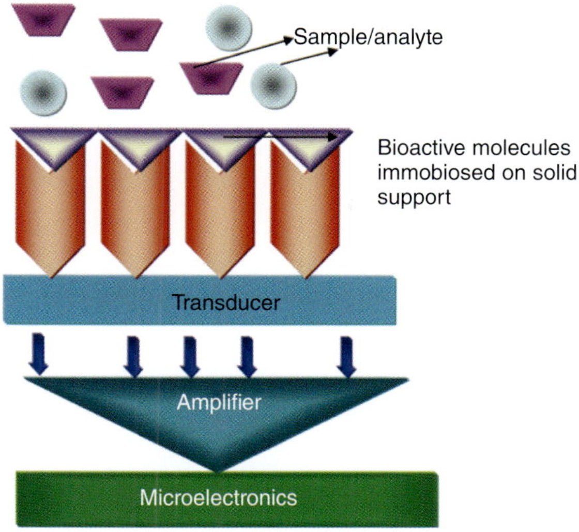

Figure 7.2 Schematic of a biosensor. (Dhawan 2009 [14]. Reproduced with permission of Elsevier.)

Biosensors are a type of chemical sensors that transform a biochemical reaction into an analytical signal. Sensors translate the information from the main domain into an output signal in order to offer the selective sensors for the analyte that can be measured. Biosensors are compact devices and analytical tools that are formed by the following two components:

- a recognition system consisting of a biological material that uses a biochemical mechanism
- a transducer from where the signal of the biosensor is obtained and changed into a meaningful signal [10, 12, 13].

According to the International Union of Pure and Applied Chemistry, a biosensor is exactly defined as a self-contained integrated device, capable of providing detailed quantitative or semiquantitative analytical information using a biological recognition element, which is retained in direct contact with a transduction element (Figure 7.2).

Biosensors can also be defined as analytical devices that combine biological materials such as cells, cell receptors, antibodies, membranes, organelles, enzymes, tissues, and microorganisms either merged within a transducer or closely in contact with a physicochemical transducer [15, 16]. Biosensors are devices that enable rapid analysis of mostly detection and monitoring of biological and chemical agents. Biosensor technology combines biological, chemical, and physical sciences together with engineering for a broad range of applications.

7.2.1 Historical Perspectives of Biosensors

The first "biosensor" was discovered by Clark in 1956 who is well-known for his invention of the Clark electrode that was later improved by Clark and Lyons in 1962 as an enzyme electrode. Updike and Hicks expanded the studies on the enzyme electrode, as they developed the first enzyme sensor in 1967 via trapping GOx in a layer of polyacrylamide gel attached to the plastic membrane of an oxygen electrode. Clark's ideas on the glucose analyzer were commercialized in 1975, which was the first of biosensor-based analyzer device to be built by companies around the world, based on the amperometric detection of hydrogen peroxide. In 1974, thermal biosensors were introduced, together with thermal enzyme probes and enzyme thermistors. Subsequently, Divi's study achieved alcohol detection via microbial electrodes with the help of bacteria. In 1975, carbon dioxide- or oxygen-based sensors were created by Lubbersand Opitz. In 1982, the *in vivo* application of glucose biosensors was presented by Shichiri *et al.* In the early 1970s, the piezoelectric or potentiometric transducer was designed. Liedberg *et al.* described the use of surface plasmon resonance to monitor affinity reactions in real time; starting from the early to the mid-1980s mainly made of modified electrodes. The first commercial lab-on-a-chip (LOC) system was announced in 1999 by Agilent, and various LOC products on glass chips exist in the literature. It is believed that the future of biosensors will depend on LOC systems due to their miniaturized biochemical analysis systems and their ability to gain miniaturized skills to biochemical analysis.

7.2.2 Parts of Biosensors: Bioreceptor and Transducer

A biosensor is composed of two parts, a bioreceptor and a transducer. Enzymes, antibodies, cell receptor, nucleic acid, tissue, and so on, can be used as a bioreceptor that selectively recognizes the analyte which is the target molecule for the receptor. The interaction between the analyte and bioreceptor produces a physical change close to the transducer surface and transducer and converts the biorecognition event into a signal by measuring the changes in the reaction. These two components, namely, the bioreceptor and transducer, are combined into one single sensor that measures the analyte under concern. The biosensor depends directly on a selective biorecognition where a conversion of biochemical information to electrical information occurs. The working principle of a biosensor depends on the diffusing of an analyte from a bulk solution to the bioreceptor immobilized sensor and on the selective and efficient reactivity with the bioreceptor. Subsequently, the biochemical reaction leads to a change depending on the transducer such as its optical, electronic properties, and so on. The alteration in the biosensor surface is converted into a signal, depending on the transducer model. Various transducers such as electrochemical, calorimetric, optical, piezoelectric are available in literature.

The transducer of a biosensor has the ability to convert a specific biological reaction (binding or catalytic) into a response that can be processed into a useable signal [17]. The working principle of the transducer is to transfer the signal from the output domain of the biorecognition system and convert a response to a signal [10]. According to the transducer type, biosensors can be classified as electrochemical (amperometric, impedimetric, potentiometric, conductimetric) optical, acoustic, thermal, and piezoelectric. Biosensing responses can be followed and converted into signals by a transducer through several ways such as oxygen consumption, hydrogen peroxide formation, changes in NADH concentration, fluorescence, absorption, pH change, conductivity, temperature, or mass [10, 17].

Electrochemical biosensors, first seen by Clark and Lyons, are the most used transducers employing amperometric glucose biosensing using the Clark oxygen electrode. Up to now, due to their high sensitivity, selectivity, inherent simplicity, ease of mass production, low cost, and availability of instrumentation, electrochemical biosensors have been used and improved now and are better in quality. In general, electrochemical transducers are used for designing a biosensor, which can have conducting, semiconducting, and ionic conducting material to coat the electrode surface combining with the biological receptor. When the electrochemical species is consumed or produced during the biological reaction, the electrochemical signal can be recorded using an electrochemical detector. The working principle of electrochemical biosensors is based on the electrochemistry of the substrate or product. Effective, simple to handle, low cost, and fast detection properties in electrochemical biosensor is related to the analytical power of the electrochemical techniques used as well as the process employed for a specified biological recognition element [13]. Amperometric, impedimetric, potentiometric, and conductimetric biosensors are the electrochemical techniques mainly used in biosensing of foods (Figure 7.3).

Amperometric electrochemical biosensors have attracted great attention in biosensor design since amperometric detection is a useful technique for substrate analysis because of its selectivity, sensitivity, rapid response, ease of construction, and reproducible performance. In amperometric biosensors, the change in the current on the working electrode upon applied constant potential is measured, which is the result of oxidation or reduction of the species in the sensing matrix due to a biological reaction. Chronoamperometry is the preferred technique in electrochemical detection systems. The main advantage of chronoamperometry is that it allows monitoring either oxygen consumption or hydrogen peroxide production. In amperometric measurements, there exists a current change that results from the reaction, which can be monitored as a function of time, which correlates with the amount of analytes in the reaction solution. The electron transfer between the analyte and the electrode surface occurs in instances where oxidation or reduction reactions are involved. Therefore, the transducer surface directly effects the functioning of amperometric biosensors. Generally, modified electrodes with nanoparticles, conducting polymer, and so on, are used in amperometric transducer design. Moreover, since constant potential is applied, maintaining the applied potential is very important to perform the process, which is correlated with the quality of the reference electrode. As a result of a biocatalytic reaction, the release of an electroactive product or consumption of reactant can easily be followed by amperometry in an inert working electrode. Due to their simple construction systems, high sensitivity, and rapid response amperometric biosensors are preferred [19]. Conductometric biosensors are based on the principle of the change of conductivity of the medium as a consequence of the biological element, designed by a two-electrode device. They are used to detect the changes in the conductance either due to consumption or to generation of charges in the species involved in the enzymatic reactions. Due to their nonspecificity and poor signal to noise ratio, conductometric biosensors have not been much used [15, 20]. The potentiometric biosensors are based on ion-selective electrodes and ion-selective field effect transistors, designed to

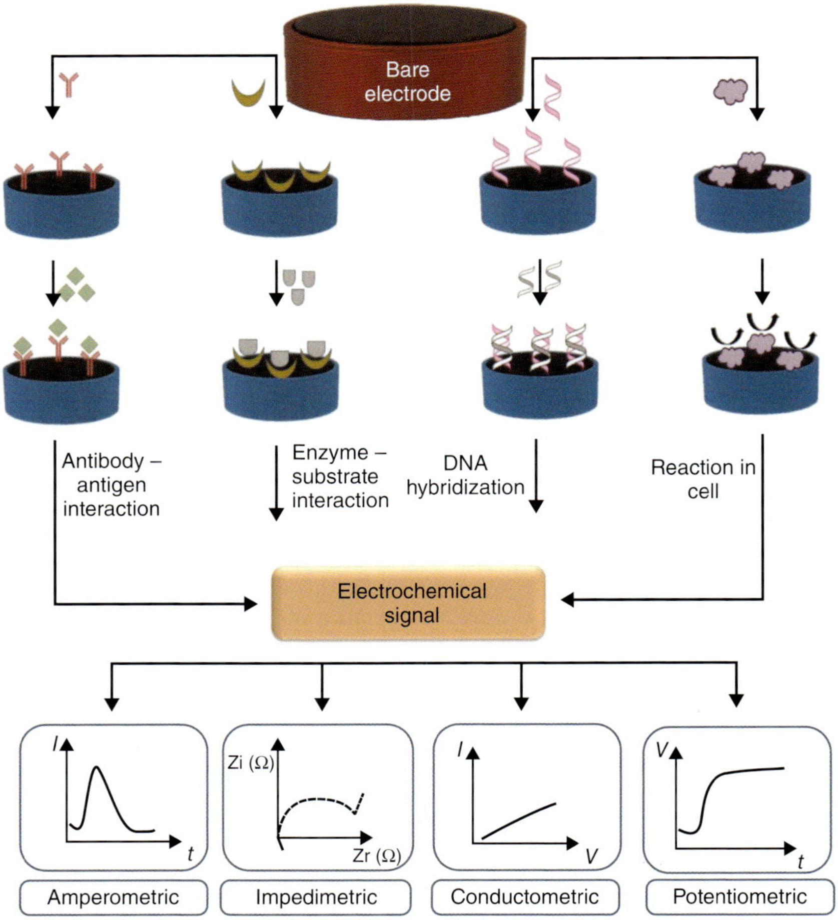

Figure 7.3 Scheme of an electrochemical biosensor. (Bahadır 2015 [18]. Reproduced with permission of Elsevier.)

measure the potentials at the working electrode with respect to the reference electrode under no current flow. They monitor electric potential related to ion concentration in the system created by selective binding of the electrode and generally possess long lifetimes and acceptable mechanical stability. The electrode measures the change in electrode potential and determines several ions in the reaction matrix due to the enzymatic reaction, while the ions accumulate and bind to a suitable ion-selective membrane. Simple instrumentation, low cost, and suitability for continuous monitoring are the advantages of potentiometric biosensors, while, on the other hand, leaching of the membrane components into the sample is a disadvantage and a major problem with potentiometric biosensors [21].

Optical biosensors are also another type of transducers that are mainly used in biosensor design. The working principle is based on the detection of changes in the refractive index. Optical biosensors contain heterogeneous group of sensors in which the interaction of light with an immobilized biorecognition material is sensed. The binding of the analyte to its specific bioreceptor produces a change in the refractive index and therefore, a response occurs at the sensing surface. Detection principle of optical biosensors is based on the changes in absorbance, fluorescence, luminescence, reflectance, light scattering, and refractive index. Thermal and acoustic transducers are adequately used in analytical applications; they have the advantage of miniaturization and the possibility of construction of arrays of sensors for simultaneous determination of several compounds. One of the disadvantages of these transducers is that they lack selectivity [15]. The change in energy is followed in thermal transducers-based biosensors but this energy can be lost and the produced heat is partly wasted by irradiation, conduction, or convection [15]. Calorimetric biosensors probe the amount of change in heat during a biochemical reaction. In general, thermistors are used to find out the modification of temperature in the solution. The recorded peak height is proportional to the enthalpy change with respect to a substrate concentration [22, 23]. A piezoelectric biosensor working principle mainly depends on measuring the decrease in frequency of vibration of an oscillating crystal by the adsorption of a foreign material on its surface. A piezoelectric biosensor uses the piezoelectric effect to probe pressure, acceleration, strain, or force by converting them to an electrical charge. The measurement of the mass changes on a piezoelectric crystal, which is sensitized by covering it with material binding or by reacting with the analyte, is the main concern. Piezoelectricity is described as a material that generates mechanical movement in response to electric potential.

In biosensor-based analyses, there exist some parameters that should/may be reported to prove that the suggested biosensor is analytically valuable. Some criteria must be satisfied in accordance with standard IUPAC protocols. Sensitivity; which is the response of the sensor to analyte concentration over a unit change, is not an obligatory to report, although reporting it is beneficial for the users. The range of biosensor where there is a linear relationship between the concentration of the substrate and the response of biosensor should be reported. This parameter is very crucial, since the measured concentration should be within the characterized linear range. Limit of detection and limit of quantification values should also be calculated, under optimized conditions of the biosensor. These parameters also show the performance of the biosensor. Selectivity of the biosensor is also important as it shows the ability of the sensor to respond only to the target analyte. IUPAC defines the response time as the time passing between the substrate added before and after or at which the activity of the enzyme starts with adding its substrate and the reaction/response has finished. Rapid response time is desired in biosensors and, therefore, response time should also be reported to support the designed biosensor. Reproducibility of the biosensor shows how accurate it is. This parameter should also be given and can be given as relative standard deviation values. Life time of the biosensor can also be reported, which is the time period of the biosensor, used without significant deterioration in performance characteristics. Shelf-life stability studies can also be shared with the readers which defines the change in its baseline or sensitivity over a fixed period of time. In addition to these parameters, a biosensor should be cheap, small and portable, biocompatible, nontoxic, nonantigenic, and stable under ambient conditions.

The major key points to note in the improvement of biosensor design involve a proper bioreceptor or a recognition molecule, suitable immobilization method, transducer to convert binding reaction into a measurable signal, range, linearity, enhancement, and sensitivity. These are crucial factors to design a biosensor and finally packaging of the biosensor into a complete device.

7.3 Enzymes

Different amino acids play important roles in metabolism; they can be found in living organisms and are the building blocks of proteins. Proteins are macromolecules containing long chains of amino acids, consisting of thousands of amino acids. Proteins are found in all forms of living organisms and perform a wide variety of tasks; for example, fibrous, elongated proteins that are insoluble in water and provide structural support, and globular, spherical proteins that are water soluble and have specific functions in the immune system and metabolism.

Many chemical reactions occur spontaneously but some of them need to be catalyzed for determined reaction speed. Catalyzers are molecules that decrease the required activation energy for chemical transformation of substrates to different materials. Thermodynamical amount of energy is defined as transversion of free energy. Catalyzers are not consumed or changed during reactions. Limitless used catalyzers can restrict the reaction that turns the substrate to product due to their stability [24]. Enzymes are the protein packages that are biological catalysts that speed biochemical reactions without being permanently changed (Figure 7.4).

A large number of biochemical reactions occurring in living organisms are catalyzed by enzymes that are protein structured biocatalysts [26]. Biomolecules

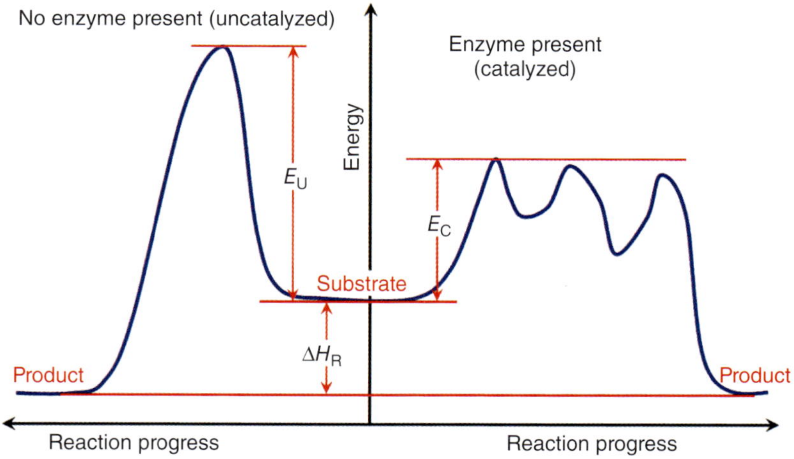

Figure 7.4 Activation energies of enzymatically catalyzed and uncatalyzed reactions. (Liu 2013 [25]. Reproduced with Permission of Elsevier.)

most used as bioreceptors are enzymes. Enzymes have high affinities toward their substrates. They select from a variety of substrates to which they are related chemically and induce the reactions. Enzymes do not change the equilibrium of a reversible reaction; they only facilitate access into balance. Characteristic properties of enzymes are correlated with the differences in their molecular structures. They are constructed by hundreds of amino acids that are bonded to each other covalently by the carbon atoms of carboxyl groups and next amino acids α-amino groups N atoms. Amino acids show either hydrophilic or hydrophobic property according to their radical groups. Enzymes are more often colorless and soluble in salt or water.

The part that directly participates in the catalytic activity is the specific binding part of enzymes called the active site which was postulated by Emil Fischer in 1860 [27]. The enzyme – substrate complex enters a catalytic reaction directed by amino acid residues in the active site. The active site is a three-dimensional niche in the protein, which constitutes a small portion of the overall protein structure. The specificity of the enzyme depends on the arrangement of atoms in the active site and the substrate – enzyme binding process involves a relatively small amount of energy.

Apoenzyme is called the protein and glycoprotein part of an enzyme. The catalytic properties of enzymes are dependent on cofactors or coenzymes, which are non-present in a peptide structure. If these structures are firmly connected to the polypeptide it is called a prosthetic group. Apoenzymes and a prosthetic group create active holoenzymes. Specificity of an enzymatic reaction is usually the result of steric harmony [28].

Emil Fischer offered "lock and key" model to fit the enzyme and substrate complex with each other.

In 1962, the "induced fit model" was offered by Daniel Koshland, suggesting that the enzyme is induced to undergo a structural rearrangement upon substrate binding due to the change in shape of the active site that facilitates the reaction [29]. As the enzymatic reaction is completed, the products are released, and the active site returns to its original state (Figure 7.5).

7.3.1 Enzyme Commission Numbers

The nomenclature of enzymes is generally based on recommendations of the Enzyme Commission in 1961, by the International Union of Biochemistry and Molecular Biology. According to the International Union of Biochemistry and Molecular Biology, every type of characterized enzymes is required to contain Enzyme Commission Numbers. The systematic name of the enzyme describes the substrate or substrates and defines the type of reaction catalyzed [29]. The Enzyme Commission Number, is associated with a recommended name, alternative names, catalytic activity, cofactors, pointers to the SWISS-PROT [31] protein sequence entries that correspond to the enzyme, pointers to the PROSITE [32] entries describing the protein families of which the enzyme is a member, pointers to human diseases associated with a deficiency of the enzyme [33].

Classification of enzymes by the International Union of Biochemistry and Molecular Biology is based on the type of chemical reaction catalyzed. The

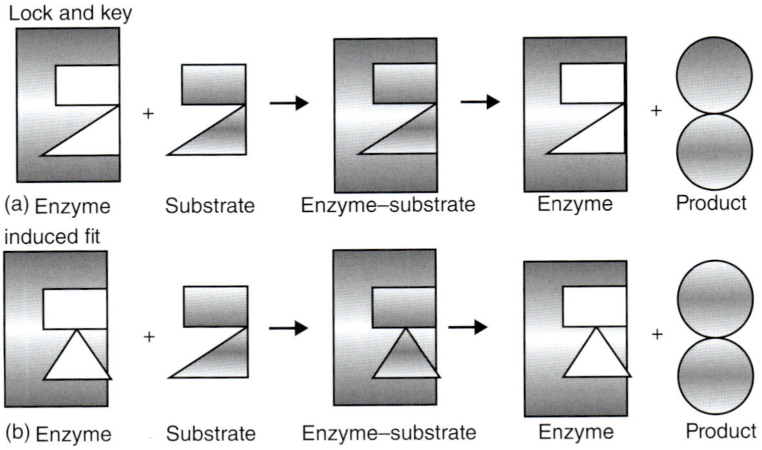

Figure 7.5 Enzyme–substrate complex with (a) lock-and-key and (b) induced fit model. (Whiteley 2006 [30]. Reproduced with permission of Elsevier.)

Enzyme Commission of the International Union of Biochemistry and Molecular Biology assigns a four digit number for each enzyme: the first number indicates the family, the second indicates the subclass of the family and also indicates its actions of chemical group type, the third indicates a subgroup within a subclass and is related to the particular chemical groups involved in the reaction, and the forth indicates the correlative number of identification within a subgroup. By using Enzyme Commission numbers, quick searching of enzymes and enzymes functions and properties can be achieved. Moreover, Enzyme Commission numbers provide original classification of enzymes [34].

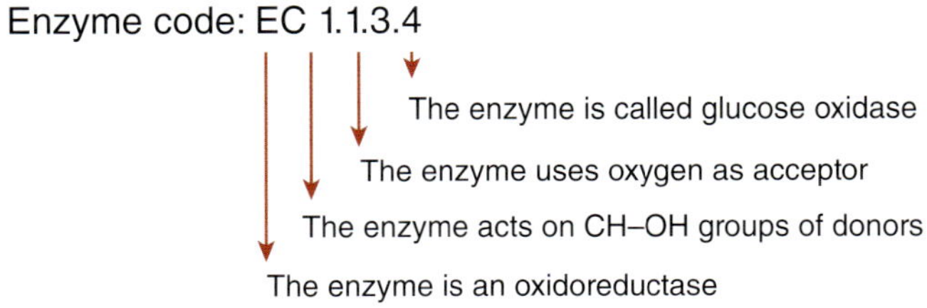

Enzymes are generally named with reference to the reaction they catalyze. The name of an enzyme is often formed by adding -ase to the name of its substrate. Suffix -ase is added to the name of the enzyme's primary substrate. Enzyme Commission has recommended nomenclature of enzymes based on the reaction being catalyzed. Enzymes are classified into six different groups. The six main types of enzyme-catalyzed reactions are as follows:

- EC1: Oxidoreductases catalyze oxidation–reduction reactions
- EC2: Transferases catalyze group transfer reactions
- EC3: Hydrolases catalyze hydrolytic reactions
- EC4: Lyases catalyze cleavage and elimination reactions
- EC5: Isomerases catalyze isomerization reactions
- EC6: Ligases catalyze synthetic reactions.

7.3.1.1 EC1 Oxidoreductases

Oxidoreductases catalyze oxidation–reduction reactions, transfer of electrons, hydrogen or oxygen atoms is the main cause of enzymes catalyzing oxidation/reduction reactions. Oxidoreductases are also called "oxidases," "dehydrogenases," or "reductases." Oxidoreductase enzymes that are also called electron donors catalyze the transfer of electrons, and are included in the central *meta*. Oxidoreductases have 22 subclasses; they all have different technological importance such as, dehydrogenases that oxidize a substrate by transferring hydrogen atoms to a coenzyme (NAD^+, $NADP^+$, FAD^+, FMN), which acts as an acceptor [29, 35].

The class Oxidoreductase includes the following:

- Dehydrogenases and oxidases (1.1–1.10)
- Peroxidases (1.11), which use H_2O_2 as the oxidant,
- Hydroxylases (1.99.1), which introduce hydroxyl groups
- Oxygenases (1.99.2), which introduce molecular O_2 in place of a double bond in the substrate.

Oxidoreductases are enzymes that catalyze oxidoreduction reactions. Simply, oxidoreductases catalyze reactions like; $A^- + B \rightarrow A + B^-$, where A is the oxidant and B is the reductant. These enzymes can be located in glycolysis, TCA cycle, oxidative phosphorylation, and amino acid metabolisms. They (dehydrogenases) use electron acceptors other than oxygen, such as the cofactor NAD^+ or $NAD(P)^+$. These cofactors have an important advantage when combined with electrochemical sensors that ensure the outwardly supplied, immobilized, and regenerated cofactor. It is reported that there are more than 250 commercially and highly specific dehydrogenases available [36].

Oxidoreductases have 22 subclasses:

- EC 1.1 contains oxidoreductases the CH—OH group of donors (alcohol oxidoreductases)
- EC 1.2 contains oxidoreductases the aldehyde or oxo group of donors
- EC 1.3 contains oxidoreductases the CH—CH group of donors (CH—CH oxidoreductases)
- EC 1.4 contains oxidoreductases the CH—NH_2 group of donors (amino acid oxidoreductases, monoamine oxidase)
- EC 1.5 contains oxidoreductases CH—NH group of donors
- EC 1.6 contains oxidoreductases NADH or NADPH
- EC 1.7 contains oxidoreductases and other nitrogenous compounds as donors
- EC 1.8 contains oxidoreductases a sulfur group of donors
- EC 1.9 contains oxidoreductases a heme group of donors
- EC 1.10 contains oxidoreductases diphenols and related substances as donors
- EC 1.11 contains oxidoreductases peroxide as an acceptor (peroxidases)
- EC 1.12 contains oxidoreductases hydrogen as donors
- EC 1.13 contains oxidoreductases single donors with incorporation of molecular oxygen (oxygenases)
- EC 1.14 contains oxidoreductases paired donors with incorporation of molecular oxygen

- EC 1.15 contains oxidoreductases superoxide radicals as acceptors
- EC 1.16 contains oxidoreductases that oxidize metal ions
- EC 1.17 contains oxidoreductases CH or CH_2 groups
- EC 1.18 contains oxidoreductases iron–sulfur proteins as donors
- EC 1.19 contains oxidoreductases reduced flavodoxin as a donor
- EC 1.20 contains oxidoreductases phosphorus or arsenic in donors
- EC 1.21 contains oxidoreductases X—H and Y—H to form an X—Y bond
- EC 1.97 contains other oxidoreductases

7.3.1.2 EC2 Transferases

Transferases catalyze functional group transfers; these are of the general form [29]. Transferase enzymes catalyze the transfer of a functional group from a donor to an appropriate receiver. According to the transferred groups' chemical properties they are classified into nine subclasses. In cell metabolisms, transferases, such as methyltransferases, acyltransferases, transaminases, phosphotransferases, and glycosyltransferases, have vital assignment. Transferases need purely intracellular coenzymes. They have limited applications but some of them are commercial enzymes of relevance in research [24].

7.3.1.3 EC3 Hydrolases

Hydrolases catalyze hydrolysis; these are of the general form [29]. Reactions of hydrolysis that is enzymes catalyzing, is the breaking of a chemical bond by water reaction. There are 12 subclasses of hydrolases according to the type of susceptible bond [37]. Hydrolases are necessary for destructive metabolism by procuring assimilable nutrients to the niche. The most incontestable hydrolase enzymes with their technological function are esterases, proteases, and glucosidases. Hydrolases can catalyze the overturn reactions of chemical bond formation with water elimination under suitable circumstances [38].

7.3.1.4 EC4 Lyases

Lyases catalyze lysis reactions, in other words nonhydrolytic removal of functional groups from substrates, often creating a double bond in the product; or the reverse reaction, that is, addition of function groups across a double bond [29]. Lyases are enzymes that catalyze the breaking of various chemical bonds nonhydrolytic, nonoxidative, and mostly forming a new double bond or a new ring structure. Lyases have seven subclasses according to the type of precise bond: C—C, C—O, C—N, C—S, C—X (halides), P—O, and other bonds. For forming double bond or a new ring, lyase enzymes act as a single substrate and eliminate a molecule. Lyase enzymes have dissimilar metabolic performance by their cell-destructive metabolism and biosynthesis by acting in reverse. Aldolases, generally acting in reverse reactions of C—C bond formation, carboxylases and decarboxylases, hydratases, and dehydratases are the most riveting lyases. Some lyases are intracellular and some do not need coenzymes. As an example pectate lyase ((1 → 4)-α-D galacturonan lyase; EC 4.2.2.2) from a separate supplier is extracellular and requires Ca^{++} as cofactor but does not need a coenzyme [39, 40].

Having different properties like, for instance, requirement of substrate is only one for one direction reactions but two for reverse direction, from other enzyme species Lyases have much more technological applications. As an example, a recognizable enzyme of lyases, nitrile hydratase (nitrile hydrolyase; EC 4.2.1.84) has a use with acrylonitrile for acrylamides industrial production [41, 42]. There are more concerned essays for lyases; production of L-aspartate with aspartase (aspartate ammonia-lyase; EC 4.3.1.1) [43–45] the production of fumarate with fumarase (fumarate hydratase; EC 4.2.1.2) [46, 47] and the production of urocanic acid with histidase (L-histidine ammonia lyase; EC 4.3.1.3) [48]. Moreover, Lyases have a role for optically active organic compounds asymmetric synthesis [49, 50].

Lyases have seven subclasses:

- EC 4.1: broken carbon–carbon bonds lyases, such as decarboxylases (EC 4.1.1), aldehyde lyases (EC 4.1.2), oxo acid lyases (EC 4.1.3), and others (EC 4.1.99)
- EC 4.2: broken carbon–oxygen bonds lyases, such as dehydratases
- EC 4.3: broken carbon–nitrogen bonds lyases
- EC 4.4: broken carbon–sulfur bonds lyases
- EC 4.5: broken carbon–halide bonds lyases
- EC 4.6: broken phosphorus–oxygen bonds, such as adenylate cyclase and guanylate cyclase lyases
- EC 4.99: other lyases, such as ferrochelatase.

7.3.1.5 EC5 Isomerases

Isomerases catalyze molecular isomerization, including racemizations and cis–trans isomerizations [29]. Isomerase enzymes convert a molecule from one isomer to another isomer. They can simplify intramolecular rearrangements in the presence of broken and formed bonds. Moreover, they can catalyze the conformational transfers. Transition of a substrate into an isomer enzyme catalyzing reactions is a substance which has the same kind and number of atoms. Isomerases have six subclasses depending on the produced kind of isomers such as racemases and epimerases, cis–trans-isomerases, intramolecular oxidoreductases, intramolecular transferases (mutases), intramolecular lyases, and other isomerases. Most of them are intracellular but some of them need cofactors but do not need organic coenzymes. Isomerases do not have much technological applications but glucose isomerase (xylose isomerase; D-xylose aldose–ketose-isomerase EC 5.3.1.5) is an impressive example of those applications [51]. This enzyme is used for the production of high fructose syrups (HFS), usually from corn starch [52].

7.3.1.6 EC6 Ligases

Ligases catalyze ligation, a synthesis reaction involving condensation of two molecules with the hydrolysis of a pyrophosphate bond [29]. Ligases are enzymes that can catalyze the covalent bonding of two large molecules by making a new chemical bond. Generally, this covalent bonding occurs by hydrolysis or a small chemical group added on one of the bigger molecules or catalyzes the bonding of two molecules with an enzyme. These enzymes have a crucial role in cell anabolism by making synthesis in the cell. Ligases have six subclasses according to the C—O, C—S, C—N, C—C, phosphoric esters, and C-metal bonds. Ligases have high molecular weight and they need intracellular coenzymes. Ligases have no wide application in technology. But they are priceless trading enzymes for research practices. For instance, T4 DNA ligase (poly deoxy ribonucleotide synthase; EC 6.5.1.1) is routinely used in genetic engineering protocols [53].

7.3.2 Enzyme Immobilization

Enzymes are biological catalyzers that speed up biochemical reactions quickly and efficiently. Due to their large application area, usages of enzyme biosensors have great attention in areas such as fine chemistry, food chemistry, therapeutics applications, protein engineering, and so on. Even though enzyme biosensors have great potential, due to economical and stabilization on storage aspects their usages are limited. Usage of enzymes has some drawbacks due to some of their characteristics that are not appropriate for industrial application such as being natively unstable, easily inhibited and working well only on natural substrates and under physiological conditions [54]. In this point, procedures to improve their stability and enable them to be reused, gains importance. The term *immobilization* refers to "restricted movement" or "making unable to move." The term *immobilized enzymes* refers to "enzymes localized in a certain defined region of space with retention of their catalytic activities, and which can be used repeatedly and continuously." Subsequently, immobilization means physical localization of biocatalysts or connection of the biocatalysts with an insoluble matrix or immobilized proteins and cells to an insoluble support. Mixing together the enzyme and the support material under optimized conditions and following a period of incubation, is the main procedure for immobilization (Figure 7.6).

Enzyme immobilization offers benefits in comparison with soluble enzymes or alternative technologies [56, 57]. Enzyme immobilization technology has become reasonable in the design of enzyme biosensors due to their many advantages such as:

- Increase in enzyme stability
- Easily separating or removing from solution without contamination by the contents of the reaction mixture
- Repetitive usage with the same amount of enzyme in solution and immobilized enzyme
- Greater control and follow of the catalytic process
- Decrease in effluent problems
- Noninfected product by the enzyme
- Developing continuous process.

The enzyme immobilization process consists of three main components such as the enzyme, the matrix, and the mode of attachment or entrapment. An ideal matrix must be inert, physically strong, stable, regenerative, and be able to increase enzyme activity and decrease product inhibition, nonspecific adsorption, and microbial contamination [58]. There can be different carriers

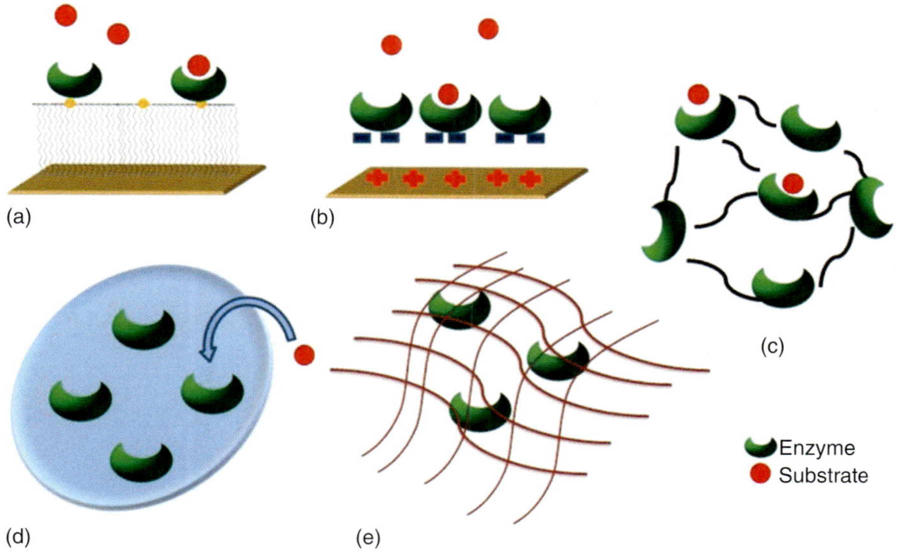

Figure 7.6 Common methods of immobilization used in biosensing: (a) covalent binding; (b) adsorption; (c) cross-linking; (d) encapsulation; and (e) entrapment. (Rodríguez-Delgado 2015 [55]. http://www.sciencedirect.com/science/article/pii/S0165993615002290 created under creative common license http://creativecommons.org/licenses/by/4.0/.)

used for enzyme immobilization in terms of organic and inorganic [59, 60]. Organic carriers are polysaccharides such as cellulose, agar, agarose, chitin, alginate dextrans; proteins such as collagen and albumin; carbon, polystyrene, and other polymers such as polyacrylate polymethacrylates, polyacrylamide, polyamides, vinyl, and allyl-polymers. Inorganic carriers are natural minerals such as bentonite, silica, sand; processed materials such as glass (nonporous and controlled pore), metals, controlled pore, and metal oxides such as ZrO_2, TiO_2, Al_2O_3. Chemical characteristics of enzymes, different properties of substrates and products, and the range of potential processes employed should be considered while selecting the immobilization methods. For enzymes, immobilization techniques can be classified based on the entrapment of enzyme or bounding of enzyme. Entrapment can be matrix entrapped or membrane entrapped. Bounded methods are classified as adsorption and covalent binding (Figure 7.7).

For food analysis applications, enzymes have to be immobilized, with very simple and cost-effective procedures [61]. In immobilization methods, how well the enzyme bonds to the sensor surface is crucial, since it directly affects the success of the enzyme biosensor. Moreover, not only the immobilization method, but also immobilization steps are important since there can be a loss of enzyme activity in the immobilization process and, therefore, it should be carefully monitored. Reuse of the enzyme is very critical because enzymes are expensive catalysts [15]. The selection of an appropriate immobilization method depends on the nature of the enzyme, type of transducer used, physicochemical properties of the analyte, and operating conditions for the enzyme biosensor

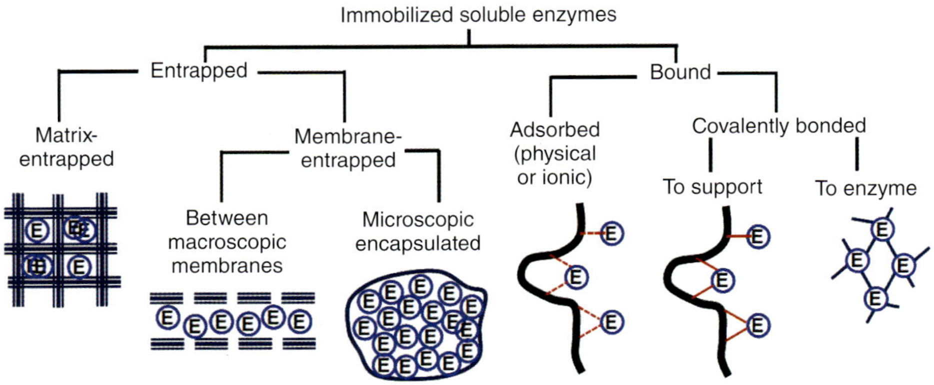

Figure 7.7 Immobilization techniques. (Liu 2013 [25]. Reproduced with permission of Elsevier.)

[15]. A successful matrix should immobilize the enzyme at a transducer surface efficiently, while maintaining the functionality of the enzyme and providing accessibility toward the target analyte with an intimate contact with the transducer surface. The development of a good biocompatible matrix for enzyme immobilization is very crucial to improving the analytical performance of a biosensor [62]. Although the most common methods for immobilization of biocomponent are adsorption and covalent bonding, the suitability of a method for a particular task is at present still being empirically elucidated.

7.3.2.1 Physical Adsorption

Physical adsorption is the simplest and oldest method made up of mixing together enzyme and support under suitable conditions which are pH, ionic strength; incubation period for immobilization of enzyme. In the physical adsorption procedure, reversible surface interactions between enzyme and support material occurs. Weak forces of electrostatic such as van der Waals forces, ionic and hydrogen bonds are involved in the physical adsorption steps. Surface chemistry between enzyme and support by these forces enable the necessary binding. In this method, there is no need for activation and chemical modification and, therefore this is a simple immobilization technique.

Since the adsorption of enzyme to a surface is a reversible process, changes in pH, ionic strength, substrate concentration, and temperature may remove the enzyme from the surface [63]. On the other hand, there exists a major disadvantage of physical adsorption such as leakages of the enzyme from the support with desorption and contamination of the solution as a result of flowrate, bubble agitation, particle–particle abrasion, and so on. Consequently, usage of physical adsorption immobilization method has its limitations. Despite having some limitations and disadvantages, immobilization by physical adsorption is still a simple procedure since it does not need physiological coupling conditions or chemicals which is an additional advantage [63]. Currently, physical entrapment is no longer used due to the short life of enzymatic activities achieved. Therefore, chemical immobilization techniques are more frequently used in enzyme sensors such as cross-linking, entrapment, and so on.

7.3.2.2 Covalent Binding

Formation of covalent bonds is the main principle of this method. Functional groups such as $—NH_2$, —COOH that are available on the surface of the enzymes mainly originate from the side chain of the amino acids attached covalently to chemically activated supports [15]. The immobilization by covalent binding is conducted by activating the carrier, coupling of the biomolecule, and removal of the adsorbed biomolecules. A disadvantage of covalent coupling is the frequently occurring loss of enzyme activity [63]. In their work, Xu *et al.* covalently immobilized the enzyme using glutaraldehyde as a spacer arm on the surface of polyvinyl alcohol (PVA)/chitosan (CS)/multiwalled carbon nanotubes (MWNTs) and composite nanofibrous membrane as shown in Figure 7.8.

7.3.2.3 Entrapment

Immobilization by entrapment differs from adsorption and covalent binding in that enzymes are free in solution but are restricted by the lattice structure of the entrapment system. Entrapment is the easiest immobilization method where there is no alteration in the structure of the enzyme.

Three general methods for entrapment are as follows:

- Entrapment behind a membrane: a solution of enzyme, a suspension of cells, or a slice of tissue is simply confined by an analyte permeable membrane as a thin film covering the detector.

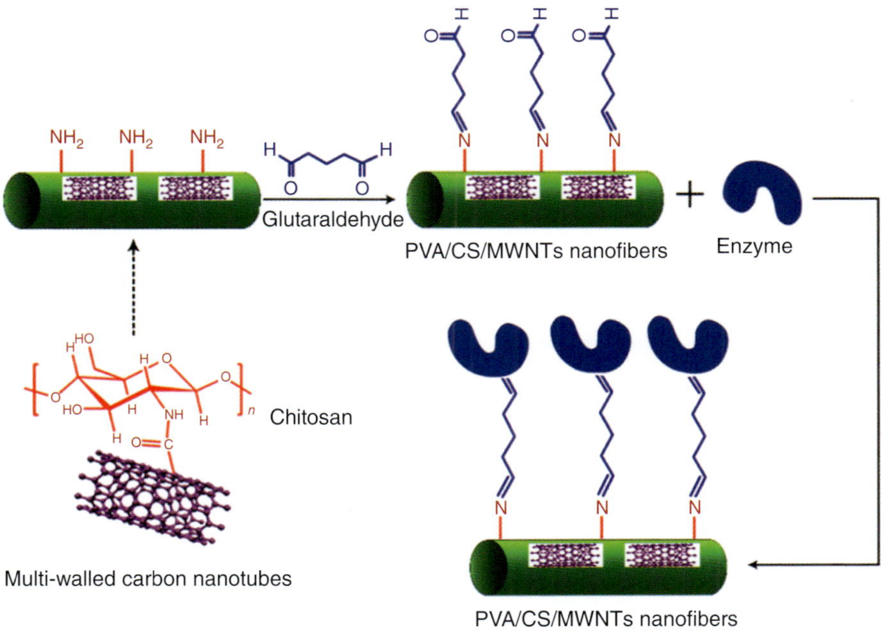

Figure 7.8 Schematic illustration of laccase immobilization on PVA/CS/MWNTs nanofibrous membranes via activation. (Xu 2015 [64]. Reproduced with permission of Elsevier.)

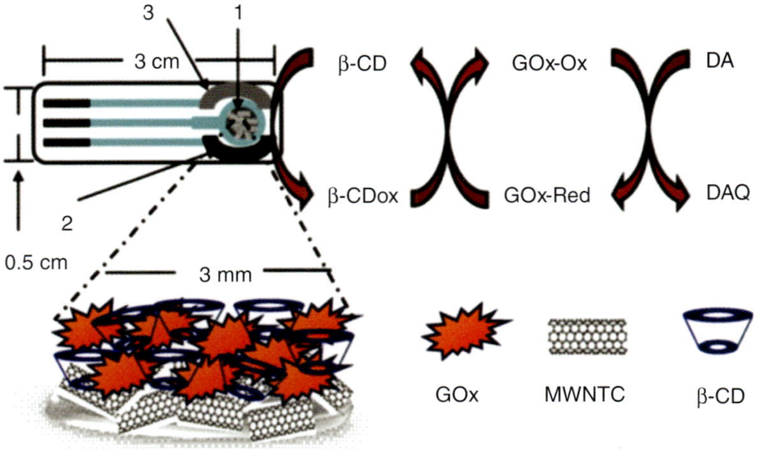

Figure 7.9 Entrapment of glucose oxidase. (Alarcón-Ángeles 2010 [65]. Reproduced with permission of Elsevier.)

- Entrapment of biological receptors within a polymeric matrix: polyacrylonitrile, agar gel, polyurethane or polyvinyl-alcohol membranes, sol gels, or redox hydrogels with redox centers.
- Entrapment of biological receptors within self-assembled monolayers or bilayer lipid membranes [10].

The entrapment method for enzyme immobilization was first carried out by Bernfeld and Wan in 1963 by entrapping them into lattices of synthetic polymers. Till date, entrapment method is the commonly used method, which is quite simple and easy. The main difference from other techniques is that the enzyme does not bind directly to the matrix. The main disadvantage of an enzyme entrapment method is that, the response time is longer compared to other methods due to difficulties in the accessibility of entrapped biomolecules. The immobilized support should allow the diffusion of substrates and products to overcome this problem.

In a recent work, Alarcón-Ángeles *et al.*, developed a novel enzyme entrapment approach based on an electropolymerization process using multi-walled carbon nanotubes (MWCNTs), ß-cyclodextrin (ß-CD), and GOx as shown in Figure 7.9.

7.3.2.4 Encapsulation

Encapsulation is the other immobilization method that can be used to immobilize enzymes within a semipermeable membrane. Similar to entrapment, in this method the enzyme is restricted in a limited space. Depending on their molecular weight, small substrates and products can be moved in and out of the semipermeable membrane. The main disadvantage of this method is that if the diffusion of the product or substrate is high, or the density of the encapsulated enzyme is high, the membrane can be damaged. As an advantage, this encapsulation method can lead to immobilization of biological materials in any desired combination.

To illustrate this method, Macario *et al.* reported an encapsulation strategy for lipase enzyme in a highly ordered mesoporous matrix by a sol–gel method. The

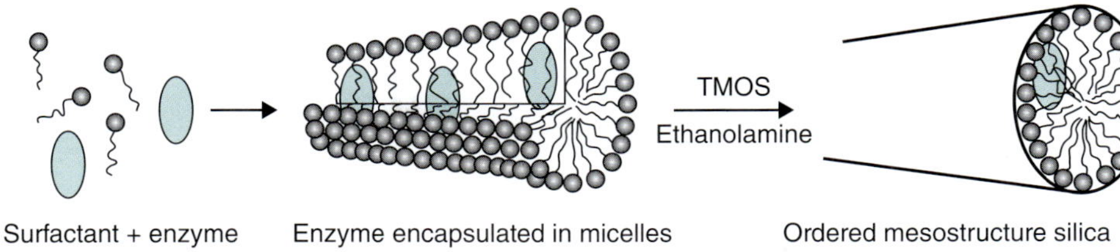

Figure 7.10 Enzyme encapsulation in micelles. (Macario 2009 [66]. Reproduced with permission of Elsevier [66].)

micellar phase of the surfactant, which is self-assembled with silica is used to encapsulate the enzyme. The authors came up with the result that an immobilization procedure saves the mobility of the enzyme and there is a sixfold increase in enzyme activity with increased stability (Figure 7.10).

7.3.2.5 Cross-Linking

One of the commonly used methods for enzyme immobilization is the cross-linking method. In this method, enzyme form cross-linked or covalent bonds with the support material and this method is a support-free procedure and involves joining the receptor molecules with each other to form a large three-dimensional complex structure. Cross-linking or co-cross-linking can be achieved by chemical or physical methods. The chemical method normally involves the formation of covalent bonds between the enzymes by means of bi- or multifunctional reagents such as bovine serum albumin, glutaraldehyde, or carbodiimide. The cross-linking method offers advantages of being a simple method and having a strong chemical binding of the enzymes. Moreover, the choice of the degree of cross-linking influences the physical properties and the particle size. The main disadvantage is that intermolecular cross-linking is difficult to control; therefore, there is a possibility of activity loss due to chemical alternations of the active site of the enzymes [63] (Figure 7.11).

7.4 Application of Enzyme-Based Biosensors in Food Industry

Poly(ethyleneoxide)/polypyrrole (PEO-*co*-PPy) and 3-methylthienyl methacrylate-*co*-*p*-vinyl benzyloxy poly(ethyleneoxide)/polypyrrole (CP-*co*-PPy) are covered with thiophene and yeast cells (*Saccharomyces cerevisiae*) and invertase enzyme were immobilized in conducting copolymers by electrochemical polymerization of pyrrole that was studied in another research [68].

Alcohol biosensor, which has a photoelectrochemical property, was constructed by using alcohol oxidase enzyme and 4-(2,5-di(thiophen-2-yl)-1*H*-pyrrol-1-yl)benzenamine (SNS-NH2) and carbon nanotubes (CNTs) modified gold slides, for detection of alcohol amount in real samples. According to research, the modified alcohol biosensor should be an alternative method for photobiocatalytic nanoscale systems based on alcohol oxidation reactions in various biotechnological, biological, and environmental analyses [69].

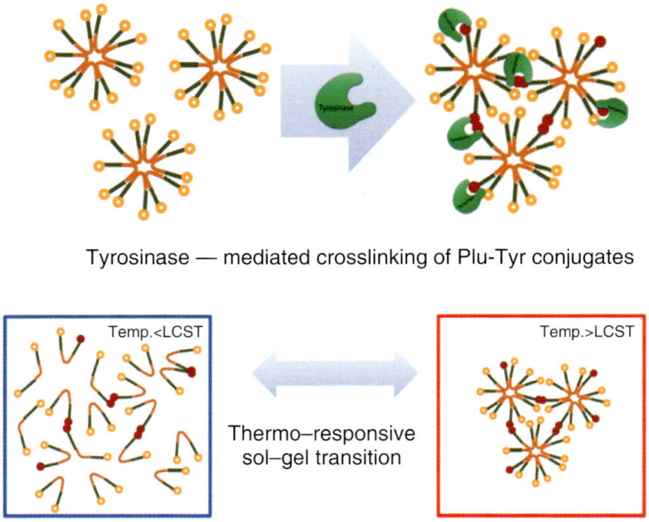

Figure 7.11 Schematic illustration of Plu-Tyr hydrogel formation by enzyme mediated cross-linking and its thermo-responsive sol–gel transition. (Lee 2011 [67]. Reproduced with permission of Elsevier.)

Another substantial leading study for biosensor application was designed by using cholesterol oxidase for cholesterol determinations. P(SNS-NH2)/ChOx/[Ru(bpy)3]2+ and ChOx/[Ru(bpy)3]2+, were covalently bonded cholesterol oxidase to modified electrode and gold slide, modified by thioaniline [70].

For cholesterol determination there has been a significant study. In this regard, thiophene capped PEO-*co*-PPy and CP-*co*-PPy were synthesized with electrochemical techniques. Cholesterol oxidase was immobilized on electrodes by electrochemical polymerization [71].

Bis(4-(3-thienyl ethylene)-oxycarbonyl)diazobenzene (TDAZO), photosensitive and thermosensitive monomer, was synthesized recently. In this study, thiophene and pyrrole were copolymerized on TDAZO, then immobilization of invertase, by pyrroles electropolymerization, was carried out for biosensor production [72].

Another important amperometric GOx biosensor was constructed by immobilization of biomolecule by thiol-containing calixarenes. For determination of the glucose amount Calix-SH was formed on a gold surface. Fabricated calix-SH/GOx biosensor was successfully used in real samples [73].

Calixaren was used in another study with modification with montmorillonite (Mt) for biosensor applications. Amino-functionalized calixarenes (Calix-NH2) were modified with Mt (Calix-NH2/Mt) and the pyronose oxidase was immobilized on the electrode for real sample tests [74].

A selenium contented polymer biosensor was improved for glucose detection. Poly(2-(2-octyldodecyl)-4,7-di(selenoph-2-yl)-2*H*-benzo[*d*][1,2,3]-triazole), poly(SBTz) was electropolymerized on graphite electrode. Immobilization was carried out with a mixture of calixaren and gold nanoparticle [75].

Another different study shows a N-functionalized dithienopyrroles (DTP-NH2) modified GOx immobilized biosensor. In addition, experiments were carried out with real samples and a reference to those experiments where commercial glucose kits shows were used shows the reliability of the biosensor [76].

A new amperometric xanthine (X) biosensor was established by immobilization of xanthine oxidase (XOx). Electopolimerized polymer film was used as a conducting polymer; DTP type polymer indicated that xanthine biosensor has high sensitivity, stability, and selectivity in real samples [77].

7.5 Conclusion

In enzyme-based biosensors, the bioreceptor takes the help of an immobilized enzyme as biocatalyst to transform the analyte to a product that can be detected by the transducer. A large number of enzymes are commercially available, but some need to be extracted in specific measurements. Some of the enzymes are naturally present in food materials, whereas others are used to enhance food processing actions or food quality. It is seen that enzyme sensors have potential applications in the food industry. The specific applications are monitoring ingredients, food additives, contaminations, and toxins.

References

1 Wang, J. (2005) Nanomaterial-based electrochemical biosensors. *Analyst*, **130**(4), 421–426.

2 Xu, Y.H. and Wang, E.K. (2012) Electrochemical biosensors based on magnetic micro/nano particles. *Electrochim. Acta*, **84**, 62–73.

3 Gerard, M., Chaubey, A., and Malhotra, B.D. (2002) Application of conducting polymers to biosensors. *Biosens. Bioelectron.*, **17**(5), 345–359.

4 Vo-Dinh, T. and Cullum, B. (2000) Biosensors and biochips: advances in biological and medical diagnostics. *Fresenius J. Anal. Chem.*, **366**(6-7), 540–551.

5 Wang, N., Zhang, N.Q., and Wang, M.H. (2006) Wireless sensors in agriculture and food industry – Recent development and future perspective. *Comput. Electron. Agric.*, **50**(1), 1–14.

6 Terry, L.A., White, S.F., and Tigwell, L.J. (2005) The application of biosensors to fresh produce and the wider food industry. *J. Agric. Food. Chem.*, **53**(5), 1309–1316.

7 Venugopal, V. (2002) Biosensors in fish production and quality control. *Biosens. Bioelectron.*, **17**(3), 147–157.

8 Patel, P.D. (2002) (Bio)sensors for measurement of analytes implicated in food safety: a review. *Trends Anal. Chem.*, **21**, 96–115.

9 Eggins, B. (2002) *Chemical Sensors and Biosensors*, John Wiley & Sons, Inc, Hoboken, NJ.

10 Thevenot, D.R., Toth, K., Durst, R.A., and Wilson, G.S. (2001) Technical report electrochemical biosensors: recommended definitions and classification. *Biosens. Bioelectron.*, **16**, 121–131.

11 Kuila, T., Bose, S., Khanra, P., Mishra, A.K., Kim, N.H., and Lee, J.H. (2011) Recent advances in graphene-based biosensors. *Biosens. Bioelectron.*, **26**, 4637–4648.

12 Gronow, M. (1991) Biosensors – a marriage of biochemistry and microelectronics. *Biotechnol. Sci. Bus.*, **20**, 355–370.

13 Freire, R.S., Pessoa, C.A., Mello, L.D., and Kubota, L.T. (2003) Direct electron transfer: an approach for electrochemical biosensors with higher selectivity and sensitivity. *J. Braz. Chem. Soc.*, **14**, 230–243.

14 Dhawan, G., Sumana, G., and Malhotra, B.D. (2009) Recent developments in urea biosensors. *Biochem. Eng. J.*, **44**, 42–52.

15 Mello, L.D. and Kubota, L.T. (2002) Review of the use of biosensors as analytical tools in the food and drink industries. *Food Chem.*, **77**, 237–256.

16 Malhotra, B.D., Singhal, R., Chaubey, A., Sharma, S.K., and Kumar, A. (2005) Recent trends in biosensors. *Curr. Appl. Phys.*, **5**, 92–97.

17 Scott, A.O. (1998) *Biosensors for Food Analysis*, The Royal Society of Chemistry, Cambridge.

18 Bahadır, E.B. and Sezgintürk, M.K. (2015) Electrochemical biosensors for hormone analyses. *Biosens. Bioelectron.*, **68**, 62–71.

19 Chaubey, A. and Malhotra, B.D. (2002) Mediated biosensors. *Biosens. Bioelectron.*, **17**, 441–663.

20 Lawrence, A.J. and Moores, G.R. (1972) Conductimetry in enzyme studies. *Eur. J. Biochem.*, **24**, 538–546.

21 Pejcic, B. and Marco, R. (2006) Impedance spectroscopy: over 35 years of electrochemical sensor optimization. *Electrochim. Acta*, **51**, 6217–6229.

22 Danielsson, B., Mattiasson, B., and Mosbach, K. (1981) Enzyme thermistor devices and their analytical applications. *Appl. Biochem. Bioeng.*, **3**, 97–143.

23 Ramanathan, K. and Danielsson, B. (2001) Principles and applications of thermal biosensors. *Biosens. Bioelectron.*, **16**, 417–423.

24 Illanes, A. (ed.) (2008) *Enzyme Biocatalysis: Principles and Applications*, Springer.

25 Liu, S. (ed.) (2013) *Process Bioengineering: Kinetics, Biosystems, Sustainability and Reactor Design*, Elsevier, Amsterdam, pp. 323–390.

26 de Carvalho, C.C.C.R. (2011) Enzymatic and whole cell catalysis: finding new strategies for old processes. *Biotechnol. Adv.*, **29**(1), 75–83.

27 Kosland, D.E. Jr., (1960) The active site and enzyme action. *Adv. Enzymol.*, **22**, 45–97.

28 Freitag, R. (1999) Utilization of enzyme–substrate interactions in analytical chemistry. *J. Chromatogr. B*, **722**(1-2), 279–301.

29 Carr, P.W. and Bowers, L.D. (1980) *Immobilized Enzymes in Analytical and Clinical Chemistry*, John Wiley & Sons, Inc., New York.

30 Whiteley, C.G. and Lee, D.-J. (2006) Enzyme technology and biological remediation. *Enzyme Microb. Technol.*, **38**, 291–316.

31 Junker, V.L., Apweiler, R., and Bairoch, A. (1999) Representation of functional information in the SWISS-PROT data bank. *Bioinformatics*, **15**(12), 1066–1067.
32 Hofmann, K., Bucher, P., Falquet, L., and Bairoch, A. (1999) The PROSITE database, its status in 1999. *Nucleic Acids Res.*, **27**(1), 215–219.
33 Bairoch, A. (2000) The ENZYME database in 2000. *Nucleic Acids Res.*, **28**(1), 304–305.
34 Tsai, C.S. (2003) *Dynamic Biochemistry: Enzyme Kinetics*, Chapter 7, Wiley-Liss, Inc.
35 Toone, E.J. (2006) *Advances in Enzymology and Related Areas of Molecular Biology*, John Wiley & Sons, Inc.
36 Ho, M. and Rechnitz, G. (1987) Highly stable biosensor using an artificial enzyme. *Anal. Chem.*, **59**(3), 536–537.
37 Holmquist, M. (2000) Alpha/beta-hydrolase fold enzymes: structures, functions and mechanisms. *Curr. Protein Pept. Sci.*, **1**(2), 209–235.
38 Davis, B.G. and Borer, V. (2001) Biocatalysis and enzymes in organic synthesis. *Nat. Prod. Rep.*, **18**(6), 618–640.
39 Nasser, W., Chalet, F., and Robertbaudouy, J. (1990) Purification and characterization of extracellular pectate lyase from bacillus-subtilis. *Biochimie*, **72** (9), 689–695.
40 Castang, S., Shevchik, V.E., Hugouvieux-Cotte-Pattat, N., Legrand, P., Haser, R., and Gouet, P. (2004) Crystallization of the pectate lyase PelI from Erwinia chrysanthemi and SAD phasing of a gold derivative. *Acta Crystallogr., Sect. D: Biol. Crystallogr.*, **60**, 190–192.
41 Yamada, H. and Kobayashi, M. (1996) Nitrile hydratase and its application to industrial production of acrylamide. *Biosci. Biotechnol., Biochem.*, **60**(9), 1391–1400.
42 Miller, J.A. and Nagarajan, V. (2000) The impact of biotechnology on the chemical industry in the 21st century. *Trends Biotechnol.*, **18**(5), 190–191.
43 Chibata, I., Sato, T., and Tosa, T. (1974) Process for the production of L-aspartic acid. Patent US3791926 A, US Patent Agency.
44 Fusee, M.C. (1987) Industrial-production of L-aspartic acid using polyurethane-immobilized cells containing aspartase. *Methods Enzymol.*, **136**, 463–471.
45 Waller, A.S. (2001) Process for the production of L-aspartic acid. Patent US6280980 B1, US Patent Agency.
46 Furui, M., Sakata, N., Otsuki, O., and Tosa, T. (1988) A bioreactor-crystallizer for L-malic acid production. *Biocatalysis*, **2**(1), 69–77.
47 Ulbert, O., Belafi-Bako, K., Tonova, K., and Gubicza, L. (2005) Thermal stability enhancement of Candida rugosa lipase using ionic liquids. *Biocatal. Biotransform.*, **23**(3-4), 177–183.
48 Shibatani, T., Nishimura, N., Nabe, K., Kakimoto, T., and Chibata, I. (1974) Enzymatic production of urocanic acid by Achromobacter liquidum. *Appl. Microbiol.*, **27**(4), 688–694.
49 van der Werf, M.J., van den Tweel, W.J., Kamphuis, J., Hartmans, S., and de Bont, J.A. (1994) The potential of lyases for the industrial production of optically active compounds. *Trends Biotechnol.*, **12**(3), 95–103.

50 Vidal, L., Ferrer, P., Álvaro, G., Benaiges, M.D., and Caminal, G. (2005) Influence of induction and operation mode on recombinant rhamnulose 1-phosphate aldolase production by *Escherichia coli* using the T5 promoter. *J. Biotechnol.*, **118**(1), 75–87.

51 Bhosale, S.H., Rao, M.B., and Deshpande, V.V. (1996) Molecular and industrial aspects of glucose isomerase. *Microbiol. Rev.*, **60**(2), 280–300.

52 Carasik, W. and Carroll, J.O. (1983) Development of immobilized enzymes for production of high-fructose corn syrup. *Food Technol.*, **37**(10), 85–91.

53 Aslanidis, C. and Dejong, P.J. (1990) Ligation-independent cloning of Pcr products (Lic-Pcr). *Nucleic Acids Res.*, **18**(20), 6069–6074.

54 Bugg, T.D.H. (2001) Enzymes: general properties. *eLS*. doi: 10.1038/npg.els.0000709

55 Rodríguez-Delgado, M.M., Alemán-Nava, G.S., Rodríguez-Delgado, J.M., Dieck-Assad, G., Martínez-Chapa, S.O., Barceló, D., and Parra, R. (2015) Laccase-based biosensors for detection of phenolic compounds. *Trends Anal. Chem.*, **74**, 21–45.

56 Costa, S.A., Azevedo, H.S., and Reis, R.L. (2005) Enzyme immobilization in biodegradable polymers for biomedical applications in *Biodegradable Systems in Tissue Engineering and Regenerative Medicine*, (eds R.L. Reis and J. San Roman), CRC Press, pp. 301–318. ISBN: 0-8493-1936-6

57 Tischer, W. and Kasche, V. (1999) Immobilized enzymes: crystals or carriers. *Trends Biotechnol.*, **17**(8), 326–335.

58 Datta, S., Christena, L.R., and Rajaram, Y.R.S. (2013) Enzyme immobilization: an overview on techniques and support materials. *Biotechnology*, **3**(1), 1–9.

59 Kennedy, J.F. and White, C.A. (1985) Principles of immobilization of enzymes, in *Handbook of Enzyme Biotechnology*, 2nd edn (ed A. Wisemann), John Wiley and Sons, New York, pp. 147–207.

60 Guisan, J.M. (ed.) (2006) *Immobilization of Enzymes and Cells*, vol. **22**, Springer.

61 Meena, K. and Raja, T.K. (2006) Immobilisation of yeast cells by gel entrapment using various metal alginates. *World J. Microbiol. Biotechnol.*, **22**, 651–653.

62 Xu, Z., Chen, X., and Dong, S. (2006) Electrochemical biosensors based on advanced bioimmobilization matrices. *Trends Anal. Chem.*, **25**, 899–908.

63 Scheller, F. and Schubert, F. (1992) *Biosensors*, Elsevier Science Publishing Company Inc., New York.

64 Xu, R., Tang, R., Zhou, Q., Li, F., and Zhang, B. (2015) Enhancement of catalytic activity of immobilized laccase for diclofenac biodegradation by carbon nanotubes. *Chem. Eng. J.*, **262**, 88–95.

65 Alarcón-Ángeles, G., Guix, M., Silva, W.C., Ramírez-Silva, M.T., Palomar-Pardavé, M., Romero-Romo, M., and Merkoci, A. (2010) Enzyme entrapment by ß-cyclodextrin electropolymerization onto a carbon nanotubes-modified screen-printed electrode. *Biosens. Bioelectron.*, **26**, 1768–1773.

66 Macario, A., Moliner, M., Corma, A., and Giordano, G. (2009) Increasing stability and productivity of lipase enzyme by encapsulation in a porous organic–inorganic system. *Microporous Mesoporous Mater.*, **118**, 334–340.

67 Lee, S.H., Lee, Y., Lee, S.-W., Ji, H.-Y., Lee, J.-H., and Lee, D.S. (2011) Tae Gwan Park Enzyme-mediated cross-linking of Pluronic copolymer micelles for injectable and in situ forming hydrogels. *Acta Biomater.*, 7, 1468–1476.
68 Yildiz, H.B., Kamaci, M., Azak, H., Secgin, O., and Suer, O. (2013) A comparative study: immobilization of yeast cells and invertase in poly(ethyleneoxide) electrodes. *J. Mol. Catal. B: Enzym.*, **91**, 52–58.
69 Yildiz, H.B., Kamaci, M., Karaman, M., Toppare, L., and Sayin, S. (2014) Photoelectrochemical biosensing approach for alcohol determination by wiring of alcohol oxidase photonically with P(SNS-NH2)/AOx/CNT/[Ru(bpy)(3)](2+) modified electrodes. *J. Macromol. Sci. Part A Pure Appl. Chem.*, **51**(5), 385–393.
70 Yildiz, H.B., Talaz, O., Kamaci, M., Caliskan, A., and Caliskan, S. (2013) Novel photoelectrochemical biosensors for cholesterol biosensing by photonic wiring of cholesterol oxidase. *J. Macromol. Sci. Part A Pure Appl. Chem.*, **50** (12), 1182–1193.
71 Yildiz, H.B., Demirkol, D.O., Sayin, S., Yilmaz, M., Koysuren, O., and Kamaci, M. (2013) New amperometric cholesterol biosensors using poly(ethyleneoxide) conducting polymers. *J. Macromol. Sci. Part A Pure Appl. Chem.*, **50**(10), 1075–1084.
72 Ak, M., Yildiz, H.B., and Toppare, L. (2014) Enzyme immobilization in a photosensitive conducting polymer bearing azobenzene in the main chain. *Polym. Bull.*, **71**(7), 1827–1841.
73 Demirkol, D.O., Yildiz, H.B., Sayin, S., and Yilmaz, M. (2014) Enzyme immobilization in biosensor constructions: self-assembled monolayers of calixarenes containing thiols. *RSC Adv.*, **4**(38), 19900–19907.
74 Sonmez, B., Sayin, S., Yalcinkaya, E.E., Seleci, D.A., Yildiz, H.B., Demirkol, D.O., and Timur, S. (2014) Calixarene modified montmorillonite: a novel design for biosensing applications. *RSC Adv.*, **4**(108), 62895–62902.
75 Gokoglan, T.C., Soylemez, S., Kesik, V., Unay, H., Sayin, S., Yildiz, H.B., Cirpan, A., and Toppare, L. (2015) A novel architecture based on a conducting polymer and calixarene derivative: its synthesis and biosensor construction. *RSC Adv.*, **5**(45), 35940–35947.
76 Azak, H., Guler, E., Can, U., Demirkol, D.O., Yildiz, H.B., Talaz, O., and Timur, S. (2013) Synthesis of an amine-functionalized naphthalene-containing conducting polymer as a matrix for biomolecule immobilization.". *RSC Adv.*, **3**(42), 19582–19590.
77 Dervisevic, M., Dervisevic, E., Azak, H., Cevik, E., Senel, M., and Yildiz, H.B. (2016) Novel amperometric xanthine biosensor based on xanthine oxidase immobilized on electrochemically polymerized 10[4H-dithieno(3,2-b:2 ',3 '-d)pyrrole-4-yl] decane-1-amine film. *Sens. Actuators, B*, **225**, 181–187.

8

Heterogeneous Catalysis from the Perspective of Surface Science

Aydin Cihanoğlu, Diego Hernán Quiñones-Murillo, and Gizem Payer

8.1 Introduction to Solid Surface

8.1.1 Historical Perspective of Surface Science and Catalysis

The history of surface science in general and surface chemistry in particular goes back more than 200 years. The historical development of surface chemistry is summarized in Figure 8.1. Catalysis first appeared in 1823 and started with the study of Johann W. Döbereiner, who discovered the platinum-surface-catalyzed reaction of H_2 and O_2. However, in 1835, the first description of catalysis was given by Jöns J. Berzelius, one of the founding father of catalysis. He presented his ideas related to catalysis as a new force: "It is, then, proved that several simple or compound bodies, soluble and insoluble, have the property of exercising on other bodies an action very different from chemical affinity. By means of this action they produce, in these bodies, decompositions of their elements and different recombinations of the same elements to which they remain indifferent." It was at about this time that the first foundation of heterogeneous catalysis was laid by Kirchhoff, Davy, Henry, Philips, and Faraday [1–4].

During the period 1860–1912, the surface chemistry of catalysis made progress as a result of many experimental observations of scientists, such as in oxidation of hydrogen chloride (Deacon process), SO_2 oxidation to SO_3, the reaction of methane with steam to form CO and H_2, the oxidation of ammonia, ethylene hydrogenation, and the synthesis of ammonia [5].

In the twentieth century, the Haber–Bosch process of ammonia synthesis was an important turning point in the understanding of the molecular basis of catalyst thanks to the explanation of the thermodynamics of this process. It was realized that a support could provide a complete picture of several catalyst reactions. Even during the 1930s, the original catalytic cracking catalysts were based on natural clays, and graded river pebbles were used as catalyst supports from 1950 through the 1970s. Before this period, metal catalyst supports were made from various materials such as activated carbon, clays, asbestos, quartz, corundum, firebrick, kieselguhr, and pumice.

Until recently, many scientific researches were rewarded with Nobel prices. In 2001, William Knowles, Ryoji Noyori, and Barry Sharpless were awarded the Nobel Prize for chirally catalyzed hydrogenation and oxidation reactions.

Surface Treatments for Biological, Chemical and Physical Applications, First Edition.
Edited by Mehmet Gürsoy and Mustafa Karaman.

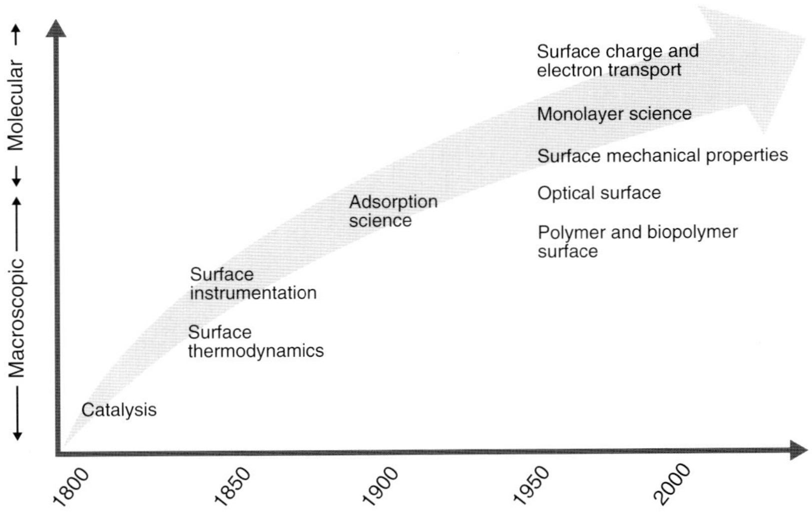

Figure 8.1 The historical development of surface chemistry.

Yves Chauvin, Robert Grubbs, and Richard Schrock received the same honor for the improvement of the metathesis process in organic synthesis in 2005. Gerhard Ertl, who was awarded the Nobel Prize for Chemistry in 2007, clarified what occurs on the surface of a catalyst [6]. With the development of molecular chemistry, and their increasing use in domestic and industrial applications, the research in catalyst technologies has been increasing steadily. The significant dates related to catalysis in relevant historical records are summarized in Table 8.1.

8.1.2 Industrial and Economical Aspects of Catalysis

Catalytic technology has to be complemented by advancements in the chemical industry. The development of catalytic technology is closely related not only to the technological advances in the chemical industry, but also to significant political concerns. Due to societal and political conflicts, catalysts, especially solid catalysts, have attracted interest both from chemical industry and scientific research. In industrial catalytic process, heterogeneous catalysts (80%) are predominantly used whereas the remaining including homogeneous catalysts (17%) and biocatalysts (3%) are widely used only in special applications. The heterogeneous catalysts or special solid catalysts are suited in both batch and continuous processes. Heterogeneous catalysts provide major advantages in the continuous process, in which these catalysts are continuously separated from products, in respect to the efficient conversion for large scales.

A sense of their histories and of developments related to chemical phenomenon and chemical engineering is of notable importance to inclusively understand current catalysis processes. The industrial revolution started with the study of heterogeneous catalytic oxidation by P. Phillips, Junior, who was granted patent "Certain Improvements in Manufacturing Sulphuric Acid commonly called Oil of Vitriol" in 1831. Industrial and society requirements

Table 8.1 Significant dates of advances in catalysis.

Year	Investigator	Phenomenon
1823	J. W. Döbereiner	The platinum-surface-catalyzed reaction of H_2 and O_2
1825	Faraday	Surface-induced combustion of hydrogen
1831	Phillips	Oxidation SO_2 to SO_3 platinum
1835	J. J. Berzelius	Description of catalysis
1860		Oxidation of hydrogen chloride (Deacon process)
1869	von Hoffmann	Methanol oxidation for the production of formaldehyde on silver
1909	Haber, Bosch, and Mittasch	Development of ammonia synthesis
1913	Bergius	First industrial ammonia synthesis
1930		Original catalytic cracking catalysts
1955	Ziegler, Natta	Polymerization catalysis
1955–1959		Synthetic zeolites
1960–1970		Graded river pebbles used as catalysts supports
1980		Production of gasoline from methanol by H-ZSM-5 zeolites
2001	Knowles, Noyori, and Sharpless	Chirally catalyzed hydrogenation and oxidation reactions
2005	Chauvin, Grubbs, and Schrock	Development of the metathesis process in organic synthesis
2007	Gerhard Ertl	Description of chemical processes on the surface of catalysts

in conjunction with the demand for raw material, new products, and the need to reduce process constraints have forced to find improvementse in catalytic processes. Table 8.2 gives the information on the chemical industry based on catalysts in the twentieth century [1, 7, 8].

Increasing the growth rate of production in chemical industry leads to an increase in the demand for catalysts. Beside this, the cost of catalysts is amplified by its usage. The demand for catalysts in 2004 was estimated around 15 billion US$, with 12 billion US$/a for solid catalysts. In the last decade, the growth rate has appeared to be about 5% per year, as shown in Figure 8.2 [9, 10].

8.2 Reaction Mechanisms and Kinetics

8.2.1 Catalysis

Catalysis is of great importance for modern energy conversion, advanced technology, and for the production of several chemical products. Since the last two centuries, especially after the Industrial Revolution, catalysts have played a vital

Table 8.2 The developments in chemical industry in the twentieth century.

Year	Process	Industry
Before 1900	Sulfuric acid, chlorine, soda	Heavy industry (inorganic chemistry)
1913–Present	Solid catalysts	Continuous process in industry
1900–1940	Ammonia synthesis, nitric acid, fat hardening, hydrodenitrogenation, methanol synthesis	Coal-based industry
1940–Present	Fuel, synthetic rubber, polymers, selective oxidation, homogeneous reaction	Petrochemical industry
Before–1970	The synthesis of fine chemicals, environmental catalysts	New process for effective use of energy and raw materials

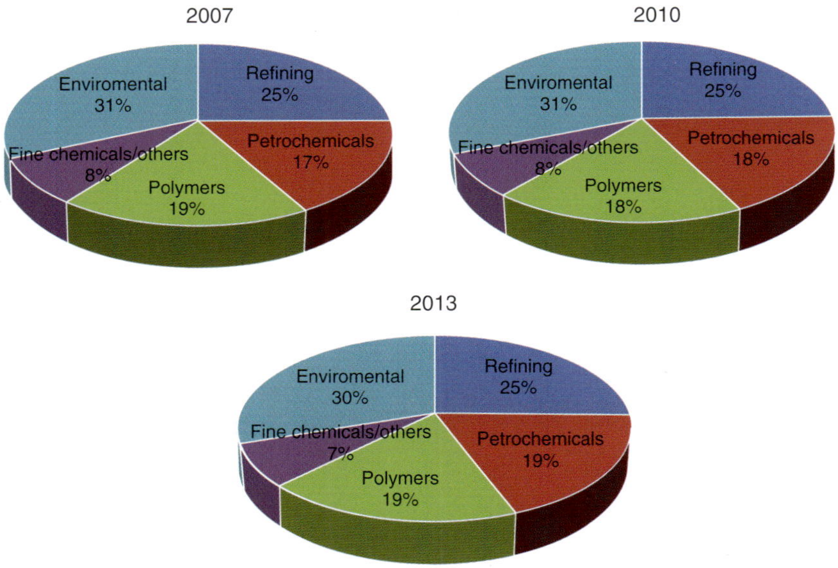

Figure 8.2 The world catalyst sales in 2007, 2010, and 2013 (billions of U.S. dollars).

role for oil refining and in the production of most chemicals. A catalyst is a substance that causes or accelerates a chemical reaction without itself being affected in the process [11, 12]. Catalysts can allow to enhance the rate of the interactions with reactants in a reaction, but will not change throughout the reaction and thereby provide control of the selectivity of a chemical reaction. Catalysts can lead the alteration of the activation energy (Ea) in a reaction, which is equated with the minimum energy possessed by the reacting molecules before a reaction, as seen in Figure 8.3 [13, 14].

Catalysis can be categorized as homogeneous and heterogeneous. Homogeneous catalysis is where the catalysts and reactants are in the same physical phase,

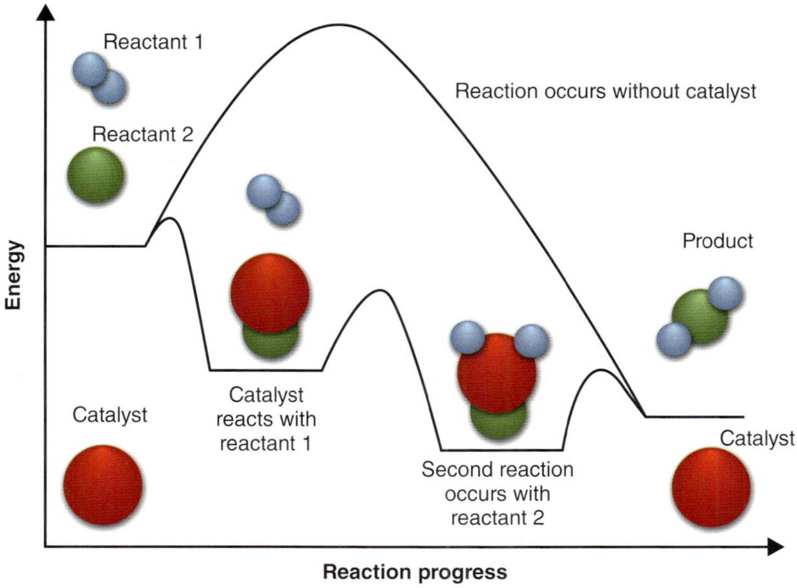

Figure 8.3 The difference between catalyzed reaction and uncatalyzed reaction.

Table 8.3 The differences between Heterogeneous catalysis and Homogeneous catalysis.

Homogeneous catalysis	Heterogeneous catalysis
In the same phase as the reactant	In bulk phase
Liquid form	Usually solid form (reaction occurs in the liquid or gas phase)
High selectivity in liquid phase	Low selectivity
Act as single active site	Act as multiple active sites
Poor thermal stability	Excellent thermal stability
Difficult recovery, expensive	Easy separation, cheap
Organometallic materials or enzymes	Inorganic materials or metal salts

called as *soluble catalysts*, usually liquids. Examples of common homogeneous catalysts are metal salts of organic acids, organometallic complexes and carbonyls of Co, Fe, and Rh, or coordination complexes, enzymes, artificial enzymes. Heterogeneous catalysis is a system in which the reaction occurs where the catalysts and reactants are in different phases, usually solids. Inorganic materials such as metals, oxides, sulfides, supported metals and metal salts, and some organic solids such as organic hydroperoxides and ion exchangers are typical heterogeneous catalysts. The adsorption of reactants from a fluid phase on a solid surface, surface reactions, and desorption of products in the fluid phase occur in a heterogeneous catalyzed-reaction, respectively [15–18]. The striking differences between homogeneous and heterogeneous catalysis are summarized in Table 8.3.

8.2.2 Individual Steps in Heterogeneous Catalysis

Heterogeneously catalyzed reactions are based on chemical and physical reaction steps. For the catalytic process to occur, the reactants must be transported from the bulk to the surface of the catalyst. Thus, diffusion, adsorption, and desorption processes play a key role for the progress of the overall reaction apart from the actual chemical reaction.

A catalytic transformation of reactants into products on the surface of the porous heterogeneous catalyst involves a sequence of elementary steps:

1) Mass transfer of reactants from the bulk fluid to the fluid–solid interface (external surface of the catalyst).
2) Diffusion of the reactant from the pore mouth through the catalyst pores to the intermediate vicinity of the internal catalytic surface (if it is porous).
3) Adsorption on the surface of the catalyst.
4) The chemical rearrangements (bond breaking, bond forming, molecular rearrangement) of the adsorbed reaction intermediates (surface reaction).
5) Desorption of the products from the surface of the catalyst.
6) Diffusion into the catalyst pores and then to the pore mouth at the external surface of the catalyst (if it is porous).
7) Mass transfer from the surface and pore mouth of the catalyst to the bulk fluid.

For a reaction (or reactions) these elementary steps occur repeatedly as long as there is a supply of reactants and no loss of catalytic activity [19, 20].

8.2.3 Rates of Reaction

For an irreversible gas-phase reaction,

$$A_{(g)} \rightarrow B_{(g)}$$

this requires a solid catalyst. Assume that the temperature is constant and that the reaction is realized on the nonporous catalyst particles. Because the catalyst is nonporous, steps 2 and 6 are not involved. The overall conversion of A to B in the bulk gas takes place according to steps 1, 3, 5, and 7. The problem may be further simplified by assuming that steps 3 and 5 may be represented by a single first-order rate equation. Then the overall reaction process may be described in three steps:

1) Gas A is transported from the bulk gas to the catalyst surface,
2) The reaction takes place at the interface,
3) Product B is transported from the catalyst surface to the bulk gas.

If the reaction is irreversible, the concentration of B at the solid surface does not influence the rate. It shows that the global rate for a catalyst particle, r_p, can be formulated by considering the first two steps. If the rates of these two steps are similar at steady state, the disappearance of A can be expressed in two ways:

1) The rate of transport of A to the catalyst surface,

$$r_p = k_m a_m (C_b - C_s)$$

2) The rate of reaction at the catalyst surface,

$$r_p = k a_m C_s$$

where k_m is the usual mass-transfer coefficient based on a unit of transfer surface, a_m is the external area per unit mass, k is the reaction rate constant per unit surface, and C_s and C_b are the concentration of the surface and bulk, respectively.

The global rate can be formulated in terms of C_b,

$$C_s = \frac{k_m}{k_m + k} C_b$$

$$r_p = \frac{k k_m a_m}{k_m + k} C_b = \frac{a_m}{1/k + 1/k_m} C_b$$

This is the expression for the global rate in terms of the bulk reactant concentration [19].

8.2.3.1 Reaction Mechanisms and Rate Laws

Reaction mechanism is a hypothetical set of elementary reactions suggesting a sequence of molecular events occurring to form products from reactants. The mechanism of a reaction is the sequence of elementary steps. In elementary steps, only one or two atomic bonds are broken or formed. Individual steps in a reaction mechanism are called elementary reactions. An elementary reaction involves a simple single step and stoichiometric coefficients are identical to powers of species concentration in the reaction. Elementary reactions must be consistent with bond energies, atomic geometry, and allow for electronic configurations.

Overall reaction consists of a series of simple steps (i.e., elementary reactions). The rate of each reaction is different and the rate of overall reaction is determined by the slowest of all the steps.

Postulated reaction mechanism must be consistent with:

1) Overall reactions and formation of all observed products
2) A rate expression (i.e., reaction rate low), derived from the proposed reaction mechanism, and must be consistent with experimentally observed kinetic data.

Reaction mechanisms are composed of a number of steps and some of those steps could involve reacting species (or called intermediates or active centers) that are not observed in the overall reaction equation. For instance, for the H_2 and O_2 reaction, the reaction mechanism contains at least 26 steps. In fact, many reaction mechanisms have a large number of elementary steps.

Derivation of reaction rate expression from a postulated reaction mechanism following assumptions could be made under certain experimental reaction conditions;

1) Neglecting reverse reaction; that is, reversible reactions are considered irreversible reactions. This assumption is valid during the initial stages of any reaction, that is, $t \to 0$ or if the reaction equilibrium constant is very large, the forward reaction rate is much greater than the reverse reaction.

2) In a sequence of reactions in a reaction mechanism, one or two elementary reactions could be assumed to be much slower than the others. Hence, this slow step(s) controls the entire overall reaction. It is also known as *rate limiting step* or *rate controlling* or *rate determining*.
3) If the concentration of one or more intermediate species in the reaction mechanism does not rapidly change with time, that is, $dC^*/dt \approx 0$, tt means no accumulation of intermediate species. This is called quasi-stationary state situation (QSS) or pseudo steady state hypothesis (PSSH). QSS (or PSSH) means that the net rate of formation of intermediate is equal to zero; that is, the rate of production of intermediates is nearly equal to the rate of consumption of intermediates. This is valid for highly reactive intermediates.
4) It is assumed that in a reaction mechanism, one or more elementary reactions are in quasi-equilibrium condition, that is, the forward and reverse reaction rates of the reversible elementary reactions are much greater than the reaction rates of other elementary reactions. This means that reversible elementary reaction stays at equilibrium during a reaction.

8.2.3.2 Microscopic Reversibility Principle

In a reaction system at equilibrium, the most probable path for forward reaction is also the most probable path for the reverse reaction, that is, the reacting molecules flowing along a particular reaction path must equally flow in the opposite direction. The microscopic reversibility indicates that reaction rate constants of forward and reverse reaction are related to equilibrium constants.

$$A \underset{k_2}{\overset{k_1}{\rightleftharpoons}} B \quad K = \frac{C_B}{C_A} = \frac{k_1}{k_{-1}}$$

8.2.3.3 Rule of Simplicity

How to choose between two alternative mechanisms is decided by rule of simplicity. It is derived from a general philosophical principle known as *Occam's razor*. It assumes that nature prefers the simplest to complex; thus, reaction follows the simplest path.

The next question will be "Which mechanism is simpler." The following information will help to determine the simpler steps.

1) Choose bimolecular steps over termolecular steps.
2) Choose steps involving smaller change in chemical structure.
3) Complicated atomic rearrangement in a single step is not a simple path.
   a) Unimolecular reactions: A → Products
   b) Bimolecular reactions: A + B → Products
   c) Termolecular reactions: A + 2B → Products.

8.2.3.4 Chain Reactions

One of the reaction intermediates in one elementary reaction reacts with another species in an earlier elementary reaction. This is called chain reaction.

Chain reactions occur through the following:

1) *Initiation reaction*: It generates species that will be chain carries or active centers for the elementary reactions after an initiation reaction step.

2) *Propagation reaction*: These are the elementary reactions responsible for the formation of products. Each reaction also forms a chain carrying species that propagates the reaction. In fact, combination of elementary reactions in propagation reactions results in the overall reaction (i.e., stoichiometric equation).
3) *Termination reaction*: Some of the chain carriers are consumed in termination reactions; that do not form species maintaining a chain reaction.

Relative rates of propagation and termination steps determine the length of the chain; for example, distribution of molecular weight in polymers, formed by chain reactions, is the result of relative rates of propagation and termination steps.

$$\text{Chain length} = \frac{\text{rate of disappearance of reactant}}{\text{rate of chain carriers formation by initiation}}$$

If a single chain carrier is formed in the propagation steps, it is called linear chain reaction. If more than one chain carrier is formed in propagation steps, it is known as *branching reaction*.

$$H_2 + Br \leftrightarrow 2HBr \quad \text{overall reaction} \quad {}^{*}\text{“Active species”}$$

Initiation step:

$$Br_2 \xrightarrow{k_1} 2Br^*$$

Propagation step:

$$Br^* + H_2 \xrightarrow{k_2} H^* + HBr$$

$$H^* + Br_2 \xrightarrow{k_3} Br^* + HBr$$

$$H^* + HBr \xrightarrow{k_4} H_2 + Br_2$$

Termination step:

$$2Br^* \xrightarrow{k_5} Br_2$$

$$r_2 = k_2 C_{Br^*} C_{H2} \quad r_3 = k_3 C_{H^*} C_{Br2} \quad r_4 = k_4 C_{H^*} C_{HBr} \quad r_{HBr\,Net} = r_2 + r_3 - r_4$$

by using QSS (i.e., PSSH)

$r_{H^*Net} \approx 0$ $\quad r_{Br^*Net} \approx 0$ $\quad C_{H^*}$ and C_{Br^*} are obtained using the rates knowledge.

8.2.3.5 Chain Transfer Reactions

Initially formed active species reacts with a reactant to produce a new active species in the propagation steps. In fact, this new active species keeps reacting with a reactant to form another new active species; that is, higher molecular weight new active species is formed every time.

Free radical polymerization reactions follow chain transfer reaction mechanism:

Initiation step:

$$A_2 \rightarrow A^* + A^*$$

Propagation step:

$$A^* + B \rightarrow BA^*$$
$$BA^* + B \rightarrow B_2A^*$$
$$B_2A^* + B \rightarrow B_3A^*$$
$$B_{n-1}A^* + B \rightarrow B_nA^*$$

Termination step:

$$BA^* + BA^* \rightarrow (BA)_2$$

8.2.3.6 Enzymatic Reactions

Enzymes are macromolecular biological catalysts. They accelerate chemical reactions. The molecules at the beginning of the reaction are called substrates and the enzyme converts the substrates into different molecules, called products. The rate of a chemical reaction is increased by means of active sites on the enzyme catalysis.

In principle, the mechanism of enzyme catalysis is the same as with other types of chemical catalysis. By ensuring an alternative reaction route, the enzyme decreases the energy required to reach the highest energy transition state of the reaction. The reduction of Ea increases the amount of reactant molecules that achieve a sufficient level of energy, such that they reach the Ea and form the product. As with other catalysts, the enzyme (catalyst) is not consumed during the reaction.

The reaction mechanism sequences are thus:

$$S \xrightarrow{E} P \quad S + E \underset{k_{-1}}{\overset{k_1}{\rightleftharpoons}} S^*E$$

S*E: intermediate (i.e., enzyme attached to reactant)

$$S^*E \xrightarrow{k_2} P + E$$

The reaction rate is, therefore, given by Michaelis–Menten kinetics

$$-r_S = \frac{k_2 \, C_S \, C_{E0}}{\frac{k_{-1}}{k_1} + C_S}$$

8.2.3.7 Inhibition of Enzymatic Reactions

This type of enzymatic reactions divides into two different ways:

1) *Inhibition by a third substrate, called inhibitor.*
   a) *Competitive inhibition*: when A (reactant) and B (inhibitor) attack the same site on enzyme.

$$A + E \underset{k_{-1}}{\overset{k_1}{\rightleftharpoons}} A^*E \xrightarrow{k_3} P + E$$

$$B + E \underset{k_{-2}}{\overset{k_2}{\rightleftharpoons}} B * E$$

b) *Non-competitive inhibition*: when B (inhibitor) attacks a different size on enzyme but it stops action of A (reactant).

$$A + E \underset{k_{-1}}{\overset{k_1}{\rightleftharpoons}} A^*E \xrightarrow{k_4} P + E$$

$$B + E \underset{k_{-2}}{\overset{k_2}{\rightleftharpoons}} B^*E$$

$$B + A^*E \underset{k_{-3}}{\overset{k_3}{\rightleftharpoons}} A^*B^*E$$

2) *Substrate inhibition*
Excess substrate causes interference at the active site of enzyme thus lowering the reaction rate.

$$A + E \underset{k_{-1}}{\overset{k_1}{\rightleftharpoons}} A^*E \xrightarrow{k_3} P + E$$

$$A + A^*E \underset{k_{-2}}{\overset{k_2}{\rightleftharpoons}} A^*E^*A$$

8.2.3.8 Heterogeneous Catalytic Reactions

Two mechanisms, Langmuir–Hinshelwood and Eley–Rideal, are often discussed for bimolecular gas phase reactions, which occur widely in heterogeneous catalysis.

$$A + B \xrightarrow{\text{cat}} C \qquad {}^*\text{: adsorbed molecules}$$

1) *Langmuir–Hinshelwood mechanism*
In this reaction mechanism, first of all, both reactants are adsorbed onto the catalyst surface, before a reaction occurs. Then surface diffusion facilitates interaction between neighboring chemisorbed molecules to give the product C, which is adsorbed on the catalyst surface [20]. Finally, the reaction product desorbs from the catalyst surface.
The reaction mechanism sequences are thus:

$$\left. A + {}^* \underset{k_{-1}}{\overset{k_1}{\rightleftharpoons}} A^* \right\} \text{adsorption steps for reactants A and B}$$

$$\left. \begin{array}{l} B + {}^* \underset{k_{-2}}{\overset{k_2}{\rightleftharpoons}} B^* \\ A^* + B^* \underset{k_{-3}}{\overset{k_3}{\rightleftharpoons}} C^* \end{array} \right\} \text{surface reaction step}$$

$$\left. C^* \underset{k_{-4}}{\overset{k_4}{\rightleftharpoons}} C + {}^* \right\} \text{desorption steps for products C and D}$$

The rate of a heterogeneous reaction is controlled by the reaction of the adsorbed molecules, and that all adsorption and desorption pressure are in equilibrium. The reaction rate is, therefore, given by,

$$r_{\text{eff}} = \frac{dpc}{dt} = k\theta_A\theta_B = \frac{kK_A p_A K_B p_B}{(1 + K_A p_A + K_B p_B)^2}$$

The Langmuir–Hinshelwood mechanism has been verified by many industrial scale reactions. For instance:

a) Oxidation of CO on Pt catalysts:

$$2CO + O_2 \rightarrow 2CO_2$$

b) Methanol synthesis on ZnO catalysts:

$$CO + 2H_2 \rightarrow CH_3OH$$

c) Reduction of N_2O with H_2 on Pt or Au catalysts:

$$N_2O + H_2 \rightarrow N_2 + H_2O$$

2) *Eley–Rideal mechanism*

In this reaction mechanism, first of all, only one of the gaseous reactants is adsorbed onto the catalyst surface. Subsequently, the other reactant reacts with the adsorbed species directly from the gas phase to give the chemisorbed product C. Finally, the reaction product desorbs from the catalyst surface [20]. The reaction mechanism sequences are thus:

$$A + {}^* \underset{k_{-1}}{\overset{k_1}{\rightleftharpoons}} A^* \quad \text{adsorption step for reactant A}$$

$$B + A^* \underset{k_{-2}}{\overset{k_2}{\rightleftharpoons}} C^* \quad \text{surface reaction between reactant B and surface adsorption } A^*$$

$$A^* \underset{k_{-3}}{\overset{k_3}{\rightleftharpoons}} C + {}^* \quad \text{desorption of product C}$$

The reaction rate is therefore given by,

$$r_{\text{eff}} = k\theta_A p_B = k\frac{K_A p_A}{(1 + K_A p_A)} p_B$$

The Eley–Rideal mechanism has been proven by many industrial scale reactions. For instance:

a) Reduction of CO_2 with H_2:

$$CO_2 + H_2^* \rightarrow CO + H_2O$$

b) Oxidation of ammonia on Pt catalyst:

$$2NH_3 + 3/2O_2^* \rightarrow N_2 + 3H_2O$$

c) Selective hydrogenation of acetylene on Ni or Fe catalysts:

$$HC \equiv CH + H_2^* \rightarrow H_2C{=}CH_2$$

8.3 Preparation of Catalysts

Catalysts can be prepared by different strategies that usually involve a series of necessary elementary steps that depend on the nature of the material to be prepared [21]. Each step is aimed at having the desirable physicochemical characteristics on the three fundamental components of a catalyst: support (used to hold and stabilize the catalytic particles), catalytic particles (which contain the active sites), and promoters (used to enhance the catalytic activity). Figure 8.4 represents schematically a typical catalyst preparation procedure (adapted from [21]).

The catalyst preparation starts with a preliminary stage of preparation of the primary solid (or components) to be used in the obtainment of the catalyst precursor; afterwards that primary solid is processed to develop in it the characteristics that will make it an optimum catalyst precursor; then the precursor is treated to have the desired texture, shape, and resistance conditions, and finally, an activation stage is carried out to produce an active catalytic material [22, 23]. Strategies most frequently used to manufacture catalysts can comprise precipitation or other deposition process, hydrothermal transformation, decantation, filtration, centrifugation, washing, crushing and grinding, forming and/or shaping operations, calcination, impregnation, reduction, mixing and activation [21–23].

8.3.1 Precipitation

During precipitation, a solid-phase product (or precipitate) is formed from a homogeneous liquid medium, due to physical changes (variation of temperature, solvent, or concentration), or chemical processes (change of pH, complexation reactions). This occurs through three elementary steps, namely supersaturation (rising of the concentration above the saturation point), nucleation (formation of small and stable solid particles), and growth (agglomeration of the formed insoluble particles) [23]. In addition, multicomponent precipitates can be prepared by co-precipitation and used as the precursors of binary or multimetallic oxidic catalysts [24]. However, co-precipitation hardly leads to good macroscopic homogeneity and the formed solid phase may be subjected to treatments such as aging, washing, filtration, hydrothermal transformation, drying, impregnation, and calcination specially to have the desired physicochemical properties, structure, and composition on the surface of the final catalyst [24]. The size of the crystals of the precipitate depends on the ratio of nucleation rate to crystal growth rate. A greater ratio leads to smaller crystallites, and vice versa [21].

Figure 8.5 summarizes the main factors influencing the properties of the final product in a precipitation procedure.

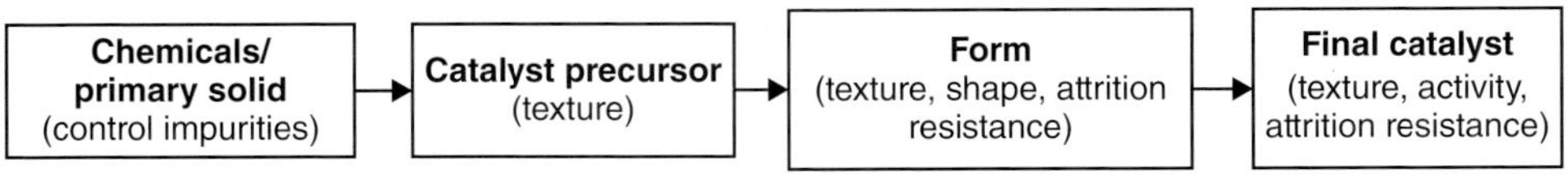

Figure 8.4 Elementary steps of the preparation of solid catalysts.

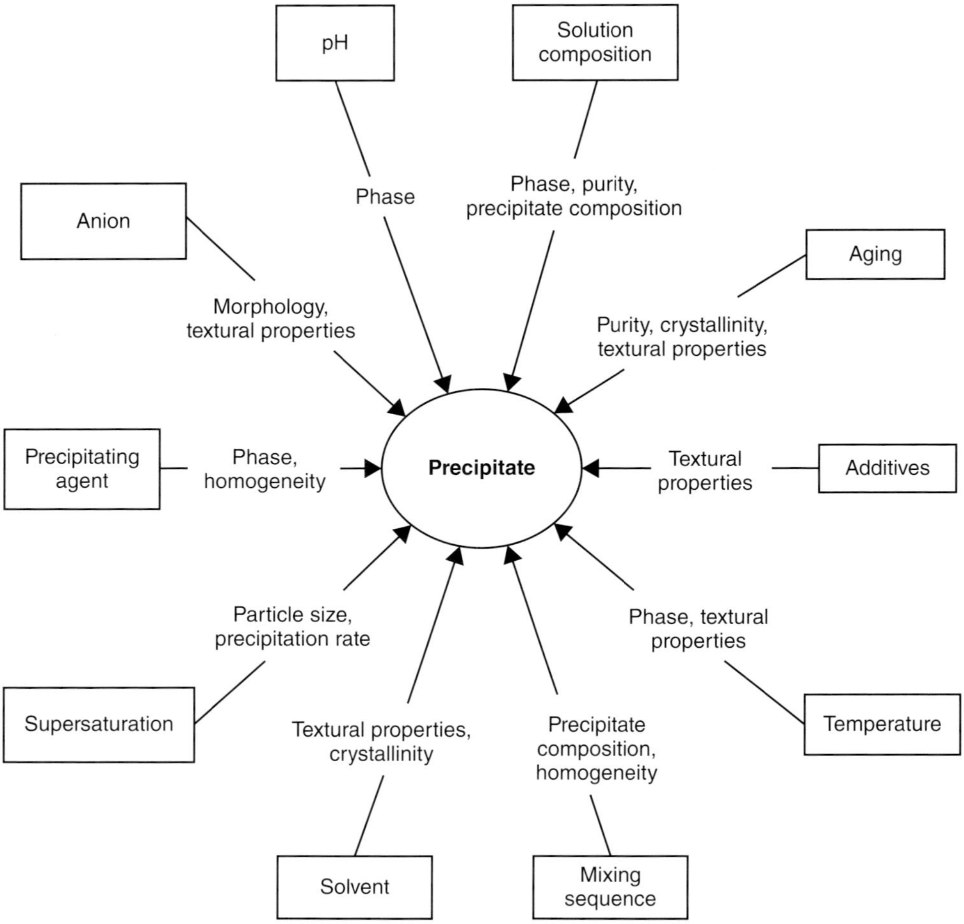

Figure 8.5 Main parameters affecting the properties of final precipitate and effects. (Hutchings 2004, [21]. Reproduced with permission of Springer.)

According to Figure 8.5, in a precipitation process the variables that influence the properties of the final product are the use of additives (such as templating agents and other miscellaneous organic molecules) or precipitating agents, the mixing procedure (which affects mass homogeneity and uniform temperature control), variation of pH during the process, the maturation process, among others [22].

The most common catalysts derived from precipitation are silica-aluminas used in fluid catalytic cracking technology, metal oxides used in Fischer–Tropsch reactions, strong acid reactions, selective oxidations, and active copper catalysts for the synthesis of hydrocarbons or methanol by hydrogenation [21, 24].

8.3.2 Gelation (Sol–Gel Process)

Gelation method (or the so-called sol–gel process) is based on the preparation of gels or solid-like substances. The method consists in continuously transforming a homogeneous solution into a three-dimensional network of micelles that encapsulate molecules of water (or liquid). Such micelles remain separated due to repulsive electrical charges on their surfaces and in the solution, thus forming a hydrophilic colloidal solution, known as *hydrogel*, which is a hydrated solid precursor that retains the active elements contained in the starting solutions [21, 23, 24].

The gel can be obtained by chemical reaction (alkoxide hydrolysis or polymerization of an anion), complexation reaction (with an acid–alcohol), freeze drying, and use of gelling agents (e.g., hydroxymethyl cellulose). During the gelation processes it is crucial to avoid segregation of the active components present in the homogeneous solution. When the gel is formed, diffusion becomes highly limited and, therefore,segregation of the active components is nonviable. The solvent and reaction by-products are eliminated through evaporation or sublimation operations and the gels are decomposed or transformed, normally into oxides [21]. Sol–gel advantages over precipitation in the preparation of catalysts are better control over surface area, volume, and pore size distribution. These parameters are controlled by reaction conditions such as pH, solvent, and amount of water [23]. This method also stands out because of its versatility since it makes possible to control parameters of the final solids such as the structure, texture, composition, and homogeneity. The method offers possibilities to prepare tailored materials, such as dispersed metals, oxidic catalysts, and chemically modified supports [21].

Typical precursors used for the preparation of oxide aerogels are organic alkoxides, acetates, acetylacetonates, and inorganic salts. Common solvents are alcohols, although other solvents such as benzene can be used for some alkoxides. In some cases, a polycondensation stage is carried out under the catalysis of some volatile acids or bases [24].

The sol–gel method used to create polymeric oxide gels can be represented by the following sequence of acid- or base-catalyzed nucleophilic additions or substitutions [24] and employs monometallic alkoxides ($M(OR)_n$) to be subjected to hydrolysis with monomer formation (Eq. (8.1)), followed by a condensation to generate M—OH—M and M—O—M bonds (Eqs. (8.2) and (8.3)) and then forms a sol through dehydration or dealcoholation operations, and cross-linking (Eqs. (8.4) and (8.5)) [21, 24].

$$M(OR)_n + xH_2O \rightarrow M(OH)_x(OR)_{n-x} + xROH \tag{8.1}$$

where M = metal or Si; R = alkyl.

Olation (condensation with formation of hydroxyl bridges):

$$M - OH + M - OHX \rightarrow M - OH - M + X - OH \tag{8.2}$$

where X = H or R.

Oxolation (condensation with formation of oxygen bridges):

$$-M - OH + XO - M- \rightarrow -M - O - M - +X - OH \tag{8.3}$$

Olation process is much faster than oxolation. As polymerization and cross-linking progresses, the viscosity of the system gradually increases. Then the formation of a gel occurs during drying or aging operations.

$$(n+1)\,(\mathrm{R-O})_x-\mathrm{M}-\mathrm{OH} \rightarrow (\mathrm{R-O})_x-\mathrm{M}\left[\mathrm{OM(O-R)}_{x-1}\right]_n\mathrm{OH}+n\mathrm{R-OH} \quad (8.4)$$

$$(\mathrm{R-O})_x-\mathrm{M}\left[\mathrm{OM(O-R)}_{x-1}\right]_n\mathrm{OH}+\mathrm{H_2O} \rightarrow (\mathrm{R-O})_x-\mathrm{M}\left[\mathrm{OM(O-R)}_{x-2}\mathrm{OH}\right]_n\mathrm{OH}+n\mathrm{R-OH} \rightarrow \mathrm{gel} \quad (8.5)$$

Through the sol–gel process, a wide number of catalytic materials can be prepared as single-metal oxides such as alumina supports with high surface area and variable crystallinity, metastable cubic-phase ZrO_2 in ZrO_2-SiO_2 system, semiconductor for photocatalytic applications [25], and platinic metal catalysts with metal precursor incorporated during the gelation step [24].

8.3.3 Impregnation

During impregnation, a solid is put into contact with a solution that contains some components that will be deposited and retained on the solid [22]. This method, useful in the preparation of supported catalysts, involves three basic steps: contacting a support with a solution containing the components to be impregnated during a period of time (typically the solution consists of an active metal precursor in water or an organic solvent), drying the support in order to drive off and evaporate the volatile components in the solution depositing the metal on the catalyst surface, and activating the catalyst by the appropriate treatment [23]. Depending on the amount of impregnating solution, this procedure can be done by two routes: either by using an excess of solution or by applying repeatedly the solution (known as *dry impregnation* or *impregnation to incipient wetness*).

When the solution is used in excess, the catalyst support is dripped with an amount of precursor-containing solution exceeding the support pore volume. The process is carried out during the necessary time for total impregnation. Afterwards, the solid is then drained to eliminate volatile components within the solution. Sometimes, a mud can be formed making it difficult to make use of the solution completely. When impregnation is carried out by incipient wetness, the support is put in contact repeatedly with an amount of solution equal to or less than the total pore volume of the support (usually drop wise). This process is repeated until the desired amount of the precursor has been impregnated [23].

The processes that take place during an impregnation procedure are: the selective adsorption of (precursor) species on the support by Coulomb force, van der Waals forces, or H-bonds; ion exchange phenomena between the support charged surface and electrolyte solution; and polymerization of species attached to the support surface [22]. Parameters such as pH, nature of the solvent, nature, and concentrations of the dissolved substances, ions containing the active elements, morphology of the solid, nature, and strength of functional groups on the solid, and the reactivity between the solid and liquid may affect the final product.

8.3.4 Chemical Vapor Deposition

In a chemical vapor deposition (CVD) process a deposit of one or more volatile precursors from gas phase is formed on the exposed surface of a solid substrate. The deposition may occur by reaction and/or decomposition of the gas-phase precursor on the substrate surface. This method leads to the preparation of catalyst supports with excellent dispersion and well-controlled distribution of active species, which enhances the activity of the catalyst. An example of the method is the preparation of zeolites with controlled pore-opening sizes. Among the advantages of the CVD method are the elimination of surface poisoning problems that may occur during liquid-phase preparations using solvents and the re-distribution of active material during drying. Thus, the CVD method can be highlighted as a promising pathway for vapor deposition or subliming of catalyst precursors, such as volatile organic complexes, metal carbonyls, halides, diketonates, alkoxides, acetates, and so on, onto a catalyst support, usually under low-pressure conditions [8, 24, 26, 27].

Most recently, metal oxide films deposited on supports were prepared using volatile alkoxides and metal oxychloride via CVD method. For preparation of TiO_2 catalysts, volatile metallo-(organic/inorganic) precursors are used (such as titanium tetra-isopropoxide, tetrabutyltitanate, titanium tetrachloride, or tetranitratotitanium) [28–30]. Metal–organic chemical vapor deposition (MOCVD), compared to the sol–gel method, for preparation of TiO_2 films supported on activated carbon offers a simple and convenient pathway for catalyst preparation since it leads to TiO_2 particles better dispersed on the porous support [31–33]. Through CVD, zeolite catalysts with controlled pore-opening size can be obtained, which may ensure shape-selective reactions and separation, as described in the work of Tominaga and co-workers [34] whereby silica was deposited on the zeolite via CVD to be used in the production of shape-selective *p*-xylene from methylation of toluene [34].

8.3.5 Solvothermal Treatments

A solvothermal process is a method of producing nanocrystalline inorganic materials through a heterogeneous chemical reaction that takes place within a solvent at temperatures higher than that of the boiling point of such a solvent and at pressures greater than 1 bar. According to the solvent used, this method can be named as hydrothermal synthesis (water), ammonothermal (ammonia), alcothermal, or glycothermal (an alcohol); likewise, other organic or inorganic solvents can be used.

During a solvothermal process, the solvent temperature-and-pressure-dependent physical properties (such as density, viscosity, heat capacity, thermal conductivity, dielectric constant) are adjusted as the system pressure and temperature is increased, and along with the use of a correct precursor type and concentration makes it possible to maintain a high degree of supersaturation leading to high rates of nucleation and avoiding nuclei growth. By adjusting system parameters such as temperature, pressure and precursor nature, and concentration the tailoring of almost any materials synthesis without the presence of further structure directing agents [35, 36] can be accomplished.

Table 8.4 Effect of temperature, pressure, and time during a solvothermal process [35, 37].

Parameter	Effects on
Temperature	• Kinetics of product formation • Solubility • Thermodynamic stability of the product phase
Pressure	• Solubility • Supersaturation • Crystallization process • Thermodynamic stability of the product phase
Time	• Kinetic of stable phases • Re-dissolution • Re-crystallization
Precursor concentration	• Control of the crystallite shape (dot, rod, etc.)
Solvent type	• Solubility • Reaction mechanism (intermediate formation, product structural orientation)

Table 8.4 shows the principal effects of temperature, pressure, and time during a solvothermal process.

Solvothermal processes usually involve five different types of reactions such as redox, hydrolysis, thermolysis, complexation, and metathesis. In all cases it is crucial to carefully control both physical and chemical parameters to obtain the desired characteristics of the final products [35]. Hydrothermal transformations are useful in the preparation of mixed oxides, zeolites, and some molecular sieves.

8.3.6 Ion Exchange

The ion exchange method is a way to introduce precursor particles (usually ionic species) from a precursor-containing solution on the surface of a catalyst support [24]. During an ion exchange procedure typically the support that contains an ion A is immersed into an amount of solution containing ion B that exceeds the support pore volume. Thus, ion B steadily penetrates into the pore space of the support displacing the ion A that passes into the solution until equilibrium is reached [8]. This method leads to perfect dispersions of precursor particles on catalyst supports by electrostatic attraction on charged sites of the catalyst surface.

This method is particularly useful in the preparation of catalytic systems, which require charge compensating ions such as some zeolites, cationic clays, and layered double hydroxides that bear electric charges in the crystalline lattice and need the compensation of such electric charges to stabilize their crystalline structure. The charge compensation is done using oppositely charged ions that are bound electrostatically in extra lattice positions [24].

Zeolites are natural cationic exchangers with unique ion-exchange capacity, crystalline structure, and uniform pore sizes. Their framework bears negative charges in the tetrahedron AlO_4 formed by one aluminum atom with three positive charges and four oxygen atoms that provide four negative charges; in this way, the resulting negative charge that is neutralized by the presence of cations such as Na^+, K^+, and so on, may be replaced by other cations during an ion-exchange procedure [8]. By this kind of process, highly selective supported metal catalysts having narrow particle-size distributions can be prepared using zeolites and noble and transition metals. Other natural exchangers are hydrotalcites that can undergo anion exchanging. Ion exchange method is also useful to prepare catalysts for organic reactions like montmorillonites, hectorites, beidellites, and smectites, by modifying the size and accessibility of the internal pores, and thus altering the chemical composition and catalytic activity [38].

8.4 Modifications and Characterizations of Solid Surface

Most of the particulate materials used as heterogeneous catalysts present some limitations such as low stability, formation of agglomerates, and little selectivity, these conditions are likely to be due to weak surface conditions rather than bulk deficiencies themselves; surface optimization can improve the performance in materials that already have the necessary bulk properties [39]. Surface modification and/or functionalization are used to change the physical, chemical, and mechanical properties of solid catalysts promoting a more efficient interfacial interaction with target substrates.

8.4.1 Modification Methods

The modification technique to be used in the preparation of materials with optimal surface properties depends on the characteristics of the substrates to be tuned. The most recurrent method is chemical modification by chemical reaction of the surface components of the catalyst with a chemical agent, or by grafting (or surface fixation) of functionalities over the catalyst surface, as schematized in Figure 8.6. These techniques will be explained through some examples below.

Carbon materials such as activated carbons, carbon nanotubes, and carbon fibers are very important in catalytic and adsorption applications thanks to their exceptional textural properties and surface chemistry. In some cases this potential needs to be improved to enhance their catalytic performance by changing their surface chemistry [40, 41]. These materials are especially useful in processes such as selective catalytic reduction of NO_x with NH_3 in which acidic and basic sites may play an important role; dehydrogenation of alcohols; NO/N_2O reduction; NO reduction with propene; oxidation of SO_2, NO, or H_2S; catalytic ozonation or catalytic wet-air oxidation which require basic sites; or hydrodehalogenation through pyridinic nitrogen sites [40, 42]. To have better activity in such applications, carbonaceous materials are chemically modified by the introduction of oxygen or nitrogen functionalities on their surface (see Figure 8.7). The amount of active oxygen or nitrogen-containing groups and

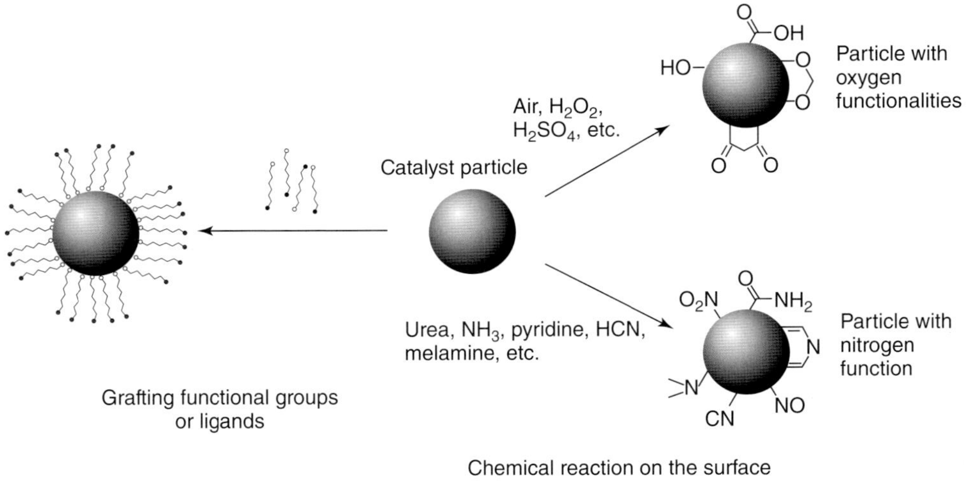

Figure 8.6 Scheme of some typical chemical oxidation and grafting of catalyst surface.

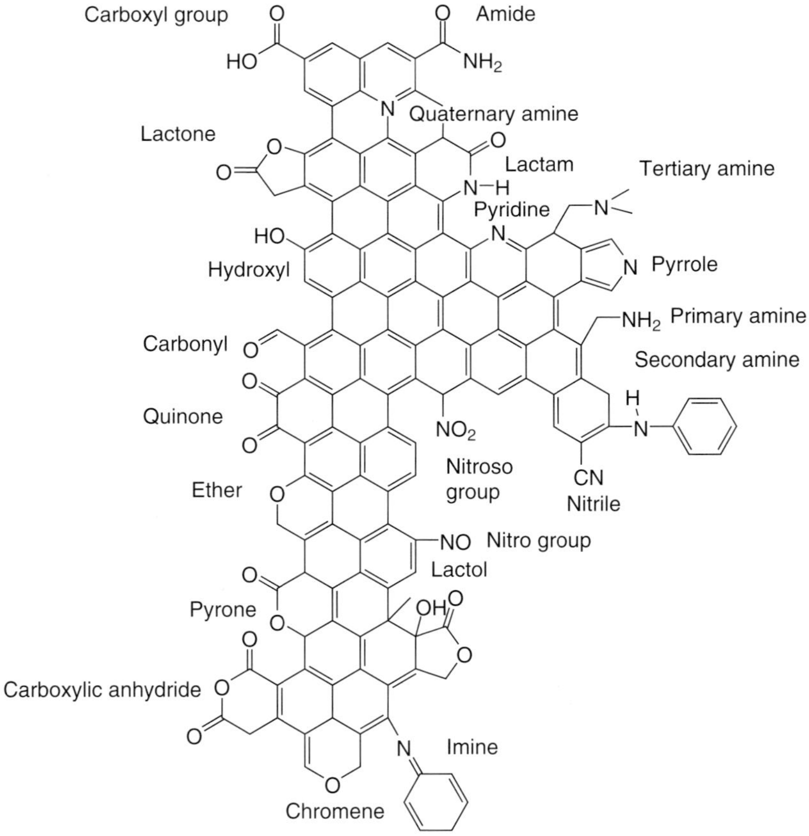

Figure 8.7 Oxygen and nitrogen functionalities that may be introduced in carbonaceous materials. (Adapted from Samojeden *et al.* [40].)

the adsorption capacity on the active sites of carbocatalysts determine their performance [43].

Oxygen surface groups can be introduced by treatment with different oxidizing agents such as concentrated HNO_3, H_2O_2, $(NH_4)_2S_2O_8$, or H_2SO_4 [44–46]. Treatments, that may vary from some minutes up to several hours, can consist of a wet-oxidation process at room or low temperature, refluxing at boiling temperature, or air/O_2 oxidation at high temperature (300–400 °C) [40]. During oxidation, a wide number of oxygen functionalities are formed on the surface of carbon materials (such as hydroxyl, lactone, lactol, ether, quinone, pyrone, chromene, carboxyl groups, carbonyl, or carboxylic anhydride), which depends on the type of oxidant used [45] and the execution of heat treatments which may selectively remove some formed surface groups [40]. On the other hand, besides increasing activity, the introduction of surface nitrogen groups also may favor stability of carbon materials toward oxidation. Treatments usually comprise a previous stage of oxidation followed by the introduction of basic N-species (which can be nitroso or nitro groups; primary, secondary, tertiary, or quaternary amines; pyridine; pyrrole; lactam; amide; imine; or nitrile) [40], such groups can be formed either by high-temperature treatment of carbon materials with molecules containing nitrogen such as urea, NH_3, pyridine, HCN, melamine, hydrazine, and so on, or by activation of carbon material precursors with N-containing compounds [40, 47–49].

Carbonaceous materials also are useful in the preparation of composites for catalytic reactions. Chemical modification of carbon supports by oxidation enriches their surface oxygen-groups content, which enhances adsorption, stability, and activity of the final catalysts. For example, carbon-supported Wacker-type catalysts for CO oxidation in which the presence of surface carboxylic, carbonyl, and phenolic groups increases the hydrophilicity and promotes formation of active sites enriched with transition metals [50–52]; acetylene hydrochlorination catalyzed by gold/activated carbon in which the introduction of a higher amount of phenol, ether, and carbonyl groups on activated carbon surface promotes a higher catalytic activity and stability of Au^{3+} active sites [53]; and (photo)catalytic ozonation reactions with carbon/Fe/TiO_2 composites prepared with magnetic iron oxides and anatase-TiO_2 particles, in which the surface chemistry of a commercial activated carbon permitted the chemisorption of Fe (by incipient wetness impregnation) and TiO_2 (by dispersion on a TiO_2–sol) and exhibited better catalytic performance than bare TiO_2 or activated carbon [54].

Inorganic nanomaterials are particles with a non-organic core that possess properties such as strong optical absorption, high electron density, photoluminescence, phosphorescence, or magnetic moment [55], which are all of importance in different fields, such as medicine, textiles, agriculture, optics, construction, catalysis, and so on [56, 57]. These materials usually tend to form agglomerates and have low solubility in water or organic solvents; thus, in some cases, it is necessary to develop some surface chemical conditions to favor the particle stability and performance. These nanomaterials can be stabilized by the introduction of inert molecular chains or ligands (such as hydrocarbon chains, polyethylene glycol, functional groups with terminating linear molecules, or polymers) by chemisorption, electrostatic adsorption, covalent binding, or non-covalent affinity with the ligand [55, 56]. The typical surface modification processes for inorganic nanoparticles consist of particle dispersion/dissolution into a matrix that contains the chemical or polymeric precursor to interact on the surface, adjustment of the treatment conditions (heating, pH, etc.), further steps to favor polymerization reactions or the addition of more functional groups on the catalyst surface if required, and also an annealing processes to stabilize the introduced ligands if needed. In water applications, nanoparticles are typically stabilized introducing carboxylic acids that stabilize it by electrostatic repulsion, and then additional functional groups can be introduced to the nanoparticle through their ligands. As an example, Au nanoparticles, with high tendency to agglomeration, can be stabilized by the introduction of a dodecanethiol monolayer on the surface; in addition, this monolayer can be modified by the introduction of bromide, ferrocene, hydroxyl, and carboxyl functional groups [55, 58] and in this way more functionalities are developed in the particle.

Surface modification of inorganic nanoparticles by silane coupling agents is another approach to improve their dispersion stability in different solvents. Silane coupling agents adsorb on the surface of nanoparticles at its hydrophilic end interacting through hydroxyl groups that must be introduced in the surface of the nanoparticle. Then, the resulting composite is better dispersed in organic solvents. Typical silane agents are 3-aminopropyltriethoxysilane, *n*-propyltriethoxysilane, or 3-methacryloxypropyltrimethoxysilane [56].

Metal oxides such as TiO_2, Al_2O_3, SiO_2, and ZrO_2 also tend to form agglomerates [57] that negatively affect their performance in catalytic applications due to a decrease in surface area [59]. Thus, they can be modified via grafting using organic molecules such as amino diacids. This kind of treatment permits the development of oxide nanoparticles with a hydrophilic surface [57] and thus prevent aggregation. An example of this kind of treatment is that described by Pei *et al.* [59] in which SiO_2 spheres were tuned by amino-functionalizing silica surface by treatment with aminopropyltriethoxysilane followed by an annealing process consisting of heating in air at 100 °C. This results in the introduction of amino diacid groups on its surface and, thereby, in improved uniformity and dispersion of the particles. These treated particles can be superficially loaded with common metals such as Pt, Rh, and Ru keeping their uniformity [59]. Another approach that has been studied to modify the surfaces of silica nanoparticles to have better dispersibility is the use of coupling agents (such as triethoxyvinylsilane (VTES)) that allow the introduction of some reactive groups, for example, poly(ethyleneglycol)methacrylate (PEGMA) or poly(propylene glycol) methacrylate (PPGMA) that are grafted on the silica surface by polymerization reactions assisted by UV radiation, as represented in Figure 8.8.

Transition metal oxides used in electrocatalysis for processes such as oxygen reduction reactions, in some cases exhibit poor electronic conductivity limiting their activity. Surface modification with carbon or conducting polymers has been reported as an effective way to overcome this limitation. Conducting polymers, such as polypyrrole, can combine advantages such as high electronic conductivity, stability, and flexibility; thus, metal–polymer hybrids can be obtained through a facial chemical oxidation polymerization approach. The resulting

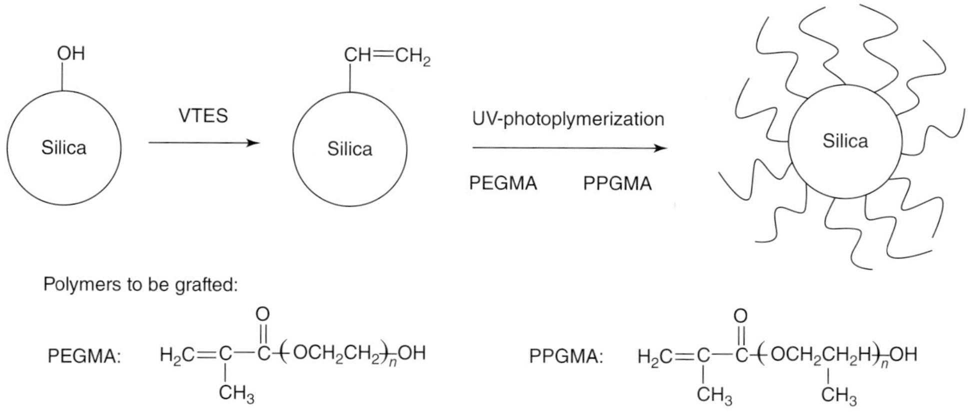

Figure 8.8 Grafting PEGMA or PPGMA onto silica nanoparticles. (Adapted from Shin *et al.* [60].)

hybrids present optimal surface conditions and enhanced electrocatalytic activity [61].

Zeolites are useful in many catalytic applications thanks to its high activity and low cost. In some cases as in the production of hydrogen using H-type of zeolites, coke is formed in the process through the strong acid sites of the zeolite, which decreases its activity. This limitation may be overcome by tuning the acidity of the zeolite to remove unnecessary strong acid sites by introducing La, Mg, Ga, B, or P as well as an alkaline treatment. Lü *et al.* [62] by the introduction of only 5 wt% P in a HZSM-5 zeolite completely covered the strong acid sites transforming them into weak acid sites, which led to better stability and activity on the catalyst [62].

Grafting of synthetic polymers constitutes another way to modify inorganic surfaces. Besides altering the chemical functionality on the surface this approach can also modify the surface topology of inorganic (and organic materials) and develop higher mechanical toughness, better electronic properties, and surface hydrophobicity; in this way, the grafted nanoparticles show better dispersion behavior. Rong *et al.* [63] describe an approach to modify the surface of nanosized alumina particles using polystyrene and polyacrylamide as grafting polymers. The method consists of an initial treatment with silane followed by radical grafting polymerization. The polymer was covalently bonded onto the surface of the alumina particles and changed their surface characteristics. According to Tang *et al.* [64], commercial ZnO nanoparticles can be superficially modified by grafting or anchoring polymethacrylic acid chains. The hydroxyl groups on the surface of nano-ZnO particles interact with carboxyl groups of the polymethacrylic acid forming poly(zinc methacrylate) complexes on the surface. This significantly improved the dispersion stabilization of modified ZnO nanoparticles in water without altering the crystalline structure of the nanoparticles.

Inorganic nanoparticles can be also modified by adsorption of polymeric anionic or cationic dispersants that generate steric repulsive forces among the polymer chains and increase the surface charge; consequently, a better dispersibility of the nanoparticles is obtained. The use of various types of polycarboxylic acids and their salts as anionic surfactants to disperse many types of metal oxide nanoparticles, such as TiO_2, Al_2O_3, and Fe_2O_3 [65] has been reported in the literature.

8.4.2 Characterizations

Typical surface characterization techniques are useful for the evaluation or monitoring of solid catalysts surface before and after a modification procedure. Transmission electron microscopy (TEM), scanning electron microscopy (SEM), temperature-programmed desorption (TPD), X-ray photoelectron spectroscopy (XPS), infrared spectroscopic methods (FTIR and DRIFT), N_2 adsorption, X-ray diffraction (XRD), and inductively coupled plasma (ICP) being the most recurring techniques.

Particle size is one of the most important characteristics of nanoparticles; this is usually related to the specific surface area and can be altered through the modification of the nanoparticle surface. The attachment of large biomolecules, functional groups, or polymeric ligands can be observed by the increase of the total particle size. Although this parameter can be determined by TEM, this technique is more suitable for inorganic particles than for organic molecules [55], as organic molecules might be stained with heavy elements due to their poor contrast in TEM. On the other hand, SEM images can be used to observe the shape and surface morphology of catalysts, revealing some structural parameters, pore formation, surface impurity segregation, and crystal shape [66]. Elemental mapping by SEM with energy dispersive X-ray spectrometry (SEM/EDX) besides providing the conventional SEM images also gives an image showing the element distribution on the surface.

XRD constitutes another way to determine particle size in case of nanocrystalline materials for which it is assumed that particles are spherical, and thus the Bragg peaks' widths are used to estimate the crystallite mean size through the Scherrer equation [67]. XRD is also useful to characterize the phase composition and crystallinity of nanoparticles.

Surface composition and oxidation state of the atoms on the catalysts' surface are fundamental parameters to follow changes on the surface composition of solids subjected to modification. In an XPS analysis the catalyst is irradiated with energetic X-rays that cause the emission of surface photoelectrons having a characteristic kinetic energy producing a spectrum that is used to identify the surface composition [68], surface groups such as C—C, C—O (alcohol), C=O (carbonyl) and COO (carboxyl) [40], or surface polymeric groups [69]. In addition, relative atomic surface composition can be determined from XPS analysis using the Wagner atomic sensitivity factors [70, 71].

In a TPD analysis, the sample is subjected to a heating rate that can be linearly programmed and the evolved gas can be analyzed to determine surface oxygen species introduced in the catalyst. This technique can differentiate between carboxyls, lactones, ethers, phenols, quinones, and so on [40]. These groups are identified by gas chromatography or mass spectrometry analysis of the evolved gas, taking into account the decomposition temperatures of such groups and the type of generated gas. For example, CO_2 is ascribed to carboxylic acid and lactones, CO to phenols and quinones, and H_2 is attributed to dissociation of CH or OH [72]. However, there is some controversy on the assignment of some peaks to some specific groups in TPD analysis, as the peak temperatures may be affected by the texture of the sample, the geometry of the system, and heating rate [73].

Oxygen groups can also be identified by infrared spectroscopic methods such as Fourier transform infrared spectroscopy (FTIR) and diffuse reflectance infrared Fourier transform spectroscopy (DRIFTS). These methods are only useful with highly oxidized samples in order to have adsorption bands with enough intensity to be clearly identified, but the spectra interpretation must be done with caution as each group can give several bands at different wavenumbers, and the observed bands may be the result of contributions of various groups [72].

Titration methods are useful to quantify and differentiate between acidic or basic groups (of strong, medium, and weak nature) [40, 74] using bases or acids of different strengths. Through the Boehm titration method, which is widely used, acidic oxygen surface functional groups on carbon samples can be determined by using bases of different strengths such as $NaHCO_3$ (the weakest, that neutralizes only carboxylic groups), Na_2CO_3 (of intermediate strength, that neutralizes carboxylic, and lactonic groups), and NaOH (the strongest, that neutralizes carboxylic, lactonic, and phenolic groups). This method may be affected by the way the different steps of the method are carried out (agitation and filtration method, use of dilute titrant, etc.) this alters the catalyst surface making it difficult to compare the results from different works [75, 76]. On the other hand, the Johnson Matthey titration method consists of a pH titration by HCl to determine the basic functions or by NaOH for the acidic functions. However, this method does not quantify the content of each type of acidic or basic function, but by a mathematical differential treatment of the experimental data the pK_a of acidic functions on the carbon surface can be estimated [73].

N-functionalities are more usually identified by XPS and FTIR. However, in the case of XPS, there is no total agreement about the assignment of binding energy of N 1 s, especially for the assignment of amine and amide groups. Other types of groups identified by XPS are pirydinic, pyrrolic/pyridone, quaternary nitrogen, and N-oxides groups, respectively [40].

ICP and N_2 adsorption–desorption analysis can provide bulk information about the catalysts. For ICP analysis, the solid samples typically are dissolved in a mixture of acids and then the overall atomic composition of the bulk catalyst is calculated and these results compared with XPS or SEM/EDX analysis of the same sample can give an idea of the degree of penetration of a dopant introduced in porous particles [77]. Nitrogen adsorption–desorption analysis may be used to determine the specific surface area of a catalyst and to estimate the pore size distribution. The determination of the total surface area can be obtained by the BET (Brunauer, Emett, Teller) method in which a N_2 adsorption isotherm is generated at different relative pressure values where the form of the obtained isotherm gives information on the type of pores contained in the solid [78].

Acknowledgment

One of the authors, Aydin Cihanoğlu, was supported by the Scientific and Technological Research Council of Turkey (TÜBİTAK) with "National Scholarship Programme for PhD Students."

References

1 Robertson, A.J.B. (1975) The early history of catalysis. *Platinum Met. Rev.*, **19**(2), 64–69.
2 Gregg, S.J. and Sing, K.S.W. (1982) *Adsorption, Surface Area and Porosity*, Academic Press, New York.
3 Moulijn, J.A., Leeuwen, P.W.N.M., and Santen, R.A. (1993) *Studies in Surface Science and Catalysis*, vol. **79**, Elsevier, New York.
4 Lloyd, L. (2011) *Handbook of Industrial Catalysts*, Springer, New York.
5 Somorjai, G.A. and Li, Y. (2010) *Introduction to Surface Chemistry and Catalysis*, John Wiley & Sons, Inc., Hoboken, NJ.
6 Beller, M., Renken, A., and Santen, R. (2012) *Catalysis: From Principles to Applications*, John Wiley & Sons, Ltd.
7 Santen, R.A. and Niemantsverdriet, J.W. (1995) *Chemical Kinetics and Catalysis*, Springer Science+Business Media, New York.
8 Ertl, G., Knözinger, H., and Weitkamp, J. (1999) *Preparation of Solid Catalysts*, John Wiley & Sons, Ltd.
9 Bartholomew, C.H. and Farrauto, R.J. (2006) *Fundamentals of Industrial Catalytic Process*, John Wiley & Sons, Ltd.
10 Krijn, P.J. (2009) *Synthesis of Solid Catalysts*, John Wiley & Sons, Ltd.
11 Richardson, J.T. (1992) *Principles of Catalyst Development*, Springer, New York.
12 Jong, K.P. (2009) *Synthesis of Solid Catalysts*, John Wiley & Sons, Ltd.
13 Fogler, H.S. (1999) *Elements of Chemical Reaction Engineering*, Prentice Hall PTR, New Jersey.
14 Schlögl, R. (2015) Heterogeneous catalysis. *Angew. Rev.*, **54**(11), 3465–3520.
15 Campbell, I.M. (1988) *Catalysis at Surfaces*, Chapman & Hall, New York.
16 Davis, M.E. and Davis, R.J. (2003) *Fundamentals of Chemical Reaction Engineering*, Dover Publications, New York.
17 Deutschmann, O., Knözinger, H., Kochloefl, K., and Turek, T. (2011) Heterogeneous catalysis and solid catalysts, 1. Fundamentals, in *Ulmann's Encyclopedia of Industrial Chemistry*, John Wiley & Sons, Ltd.
18 Farnetti, E., Monte, R.D., and Kašpar, J. (2014) *Encyclopedia of Inorganic and Bio-Inorganic Chemistry*, vol. **2**, John Wiley & Sons, Ltd.
19 Smith, J.M. (1981) *Chemical Engineering Kinetics*, McGraw-Hill.
20 Hagen, J. (2006) *Industrial Catalysis*, John Wiley & Sons, Ltd.
21 Hutchings, G.J. and Vedrie, J.C. (2004) *Heterogeneous Catalyst Preparation in: Basic Principles in Applied Catalysis*, Springer, Heidelberg.

22 Haber, J., Block, J.H., and Delmon, B. (1995) Manual of methods and procedures for catalyst characterization (Technical Report). *Pure Appl. Chem.*, **67**(8-9), 1257–1306.

23 Perego, C. and Villa, P. (1997) Catalyst preparation methods. *Catal. Today*, **34**(3-4), 281–305.

24 Schwarz, J.A., Contescu, C., and Contescu, A. (1995) Methods for preparation of catalytic materials. *Chem. Rev.*, **95**(3), 477–510.

25 Macwan, D.P., Dave, P.N., and Chaturvedi, S. (2011) A review on nano-TiO_2 sol–gel type syntheses and its applications. *J. Mater. Sci.*, **46**(11), 3669–3686.

26 Pierson, H.O. (1999) *Handbook of Chemical Vapor Deposition (CVD) Preinciples, Technology, and Applications*, William Andrew Publishing, New York.

27 Khare, R. and Bose, S. (2005) Carbon nanotube based composites-a review. *J. Miner. Mater. Charact. Eng.*, **4**(1), 31–46.

28 Zhang, X., Zhou, M., and Lei, L. (2005) Enhancing the concentration of TiO_2 photocatalyst on the external surface of activated carbon by MOCVD. *Mater. Res. Bull.*, **40**(11), 1899–1904.

29 Zhang, X., Zhou, M., Lei, L., and Xu, S. (2004) Synthesis of TiO_2 supported on activated carbon by MOCVD: operation parameters study. *J. Zhejiang Univ. Sci.*, **5**(12), 1548–1553.

30 Duminica, F.D., Maury, F., and Senocq, F. (2004) Atmospheric pressure MOCVD of TiO_2 thin films using various reactive gas mixtures. *Surf. Coat. Technol.*, **188-189**, 255–259.

31 Puma, G.L., Bono, A., Krishnaiah, D., and Collin, J.G. (2008) Preparation of titanium dioxide photocatalyst loaded onto activated carbon support using chemical vapor deposition: a review paper. *J. Hazard. Mater.*, **157**(2-3), 209–219.

32 Zhang, X. and Lei, L. (2008) Effect of preparation methods on the structure and catalytic performance of TiO_2/AC photocatalysts. *J. Hazard. Mater.*, **153**(1-2), 827–833.

33 Shan, A.Y., Ghazi, T.I.M., and Rashid, S.A. (2010) Immobilisation of titanium dioxide onto supporting materials in heterogeneous photocatalysis: a review. *Appl. Catal., A*, **389**(1-2), 1–8.

34 Tominaga, K., Maruoka, S., Gotoh, M., Katada, N., and Niwa, M. (2009) HZSM-5 modified by silica CVD for shape-selective production of *p*-xylene: influence of in situ and *ex situ* preparation conditions of the zeolite. *Microporous Mesoporous Mater.*, **117**(3), 523–529.

35 Schäf, O., Ghobarkar, H., and Knauth, P. (2002) Hydrothermal synthesis of nanomaterials, in *Nanostructured Materials*, vol. **8** (eds P. Knauth and J. Schoonman), Springer, pp. 23–41.

36 Dalvand, P. and Mohammadi, M.R. (2011) Controlling morphology and structure of nanocrystalline cadmium sulfide (CdS) by tailoring solvothermal processing parameters. *J. Nanopart. Res.*, **13**(7), 3011–3018.

37 Wang, Q., Pan, D., Jiang, S., Ji, X., An, L., and Jiang, B. (2006) A solvothermal route to size- and shape-controlled CdSe and CdTe nanocrystals. *J. Cryst. Growth*, **286**(1), 83–90.

38 Morikawa, Y. (1993) Catalysis by metal ions intercalated in layer lattice silicates. *Adv. Catal.*, **39**, 303–327.

39 Govindarajan, T. and Shandas, R. (2014) A survey of surface modification techniques for next-generation shape memory polymer stent devices. *Polymers*, **6**(9), 2309–2331.

40 Samojeden, B., Motak, M., and Grzybek, T. (2015) The influence of the modification of carbonaceous materials on their catalytic properties in SCR-NH_3. A short review. *C.R. Chim.*, **18**(10), 1049–1073.

41 Serp, P. and Figueiredo, J.L. (2008) *Carbon Materials for Catalysis*, John Wiley & Sons, Inc., Hoboken, NJ.

42 Tascón, J.M.D. (2012) *Novel Carbon Adsorbents*, Elsevier, UK.

43 Xue, Y., Guo, Y., Zhang, Z., Guo, Y., Wang, Y., and Lu, G. (2008) The role of surface properties of activated carbon in the catalytic reduction of NO by carbon. *Appl. Surf. Sci.*, **255**(5), 2591–2595.

44 Pradhan, B.K. and Sandle, N.K. (1999) Effect of different oxidizing agent treatments on the surface properties of activated carbons. *Carbon*, **37**(8), 1323–1332.

45 Zhang, J., Zou, H., Qing, Q., Yang, Y., Li, Q., Liu, Z., Guo, X., and Du, Z. (2003) Effect of chemical oxidation on the structure of single-walled carbon nanotubes. *J. Phys. Chem. B*, **107**(16), 3712–3718.

46 Teng, H., Tu, Y.T., Lai, Y.C., and Lin, C.C. (2001) Reduction of NO with NH_3 over carbon catalysts: the effects of treating carbon with H_2SO_4 and HNO_3. *Carbon*, **39**(4), 575–582.

47 Jansen, R.J.J. and Bekkum, H. (1994) Amination and ammoxidation of activated carbons. *Carbon*, **32**(8), 1507–1516.

48 Zhao, Z., Dai, Y., Ge, G., and Wang, G. (2015) Efficient tuning of microstructure and surface chemistry of nanocarbon catalysts for ethylbenzene direct dehydrogenation. *AlChE J.*, **61**(8), 2543–2561.

49 Chen, Y., Lin, B., Wang, H., Yang, Y., Zhu, H., Yu, W., and Basset, J.-M. (2016) Surface modification of g-C_3N_4 by hydrazine: simple way for noble-metal free hydrogen evolution catalysts. *Chem. Eng. J.*, **286**, 339–346.

50 Park, E.D. and Lee, J.S. (2000) Effect of surface treatment of the support on CO oxidation over carbon-supported Wacker-type catalysts. *J. Catal.*, **193**(1), 5–15.

51 Wang, S., Wei, L., Yuanyuan, D., Zhao, Y., and Ma, X. (2015) Dimethyl carbonate synthesis from methyl nitrite and CO over activated carbon supported Wacker-type catalysts: the surface chemistry of activated carbon. *Catal. Commun.*, **72**, 43–48.

52 Wang, L., Feng, Y., Zhang, Y., Lou, Y., Lu, G., and Guo, Y. (2012) Effect of original activated carbon support and the presence of NO_x on CO oxidation over supported Wacker-type catalysts. *Fuel*, **96**, 440–445.

53 Xu, J., Zhao, J., Xu, J., Zhang, T., Li, X., Di, X., Ni, J., Wang, J., and Cen, J. (2014) Influence of surface chemistry of activated carbon on the activity of gold/activated carbon catalyst in acetylene hydrochlorination. *Ind. Eng. Chem. Res.*, **53**(37), 14272–14281.

54 Quiñones, D.H., Rey, A., Álvarez, P.M., Beltrán, F.J., and Plucinski, P.K. (2014) Enhanced activity and reusability of TiO_2 loaded magnetic activated carbon for solar photocatalytic ozonation. *Appl. Catal., B*, **144**, 96–106.

55 Sperling, R.A. and Parak, W.J. (2010) Surface modification, functionalization and bioconjugation of colloidal inorganic nanoparticles. *Philos. Trans. R. Soc. London, Ser. A*, **368**, 1333–1383.
56 Kango, S., Kalia, S., Celli, A., Njuguna, J., Habibi, Y., and Kumara, R. (2013) Surface modification of inorganic nanoparticles for development of organic–inorganic nanocomposites—A review. *Prog. Polym. Sci.*, **38**(8), 1232–1261.
57 Mallakpour, S. and Nikkhoo, E. (2014) Surface modification of nano-TiO_2 with trimellitylimido-amino acid-based diacids for preventing aggregation of nanoparticles. *Adv. Powder Technol.*, **25**(1), 348–353.
58 Templeton, A.C., Wuelfing, W.P., and Murray, R.W. (2000) Monolayer-protected cluster molecules. *Acc. Chem. Res.*, **33**(1), 27–36.
59 Pei, Y., Xiao, C., Goh, T.W., Zhang, Q., Goes, S., Sun, W., and Huang, W. (2016) Tuning surface properties of amino-functionalized silica for metal nanoparticle loading: the vital role of an annealing process. *Surf. Sci.*, **648**, 299–306.
60 Shin, Y., Lee, D., Lee, K., Ahn, K.H., and Kim, B. (2008) Surface properties of silica nanoparticles modified with polymers for polymer nanocomposite applications. *J. Ind. Eng. Chem.*, **14**(4), 515–519.
61 Cao, X., Yan, W., Jin, C., Tian, J., Ke, K., and Yang, R. (2015) Surface modification of $MnCo_2O_4$ with conducting polypyrrole as a highly active bifunctional electrocatalyst for oxygen reduction and oxygen evolution reaction. *Electrochim. Acta*, **180**, 788–794.
62 Lü, J., Zhou, S., Ma, K., Meng, M., and Tian, Y. (2015) The effect of P modification on the acidity of HZSM-5 and P-HZSM-5/CuO-ZnO-Al_2O_3 mixed catalysts for hydrogen production by dimethyl ether steam reforming. *Chin. J. Catal.*, **36**(8), 1295–1303.
63 Rong, M.Z., Ji, Q.L., Zhang, M.Q., and Friedrich, K. (2002) Graft polymerization of vinyl monomers onto nanosized alumina particles. *Eur. Polym. J.*, **38**(8), 1573–1582.
64 Tang, E., Cheng, G., Ma, X., Pang, X., and Zhao, Q. (2006) Surface modification of zinc oxide nanoparticle by PMAA and its dispersion in aqueous system. *Appl. Surf. Sci.*, **252**(14), 5227–5232.
65 Palmqvist, L. and Holmberg, K. (2008) Dispersant adsorption and viscoelasticity of alumina suspensions measured by quartz crystal microbalance with dissipation monitoring and in situ dynamic rheology. *Langmuir*, **24**, 9989–9996.
66 Parker, S.C., Lawrence, P.J., Freeman, C.M., Levine, S.M., and Newsam, J.M. (1992) Information on catalyst surface structure from crystallite morphologies observed by scanning electron microscopy (SEM). *Catal. Lett.*, **15**(1), 123–131.
67 Holzwarth, U. and Gibson, N. (2011) The Scherrer equation versus the Debye-Scherrer equation. *Nat. Nanotechnol.*, **6**(9), 534.
68 Venezia, A.M. (2003) X-ray photoelectron spectroscopy (XPS) for catalysts characterization. *Catal. Today*, **77**(4), 359–370.
69 Süzer, Ş., Birer, Ö., Sevil, U.A., and Güven, O. (1998) XPS investigations on conducting polymers. *TUrk. J. Chem.*, **22**(1), 59–65.

70 Wagner, C.D., Davis, L.E., Zeller, M.V., Taylor, J.A., Raymond, R.H., and Gale, L.H. (1981) Empirical atomic sensitivity factors for quantitative analysis by electron spectroscopy for chemical analysis. *Surf. Interface Anal.*, **3**(5), 211–225.

71 Quiñones, D.H., Rey, A., Álvarez, P.M., Beltrán, F.J., and Li Puma, G. (2015) Boron doped TiO_2 catalysts for photocatalytic ozonation of aqueous mixtures of common pesticides: Diuron, *o*-phenylphenol, MCPA and terbuthylazine. *Appl. Catal., B*, **178**, 74–81.

72 Steele, W.A., Zgrablich, G., and Rudzinski, W. (1997) *Equilibria and Dynamics of Gas Adsorption on Heterogeneous Solid Surfaces*, Elsevier, Amsterdam.

73 Figueiredo, J.L., Pereira, M.F.R., Freitas, M.M.A., and Órfão, J.J.M. (1999) Modification of the surface chemistry of activated carbons. *Carbon*, **37**(9), 1379–1389.

74 Dubois, V., Dal, Y., and Jannes, G. (2002) Active carbon surface oxidation to optimize the support functionality and metallic dispersion of a Pd/C catalyst, in *Scientific Bases for the Preparation of Heterogeneous Catalysts* (eds E. Gaigneaux, D.E. De Vos, P. Grange, P.A. Jacobs, J.A. Martens, P. Ruiz, and G. Poncelet), Elsevier Science, Amsterdam, pp. 993–1002.

75 Oickle, A.M., Goertzen, S.L., Hopper, K.R., Abdalla, Y.O., and Andreas, H.A. (2010) Standardization of the Boehm titration: part II. Method of agitation, effect of filtering and dilute titrant. *Carbon*, **48**(12), 3313–3322.

76 Xiao, X., Liu, D., Yan, Y., Wu, Z., Wu, Z., and Cravotto, G. (2015) Preparation of activated carbon from Xinjiang region coal by microwave activation and its application in naphthalene, phenanthrene and pyrene adsorption. *J. Taiwan Inst. Chem. Eng.*, **53**, 160–167.

77 Schmidt, S.R. and Tanielyan, S.K. (2003) in *Catalysis of Organic Reactions* (ed. D.G. Morrell), Marcel Dekker Inc., New York, pp. 247–262.

78 Magee, J.S. and Mitchell, M.M. (1993) *Fluid Catalytic Cracking: Science and Technology*, Elsevier Science Publishers B.V., Amsterdam.

Index

Surface Treatments for Biological, Chemical and Physical Applications, First Edition.
Edited by Mehmet Gürsoy and Mustafa Karaman.